Climate Variability of Southern High Latitude Regions

Climate Variability of Southern High Latitude Regions

Sea, Ice, and Atmosphere Interactions

Edited by Neloy Khare

CRC Press
Taylor & Francis Group
Boca Raton London New York

CRC Press is an imprint of the
Taylor & Francis Group, an **informa** business

First edition published 2022
by CRC Press
6000 Broken Sound Parkway NW, Suite 300, Boca Raton, FL 33487-2742

and by CRC Press
4 Park Square, Milton Park, Abingdon, Oxon, OX14 4RN

© 2022 Taylor & Francis Group, LLC

CRC Press is an imprint of Taylor & Francis Group, LLC

Library of Congress Cataloging-in-Publication Data
Names: Khare, Neloy, editor.
Title: Climate variability of southern high latitude regions : sea, ice, and atmosphere interactions / edited by Neloy Khare.
Description: First edition. | Boca Raton : CRC Press, 2022. | Includes bibliographical references and index.
Identifiers: LCCN 2021047851 (print) | LCCN 2021047852 (ebook) |
ISBN 9781032061597 (hardback) | ISBN 9781032067629 (paperback) | ISBN 9781003203742 (ebook)
Subjects: LCSH: Climatic changes--Antarctica. | Climatic changes--Antarctic Ocean. | Antarctica--Environmental conditions. | Antarctic Ocean--Environmental conditions.
Classification: LCC GE160.A6 C55 2022 (print) | LCC GE160.A6 (ebook) |
DDC 551.69167--dc23/eng/20211216
LC record available at https://lccn.loc.gov/2021047851
LC ebook record available at https://lccn.loc.gov/2021047852

ISBN: 978-1-032-06159-7 (hbk)
ISBN: 978-1-032-06762-9 (pbk)
ISBN: 978-1-003-20374-2 (ebk)

DOI: 10.1201/9781003203742

Typeset in Times
by MPS Limited, Dehradun

Dedication

Dedicated to Late Prof Anil Kumar Jauhri

Late Prof Anil Kumar Jauhri (28.12.1947 – 08.04.2017)

Prof. Anil Kumar Jauhri was born in Lucknow (Uttar Pradesh) on 28/12/1947. He obtained his BSc in 1967 and MSc (Geology) in 1969 from the University of Lucknow. Prof. Jauhri completed his doctorate under the guidance of Prof. K.P.Vimal and later joined the department as a Lecturer in the year 1979 and become Reader in 1992. He rose to the position of Professor of Palaeontology in the year 2000. He superannuated in the year 2012 but never retired in his professional career. He continued reaching and supervising students. He maintained a rigorous research, teaching and editorial schedule throughout his academic career. Even during times of heavy administrative responsibilities as Head of the Department of Geology.

Prof. Jauhri was highly rated for his teaching ability, and commitment for research. His ability for editing was manifested when he joined the editorial board of the journal of the Paleontological

Society of India. It was his efforts that the Journal got the citation index number in 2013 and an international acclaim for its contents. Prof. Jauhri is known for his outstanding scientific contribution on Cenozoic foraminiferal biostratigraphy of Kachchh and Meghalaya, India. His long collaboration with Prof. P.K Misra, developed his interest in Coralline algae and later together they published more than 30 research papers, completed 7 research projects and guided 3 PhD theses on integrated aspects of coralline algae and larger foraminifera form the different parts of India. At the time of his death, he was working on the manuscripts of Coralline Algae form the Prang and Kopili formations of the Meghalaya and Hut Bay Formation of Little Andaman, India.

Prof Jauhri breathed his last on April 8, 2017. The void created with the demise of Prof. Jauhri in the scientific community in general and Paleontological research community in specific is difficult to be filled. All his students, colleagues and friends will always miss him and remember him for his best qualities like kind hearted, polite, soft spoken excellent human being with complete commitment and devotion to his duties. It is a great loss to all his student. Prof. Jauhri will be remembered by generations of students for advancing their careers.

As a small tribute to my teacher Prof Jauhri, I dedicate this book to him.

Contents

vii

Foreword

The Antarctic and its surrounding ocean is a highly coupled system and plays a central role in global climate variability and change due to non-linear interactions between the atmosphere, ocean, ice, and complex links to the rest of the Earth system. Nevertheless, the sustained efforts to illuminate its critical linkages to lower latitudes are lacking. Conjunction of new observational capabilities, advances in scientific understanding, and improving numerical models highlight the global relevance of Antarctica. The Indian Ocean sector is an area that is closely coupled to the global atmosphere on a variety of time scales. The coupled air-atmosphere-ice-ocean numerical models are required for global simulations and that realistically incorporate Antarctica.

The Antarctic region is a hotspot of climate change assessment and a barometer of global climate variability. Despite such a high significance, scientific understanding about the Indian Ocean sector of the Southern Ocean and atmospheric processes over southern high-latitude regions are yet to be further augmented, and results to be collated as a ready reference to the budding researchers in the contemporary fields of high-latitude research. At the same time, significant scientific endeavours by Indian researchers in these areas need proper documentation.

The present book, *Climate Variability of Southern High-Latitude Regions: Sea, Ice, and Atmosphere Interactions,* aptly provides a comprehensive account of Indian efforts to help understand the impact of climate change on the polar atmosphere and Southern Ocean vis-à-vis the influence of the changing Antarctica and its surrounding oceans, the polar atmosphere, and sea ice on global climate change in twelve chapters.

This book begins with the assessment of ozone variability in the Antarctic stratosphere over many decades by Soni. Interestingly, the significance of "Stratospheric Dynamics in Climate over Antarctica" has been highlighted by Kishore Kumar and Koushik. On the other hand, Sunitha Devi and Maheskumar have detailed "Antarctic Weather and Climate Patterns". Significant attempts have been made to study aerosols over the Antarctic region. Pant et al. measured Antarctic aerosols and linked them with climate change. Similarly, Sonbawne et al. detailed the Antarctic aerosol characteristics and their role in climate variability coupled with Goel and Sharma's physicochemical characterization of Antarctic particles. The "Impact of Near-Earth Space Environmental Condition to the Antarctic Sub-Auroral Upper Atmospheric Region" has been studied by Das.

The ocean's response to the ongoing climate variability and the associated feedback mechanisms are dealt with by Prabhu et al., who point out an "Intriguing Relationship Between Antarctic Sea Ice, ENSO, and Indian Summer Monsoon", whereas Dwivedi and Pandey attempt to quantify the predictability of southern Indian Ocean sea-ice concentration in a changing climate scenario.

Sea ice is a critical element of the climate system, which regulates heat, mass, and momentum exchanges between the atmosphere and the oceans at high latitudes, using the sea-ice extent (SIE) data generated from a series of passive microwave sensors. The variability of SIE trends for the Weddell Sea (WS), Indian Ocean (IO), western Pacific Ocean (PO), Ross Sea (RS), and Bellingshausen and Amundsen Seas (BAS) during the past decades, along with other aspects, have been detailed by Luis. On the other hand, Chattopadhyay and Sahai explored interhemispheric teleconnection. They noticed a connection between the Southern Hemispheric climate change and the decreasing trends in the Seasonal Mean Monsoon Rainfall over the Indian region in the last century (1871–2004). At the same time, spatial and temporal variability of physical parameters in Prydz Bay, due to climate change, has been assessed by Pednekar. Altogether, this book provides a comprehensive, up-to-date account of how the physical environment of the Antarctic continent and the Southern Ocean has changed with time. Some of the findings in this volume may prepare the Antarctic environment for change over the next century due to the accelerated greenhouse gas concentrations. This book is ready with a highly cross-disciplinary approach to reflect the continent's importance in global issues.

It will be of immense value to all scientists interested in the Antarctic continent and the Southern Ocean. It will also help the policymakers and those concerned with observing systems and the development of climate models.

Shishir Kumar Dube
New Delhi, December 2021

Preface

Climate variability includes all the variations in the climate that last longer than individual weather events. In contrast, climate change only refers to those variations that persist for a more extended period, typically decades or more. In other words, climate variability refers to the climatic parameter of a region varying from its long-term mean. Sea-ice loss, accelerated sea-level rise, and longer, more intense heat waves, besides others, are consequences we have started experiencing now.

Temperature is one of the significant measures of climate variability; this is primary and can be measured or reconstructed for the Earth's surface and sea surface temperature (SST). Precipitation (rainfall, snowfall, etc.) offers another indicator of relative climate variation, including humidity or water balance and water quality.

The ocean, which covers 70% of the global surface, has a significant influence on Earth's weather and climate. The ocean acts as a great reservoir that continuously exchanges heat, moisture, and carbon with the atmosphere, driving weather patterns and subtly influencing global climates.

The atmospheric circulation is derived by absorbing solar radiation and releasing heat from the ocean. The ocean also influences climate by way of releasing aerosols. It controls cloud cover, mainly by emitting most rainwater, absorbing atmospheric carbon dioxide, and storing it for years to millions of years and thus, the oceans influence climate. The oceans absorb the maximum solar energy that reaches Earth. During the past century alone, the global temperature has increased by 0.6 degrees Celsius. Similarly, as per estimates, the average global sea level over the past decade has risen steadily.

We need to understand and recognize the warming pattern. If global warming continues, it may not be uniform. In the context of "global warming," we must address the more significant issue of "global climate variability". High-latitude regions' unique and distinctive physical features enhance change in mean surface temperature for a given perturbation of planetary heat balance. Such a physical oceanographic setting also enhances regional and seasonal environmental response due to non-uniformity in poleward heat flux and the energy relationships of phase and albedo changes connected with ice and snow cover.

India's science pursuits in climate variability over the southern high-latitude regions have been long and diversified covering, air, sea, ice, and atmosphere over the Antarctic region. Such valuable data has been collated to provide insight into the most significant topic of climate variability vis-à-vis global warming in the present book titled *Climate Variability of Southern High-Latitude Regions: Sea, Ice, and Atmosphere Interactions*. The book consists of a total of 12 chapters intensely focused on thematic topics. The book begins with a chapter on the ozone measurements over the Antarctic region by Soni et al. The atmospheric ozone is the most critical trace gas that is beneficial and harmful to human beings and the ecosystem, depending on its abundance in the stratosphere and troposphere. It plays an essential role by absorbing solar radiation and determining the stratosphere's temperature profile and atmospheric circulation. The factors controlling spatiotemporal variations of ozone in different regions of the atmosphere over the Antarctic environment provide precious information about climate change and play a crucial role in influencing weather systems around the globe. The meteorological conditions in Antarctica prompt halogenated gases to be more effective. These gases deplete stratospheric ozone. It signifies the importance of ozone monitoring in Antarctica. Ground and satellite observations show that halogen levels in the stratospheric atmosphere are declining. Several model projections suggest that stratospheric ozone will recover to 1980 levels around the middle of this century.

The next chapter discusses the "Stratospheric Dynamics and Its Role in Climate Over Antarctica" by Kishore Kumar and Koushik. This chapter provides a comprehensive view of the state of the Antarctic stratosphere and its role in modulating the Antarctic climate. Modern-Era Retrospective Analysis for Research and Applications, version 2 (MERRA-2), was used. Reanalysis of the data set during 1980–2020 for three crucial climate variables, ozone, temperature, and zonal winds, are constructed. Besides discussing the meridional cross-section of these variables in the Southern Hemisphere, their mean annual cycle in the polar cap (60–90°S) region is discussed comprehensively. Forty-one years of deseasonalized perpetrations are obtained. The same is employed to estimate the trends in the troposphere and stratosphere. Height-month sections of the annual cycle of trends in ozone, temperature, and zonal winds are discussed. In light of the present understanding of the Antarctic stratospheric ozone depletion, the degree of covariability of trends in these three parameters is also discussed. A brief discussion on the role of polar stratospheric clouds and their signatures in the space-based lidar observations is also provided. Finally, a discussion on the potential pathways through which the stratospheric structure and dynamics interact with the troposphere is discussed in detail. The results discussed in this chapter suggest that the stratospheric processes over the Antarctic impact the troposphere through both chemical and dynamical processes and play a much more vital role than anticipated.

On the contrary, Sunitha Devi and Maheskumar discussed the salient features of Antarctic weather and climate, climate change impacts, and also briefly summarized the Indian Antarctic Program. This chapter starts with the geography and seasons of the Antarctic sub-continent. Weather patterns and the variation of state parameters and significant weather producing systems are discussed later. It also summarizes the climatological aspects of all the parameters over the South Pole.

Aerosols exert direct and indirect impacts on the climate via their interaction with the incoming solar radiation and participation in cloud microphysics. These aerosol effects depend on the size range of these particles. The remote oceanic regions and Antarctica serve as background sites to assess the climatic impacts of aerosols. Pant et al. present a review of literature on the Antarctic aerosol measurements and their climatic effects. Also reported are the measurements of aerosol concentrations and number size distributions made at a coastal Antarctic station, Maitri, during January–February 2005. Some ship-based measurements of aerosols from coastal Antarctic waters near the ice-shelf region are also reported. The high-resolution aerosol size distributions were measured over a wide size range of 3 nm–20 μm. The variations in number size distributions in Aitken, accumulation, and coarse mode particles are discussed. At the Maitri station, the total concentrations of coarse particles (0.5–20 μm diameter) remained below 1.0 cm^{-3}, with an accumulation mode between 0.72 and 0.77 μm diameter. However, these particles were found to be in the range of 2–40 cm^{-3} near the ice-shelf region at Antarctic coastal waters. The total concentration of submicrometer (0.003–0.7 μm) particles varied in the range of 100–800 particles cm^{-3} in January and between 100 and 2,000 particles cm^{-3} in February at Maitri. Considerable variability was found in the magnitude and size range of different modes in Aitken mode at Maitri and ice-shelf region. The total concentration of particles over the coastal Antarctic waters near the ice shelf was double that of the Maitri station. This high concentration of ultrafine particles in the coastal Antarctic environment could be attributed to the gas-to-particle conversion in the sub-polar oceanic region around Antarctica. The cyclonic storms revolving around the Antarctic continent enhanced the aerosol number concentration at the Maitri station. The observed size distributions of aerosols at Maitri and ice-shelf locations are discussed in their generation, transformation, and climatic impacts.

Similarly, Sonbawne et al. have detailed "Transient Variations in Enroute Southern Indian Ocean Aerosols, Antarctic Ozone Climate and its Relationship with HO$_x$ and NO$_x$". Recent advances in field instrumentation and remote sensing technologies have paved the way for a variety of novel approaches to study the polar atmosphere, especially aerosols and precursor gases of both land and marine origin, influencing the Antarctica climate differently. Extensive field observations have been carried out over the southern Indian Ocean on the transit journey between Cape Town to Antarctica during the 24th Indian Antarctic Expedition voyage. We used ground-based total column ozone measurements (Micro tops sunphotometer, Dobson spectrophotometer, and Brewer spectrometer) and satellites (TOMS, GOES, and SCIMACHY) data during January 2005 and December 2006 over the Indian Antarctic station Maitri (70.76°S, 11.74°E). The results revealed a short-lived ozone depletion and heterogeneous chemical effect of Antarctic aerosols. The study also points out that in addition to the ozone loss in polar regions on a seasonal time scale, the short-term ozone depletion caused by Solar Proton Events (SPEs) produced nitrate aerosols that could also have a significant impact on the Earth's biosphere. Such studies are very sparse and almost non-existint in this region, and hence the proposed measurements would help bridge this gap to a certain extent. At the same time, Goel and Mishra have studied the "Physicochemical Properties of Antarctic Aerosol Particles". They attempted to look at Antarctic aerosols' physicochemical properties (shape, size, mixing state, and chemical composition) at the individual particle level. The frequency distribution of aspect ratio and circulatory factor of the Antarctic aerosols was observed to be bimodal with their respective mode peaks at 1.3 and 1.9 and 0.3 and 0.7. The particles were rich in Al, Mg, Si, Fe, Ti, Ca, and Cr. The particles are mainly from the crustal origin with variable shapes e.g. triangular, layered, flattened, aggregated, and glass-like structure. The spectral variation of Single Scattering Albedo (SSA) shows that the particles rich in Fe$_2$O$_3$ and Cr$_2$O$_3$ are more efficient solar radiation absorbers, whereas particles rich in Al$_2$O$_3$ exhibit high scattering.

Das studied the "Impact of Near-Earth Space Environmental Condition to the Antarctic Sub-Auroral Upper Atmospheric Region". His chapter addresses the scientific interest of high-latitudinal ionospheric consequences caused by the modulation of near-Earth space environmental conditions. For better understating, this chapter is divided into three parts based on different scientific investigations. It deals with the Earth's geomagnetic perturbations due to the solar wind–magnetospheric coupling process and explores the response of the sub-auroral high-latitude ionosphere to the geomagnetic disturbances. It further explains the longitudinal ionospheric response of the Southern Hemispheric high-latitude region. This chapter elaborated that the cumulative effect of consecutive three sub-storms has been responsible for a significant increase in ring current, which triggered a moderate-type geomagnetic storm. Further, the consequences of such geomagnetic perturbations on sub-auroral as well as polar longitudinal ionospheric impact and associated phenomenon have been described.

Sea ice is a critical element of the climate system that regulates the heat exchanges between the atmosphere and the high-latitude oceans. The changed concentration of the sea ice can affect ocean circulation. It leads to changes in global climate. Sea ice also plays an intrinsic role in maintaining the energy balance of the Earth. It helps keep polar regions cool due to its ability to reflect more sunlight into space. Sea ice also keeps air cool by forming an insulating barrier between the cold air above it and the warmer water below it. Due to melting glaciers, sea levels increase. Many authors ably cover such vital aspects in this book.

While Prabhu et al. discussed the "Intriguing Relationship Between Antarctic Sea Ice, ENSO, and Indian Summer Monsoon", they demonstrated a robust relationship between Antarctic sea ice and Indian summer monsoon rainfall (ISMR) using microwave satellite data for the period 1983–2015. An in-phase significant relationship is observed between sea ice over the Western Pacific Ocean (WPO) sector and ISMR. In contrast, for the same period, an out-of-phase relationship is observed between sea ice over the Bellingshausen and Amundsen Sea (BAS) sector with that of ISMR. The underlying physical mechanism that relays southern polar variability signal to the Indian monsoon region is through the Pacific Ocean marked by El Niño Southern Oscillation (ENSO). The sea-ice variability over the BAS (WPO) sector is associated with concurrently occurring equatorial central (western) Pacific Sea Surface Temperature (SST). Anomalous meridional circulations supplemented by BAS (WPO) sea-ice variability are accompanied by an ascending (descending) motion over the central (western) equatorial Pacific. It contemporaneously impacts summer monsoon rainfall over the Indian region adversely (favorably). Though ENSO is a prime factor simultaneously modulating signatures of sea ice and precipitation, two-way interaction between sea ice over the Antarctic and SST over the Pacific is also suggested.

Furthermore, it is verified that Antarctic sea-ice variability in conjunction with ENSO could have an opposite impact on rainfall variability over central, northern, and southern parts of India. It appears that ISMR variability is linked with Antarctic sea-ice variability through large-scale atmospheric circulations. Thus, Antarctic sea ice–ENSO–Walker cell–Hadley cell–ISMR is a new channel proposed in this study.

Dwivedi and Pandey focused on the predictability of the southern Indian Ocean sea-ice concentration in a changing climate scenario using CMIP6 models. The performance of CMIP6 models in simulating the sea-ice concentration (SIC) of the southern Indian Ocean region around (10E–100E; 55S–75S) covering both the Indian Antarctic Stations Maitri and Bharati is evaluated against the corresponding satellite observations. Out of 33 CMIP6 models used for the analysis, 25 models are categorized as good and 8 as poor models. A large inter-model spread is noticed in the seasonal variability of SIC of the region, but the multi-model mean matches very well with the observed satellite data. It is found that the multi-model mean SIC time series of historical data for the period 1900–2014 as well as high greenhouse gas (GHG) concentration future projection SSP5-8.5 data for the period 2015–2100 shows a decreasing trend. The SIC will decrease at an alarming rate of nearly 0.2% per year in the SSP5-8.5 scenario. The effect of climate change on the predictability of southern Indian Ocean SIC is quantified in terms of a generalized Hurst exponent and Predictability Index. It is shown that the SIC of the south Indian Ocean is predictable. With the increase in the GHG concentration, the predictability of southern Indian Ocean SIC will decrease. The SIC of the Maitri region is more predictable compared to the Bharati region. The predictability of the SIC of the Maitri and Bharati regions shall decrease in the SSP5-8.5 projection scenario during the years 2015–2100.

Luis highlighted Decadal Sea-Ice Variability over the Antarctic region, using the sea-ice extent (SIE) data generated from a series of passive microwave sensors. The variability of SIE trends is discussed for the Weddell Sea (WS), Indian Ocean (IO), western Pacific Ocean (PO), Ross Sea (RS), and Bellingshausen and Amundsen Seas (BAS), highlighting their magnitude for each of the four decades: 1979–1988, 1989–1998, 1999–2008, 2009–2018, and the role of the atmosphere/ocean and climate indices such as Pacific Decadal Oscillation (PDO), Atlantic Meridional Oscillation (AMO), Southern Oscillation Index (SOI), and Southern Annular Mode (SAM). The WS exhibited a negative SIE trend for all seasons during 2009–2018 and 1979–1988 (except for spring). It suggests positive trends during 1999–2008 and 1989–1998 (except for summer). The IO sector exhibited negative trends for all seasons during 1979–1988 (except for spring), 1989–1998, and 2009–2018, while positive trends were observed during 1999–2008. A significant SIE trend was detected for IO in the autumn (−20.12%/decade). The negative trends during 1979–1988 were noticed in the SIE in the PO sector. It showed positive trends during 1989–1998, 1999–2008 (except for autumn and winter), and 2009–2018 (except for spring). He encountered a significant SIE trend in winter during 1989–1998 (16.38%/decade). During 1979–1988 and 1989–1998, the SIE showed positive trends in the RS sector and negative trends during 1999–2008 (except for spring) and 2009–2018. Trends significant were encountered in the RS in summer (74.48%/decade) and spring (21.26%/decade) during 1979–1988, and in winter (−11.89%/decade) during 2009–2018. With negative trends during 1979–1988 and 1999–2008, we detected positive trends during 1989–1998 (except for spring) and 2009–2018 (except for autumn and spring). He consolidates the inferences that explain the interconnection between local and remote drivers for explaining the SIE variability.

On the contrary, Chattopadhyay and Sahai detailed the "Southern Hemispheric Climate Change, Interhemispheric Teleconnection, and the Observed Trends in the Seasonal Mean Monsoon Rainfall over the Indian Region in the Last Century (1871–2004)". The Southern Hemisphere has shown strong signatures of anthropogenic warming in recent decades. This study utilizes the available climate data from 1871 to 2004 to understand and explore the role of Southern

Hemispheric climate change on the monsoonal variability based on the definition of regional inter-hemispheric gradient indices. It is already known that inter-hemispheric gradients represent inter-hemispheric teleconnections and show strong temperature anomaly asymmetry as a result of global warming. This chapter shows that the monsoon rainfall has not increased in the past hundred years in response to global warming, as a simple theory using an increase in moisture availability (e.g. perceptible water) and warming ocean (the Bay of Bengal and the Arabian Sea) would suggest. This chapter presents how the Southern Hemispheric climate change can explain this decreasing trend in a simplistic framework. They assume that the monsoon flow, to a first-order, is a land-sea breeze circulation with the inter-hemispheric link, which is known for a long time (for example, Mascarenes high and Findlater jet induced monsoonal flow). This interhemispheric temperature gradient is inversely correlated with the Southern Annular Mode (SAM) index. SAM shows positive trends in recent decades. Thus, an increase (i.e. positive trend) in the SAM index would weaken the land-sea temperature gradient and impact the monsoon. This SAM monsoon linkage can explain the reducing trend of all India area-averaged rainfall during the peak monsoon months.

Significantly the spatial and temporal variability of physical parameters in the Prydz Bay for climate change has been dealt with by Pednekar. This chapter presents the variability of Prydz Bay's physical parameter, which is bounded by the open sea to the north surrounded by clockwise Prydz gyre and a vast polar ice sheet to the south. Instrumental Seals data supported scientific communities to describe the variability in the Prydz Bay south of the polar frontal zone. An attempt has been made to highlight the changes in space and time in Prydz Bay based on previous studies. The major water masses in Prydz Bay are briefly explained and demonstrated using a potential temperature and salinity diagram. The vertical sections of potential temperature shown the occurrence of CDW below 200 m between 65°E and 75°E near to slope of the Prydz Bay along the 66.3°S transect. The presence of mCDW flows onto Prydz Bay occurs further south below 100 m in pockets in the transect of 67°S between 72°E to 78.5°E and 74°E. Time and space analyses have shown the entrance of CDW into the Prydz Bay near the shelf break during summer having warmer and saltier water characteristics as identified. The variability in the extension of mCDW each year in both isopycnal surfaces on a spatial and temporal scale exists. The distribution of $-1.7°C$ isotherm on potential temperature and salinity highlights the annual variability in space and time. The extent of the mCDW signal to the interior of the bay as isotherm $-1.7°C$ extended more southward 68.5°S with small pockets up to 69°S. The continental shelf region of the bay is influenced by the signature of warm mCDW responsible for the climate change in Prydz Bay due to the strong winds blowing from northeast to southwest in the Larsemann Hills region.

This book will serve its purpose to disseminate information on the vital aspects of climate variability over southern high-latitude regions.

Neloy Khare
New Delhi, December 2021

Acknowledgments

It is my great pleasure to express my gratitude and deep appreciation to all contributing authors. Without their valuable inputs on various facets of climate variability over the Antarctic and surrounding Southern Ocean region, the book would not have been possible. Various learned experts who have reviewed different chapters are graciously acknowledged for their timely, constructive, and critical reviews.

I sincerely thank the Ministry of Earth Sciences, Government of India, New Delhi (India) for various inputs, support, and encouragements. Secretary, Ministry of Earth Sciences, and Government of India and Prof. Govardhan Mehta, FRS have always been the source of inspiration and are acknowledged for their kind support. Dr. O.P. Mishra, Director, National Center of Seismology (NCS), New Delhi and Dr. K.J. Ramesh, former Director-General, India Meteorological Department (IMD), New Delhi (India), have always been supportive as true well-wishers.

Prof. Anil Kumar Gupta, Indian Institute of Technology, Kharagpur (India), Prof. Devesh Kumar Sinha, Department of Geology, Delhi University and Dr. Rajiv Nigam former adviser at the National Institute of Oceanography, Goa (India) are deeply acknowledged for providing many valuable suggestions to this book. This book has received significant support from Akshat Khare and Ashmit Khare, who have helped me during the book preparation. Dr. Rajni Khare has unconditionally supported enormously during various stages of this book. Shri Hari Dass Sharma from the Ministry of Earth Sciences, New Delhi (India) has helped immensely in formatting the text and figures of this book and bringing it to its present form. The publishers (Taylor & Francis) have done a commendable job and are sincerely acknowledged.

Neloy Khare
New Delhi, December 2021

Editor

Neloy Khare, PhD, is an adviser/scientist to the government of India at the Ministry of Earth Sciences (MoES). He has a distinct understanding of administration and quality science and research in his areas of expertise, covering a large spectrum of geographically distinct locations such as Antarctic, Arctic, Southern Ocean, Bay of Bengal, Arabian Sea, Indian Ocean, etc. Dr. Khare has almost 30 years of experience in the field of paleoclimate research using paleobiology (paleontology), teaching, science management, administration, coordination for scientific programs (including Indian Polar Program), etc.

He earned a PhD in tropical marine region and a DSc on southern high-latitude marine regions toward environmental and climatic implications. He used various proxies, including foraminifera (micro-fossil), to understand palaeoclimatology of southern high-latitude regions (the Antarctic and the Southern Ocean). These studies, coupled with his palaeoclimatic reconstructions from tropical regions, helped understand causal linkages and teleconnections between the processes in southern high latitudes and that of climate variability occurring in tropical regions. Dr. Khare is an honorary professor and adjunct professor at many Indian universities. He has an impressive list of publications to his credit (125 research articles in national and international scientific journals; 3 special issues of national scientific journals as guest editor; and edited a special issue of *Polar Sciences* as its managing editor). Dr. Khare had authored and edited many books, 130 abstracts have been contributed to various seminars, 23 popular science articles, and 5 technical reports. The government of India and many professional bodies have bestowed him with many prestigious awards for his humble scientific contributions to past climate changes, oceanography, polar science, and southern oceanography. The most coveted award is the 2013 Rajiv Gandhi National Award conferred by the honorable president of India. Others include ISCA Young Scientist Award, BOYSCAST Fellowship, CIES French Fellowship, Krishnan Gold Medal, Best Scientist Award, Eminent Scientist Award, ISCA Platinum Jubilee Lecture, IGU Fellowship, and many more. Dr. Khare has made tremendous efforts to popularize ocean science and polar science across the country by delivering many invited lectures, radio talks, and published popular science articles. He has many authored and edited books on thematic topics and has been published by reputed international publishers, which are testimony to his commitment to popularize science among the masses.

Dr. Khare has sailed the Arctic Ocean as a part of Science PUB in 2008 during the International Polar Year campaign for scientific exploration and became the first Indian to sail the Arctic Ocean.

Contributors

Sanjay Bist
India Meteorological Department
New Delhi, India

Rajib Chattopadhyay
Indian Institute of Tropical Meteorology
Pune, India

K.K. Dani
Indian Institute of Tropical Meteorology
Pune, India

Rupesh M. Das
Environmental Sciences and Biomedical
    Metrology Division
CSIR-National Physical Laboratory
New Delhi, India

P.C.S. Devara
Amity University Haryana
Gurugram, India

S Sunitha Devi
India Meteorological Department
New Delhi, India

Suneet Dwivedi
K Banerjee Centre of Atmospheric and
    Ocean Studies
University of Allahabad
Allahabad, India

Vikas Goel
Environmental Sciences and Biomedical
    Metrology Division
CSIR-National Physical Laboratory
New Delhi, India

A.K. Kamra
Indian Institute of Tropical Meteorology
Pune, India

N. Koushik
Space Physics Laboratory
Vikram Sarabhai Space Centre
Thiruvananthapuram, India

R.H. Kripalani
Indian Institute of Tropical Meteorology
Pune, India

Karanam Kishore Kumar
Space Physics Laboratory
Vikram Sarabhai Space Centre
Thiruvananthapuram, India

Alvarinho J. Luis
Earth System Science Organization
National Centre for Polar and Ocean
    Research
Goa, India

R.S. Maheskumar
Ministry of Earth Sciences
New Delhi, India

Kirtiranjan Mallick
Department of Geology
Utkal University
Odisha, India

Sujata K. Mandke
Indian Institute of Tropical Meteorology
Pune, India

Sumit Kumar Mishra
Academy of Scientific and Innovative
    Research (AcSIR)
Uttar Pradesh, India

Lokesh Kumar Pandey
K Banerjee Centre of Atmospheric and
    Ocean Studies
University of Allahabad
Allahabad, India

G. Pandithurai
Indian Institute of Tropical Meteorology
Pune, India

Vimlesh Pant
Indian Institute of Technology Delhi
New Delhi, India

S.M. Pednekar
National Centre for Polar and Ocean
    Research
Goa, India

Amita Prabhu
Indian Institute of Tropical Meteorology
Pune, India

P.R.C. Rahul
Indian Institute of Tropical Meteorology
Pune, India

A.K. Sahai
Indian Institute of Tropical Meteorology
Pune, India

Devendraa Siingh
Indian Institute of Tropical Meteorology
Pune, India

Jagvir Singh
Ministry of Earth Sciences
New Delhi, India

S.M. Sonbawne
Indian Institute of Tropical Meteorology
Pune, India

V.K. Soni
India Meteorological Department
New Delhi, India

Glossary

Aerosol Optical Thickness: The degree to which aerosols prevent the transmission of light by absorption or scattering of light. The aerosol optical depth (AOD) or optical thickness (τ) is defined as the integrated extinction coefficient over a vertical column of a unit cross-section.

Air masses: It is a body of air extending hundreds or thousands of miles horizontally and sometimes as high as the stratosphere and maintaining as it travels nearly uniform conditions of temperature and humidity at any given level.

Air-sea fluxes: The ocean and atmosphere interact through air-sea oscillations. These fluxes, or exchanges, are the most direct ocean climate indicator of how the ocean influences climate and weather and their extremes and how the atmosphere forces ocean variability. Map of the mean net surface heat flux into the sea.

Amundsen Sea Low (ASL): It is a climatological low-pressure centr located over the extreme southern Pacific Ocean, off the coast of West Antarctica. The ASL plays a significant role in the climate variability of West Antarctica and the adjacent oceanic environment.

Aerosols: It is a suspension of fine solid particles or liquid droplets in air or another gas. Aerosols can be natural or anthropogenic. Diseases can also spread through tiny droplets in the breath, also called aerosols.

Antarctic Coastal Current: It is also known as the East Wind Drift Current. It is the world's southernmost current. This current is the countercurrent of the largest ocean current in the world, the Antarctic Circumpolar Current. On average, it flows westward and parallels the Antarctic coastline.

Antarctic sea ice: It is the sea ice of the Southern Ocean. It extends from the far north in the winter and retreats to almost the coastline every summer, getting closer and closer to the coastline every year due to sea ice melting. Sea ice is frozen seawater that is usually less than a few meters thick.

Antarctic Slope Current (ASC): It is a coherent circulation feature that rings the Antarctic continental shelf and regulates water flow toward the Antarctic coastline.

Asymmetry parameter: It is the first Legendre moment of a phase function. Anisotropy of the phase function is characterized by all moments rather than the first one. In many cases, utterly different habit mixtures return the same or very close to asymmetry parameters.

Atlantic Multidecadal Oscillation (AMO): Also known as Atlantic Multidecadal Variability (AMV), it is the theorized variability of the sea surface temperature (SST) of the North Atlantic Ocean on the timescale of several decades.

Atmospheric science: It is the study of the dynamics and chemistry of the gas layers surrounding the Earth, other planets, and moons. It encompasses the interactions between various parts of the atmosphere and interactions with the oceans and freshwater systems, the biosphere, and human activities.

Auroral Electrojet Index (AE): It is designed to provide a global, quantitative measure of auroral zone magnetic activity produced by enhanced Ionospheric currents flowing below and within the auroral oval. Ideally, it is the total range of deviation at an instant of time from quiet day values of the horizontal magnetic field (h) around the auroral oval.

Black carbon: It consists of pure carbon in several linked forms. It is formed through the incomplete combustion of fossil fuels, biofuel, and biomass and is one of the main types of particle in both anthropogenic and naturally occurring soot. Black carbon causes human morbidity and premature mortality.

Boreal summer intra-seasonal oscillation (BSISO): Asian summer monsoon (ASM) is one of the most prominent sources of short-term climate variability in the global monsoon system.

Circumpolar Deep Water (CDW): It is a designation given to the water mass in the Pacific and Indian Oceans that essentially characterizes a mixing of other water masses in the region. Circumpolar deep water is between $1°C$ and $2°C$ ($34°F$ and $36°F$) and has a salinity between 34.62 and 34.73 practical salinity units (PSU).

Climatology: Climatology or climate science is the scientific study of climate, scientifically defined as weather conditions averaged over a period. This modern field of study is regarded as a branch of the atmospheric sciences and a subfield of physical geography, one of the Earth sciences. Climatology now includes aspects of oceanography and biogeochemistry.

Cloud condensation nuclei or CCNs: These are small particles typically $0.2\ \mu m$, or 1/100 the size of a cloud droplet on which water vapor condenses. Water requires a non-gaseous surface to transition from a vapour to a liquid; this process is called condensation.

Cloud microphysics: The branch of the atmospheric sciences concerned with the many particles that make up a cloud. Relative to the cloud, the individual particles are microscopic and so exist on the 'microscale'; that is, over distances from fractions of a micrometer to several centimeters.

Coronal Mass Ejections (CMEs): These are large expulsions of plasma and magnetic field from the sun's corona. They can eject billions of tons of coronal material and carry an embedded magnetic field (frozen in flux) that is stronger than the background solar wind interplanetary magnetic field (IMF) strength.

Correlation Coefficient (CC): It is used in statistics to measure the correlation between two data sets. The correlation between two financial instruments, simply put, is the degree to which they are related.

Counter-equatorial electrojet (CEJ): Sometimes the flow of the overhead current system reverses its direction temporarily and flows westward, producing depressions in the H-field at equatorial stations in the morning/afternoon hours during magnetically quiet days. It is known as a counter-equatorial electrojet (CEJ).

Coupled Model Inter-comparison Project (CMIP): The CMIP is a standard experimental framework for studying the output of coupled atmosphere-ocean general circulation models. It facilitates assessing the strengths and weaknesses of climate models, which can enhance and focus the development of future models.

Doppler interferometry (IDI): This technique estimates the location of scattering centres and the corresponding radial velocities.

Empirical orthogonal function (EOF): In statistics and signal processing, the method of practical orthogonal function (EOF) analysis is a decomposition of a signal or data set in terms of orthogonal basis functions determined from the data. This term is also interchangeable with the geographically weighted PCAs in geophysics.

ENSO: El Niño and the Southern Oscillation, also known as ENSO, is a periodic fluctuation in sea surface temperature (El Niño) and the air pressure of the overlying atmosphere (Southern Oscillation) across the equatorial Pacific Ocean.

Equatorial electrojet (EEJ): It is a narrow ribbon of current flowing eastward in the daytime tropical region of the Earth's ionosphere.

Extreme precipitation events (EPEs): These are defined as days with precipitation in the top 1% of all days. Increases in the intensity or frequency of heavy rainfall are vital factors that affect the risk of floods and flash floods.

Fractal and multifractal dimension analysis: A multifractal system generalizes a fractal method in which a single exponent (the fractal dimension) is not enough to describe its dynamics; instead, a continuous spectrum of exponents (the so-called singularity spectrum) is needed. Multifractal systems are standard.

Geographic location: This term refers to a position on the Earth. Two coordinates, longitude and latitude, define your absolute geographic location. These two coordinates can be used to give specific areas independent of an outside reference point.

Global conveyor belt: It is a system of ocean currents that transport water around the world. While wind primarily propels surface currents, deep currents are driven by differences in water densities in a process called thermohaline circulation.

Global Positioning System (GPS): It is a U.S.-owned utility that provides users with positioning, navigation, and timing (PNT) services. This system consists of three segments: the space segment, the control segment, and the user segment.

Greenhouse gases (GHGs): Compound gases trap heat or longwave radiation in the atmosphere. Their presence in the atmosphere makes the Earth's surface warmer. The principal GHGs, also known as heat-trapping gases, are carbon dioxide, methane, nitrous oxide, and fluorinated gases.

Horizontal component: It stretches from the start of the vector to its furthest x-coordinate. The vertical part extends from the x-axis to the most vertical point on the vector. Together, the two components and the vector form a right triangle.

Hurst exponent: It is used as a measure of long-term memory of time series. It relates to the autocorrelations of the time series and the rate at which these decrease as the lag between pairs of values increases. Studies involving the Hurst exponent were initially developed in hydrology for the practical matter of determining optimum dam sizing for the Nile river's volatile rain and drought conditions that had been observed over a long period.

Indian Ocean Dipole (IOD): It is defined by the difference in sea surface temperature between two areas (or poles, hence a dipole) – a western pole in the Arabian Sea (western Indian Ocean) and an eastern pole in the east of the Indian Ocean south of Indonesia.

Indian summer monsoon rainfall (ISMR) or southwest monsoon rainfall: During June to September is a component of the Asian monsoon system, accounting for 70–90% of annual precipitation in India. The ISMR exhibits high temporal as well as spatial variations.

Interplanetary magnetic field (IMF): Now more commonly referred to as the heliospheric magnetic field (HMF), is the component of the solar magnetic field that is dragged out from the solar corona by the solar wind flow to fill the solar system.

Intertropical Convergence Zone, or **ITCZ:** It is the region that circles the Earth, near the equator, where the trade winds of the Northern and Southern Hemispheres come together.

Katabatic winds: A katabatic wind is a drainage wind that carries high-density air from a higher elevation down a slope under the force of gravity. Such winds are sometimes also called fall winds; the spelling catabatic winds is also used.

Lidar: Light Detection and Ranging (Lidar) is a remote sensing method used to examine the surface of the Earth. These light pulses combined with other data recorded by the airborne system generate precise, three-dimensional information about the shape of the Earth and its surface characteristics.

Mean sea-level pressure (MSLP): This is the atmospheric pressure at mean sea level (PMSL). The atmospheric pressure is normally given in weather reports on radio, television, newspapers, or the Internet. Average sea-level pressure is 1013.25 mbar (101.325 kPa; 29.921 in Hg; 760.00 mm Hg).

Microwave radiometer (MWR): This provides time-series measurements of column-integrated amounts of water vapor and liquid water. The instrument itself is a sensitive microwave receiver that detects the microwave emissions of the smoke and liquid water molecules in the atmosphere at two frequencies: 23.8 and 31.4 GHz.

Mid-latitude: The middle latitudes (also called the mid-latitudes, sometimes midlatitudes, or moderate latitudes) are a spatial region on Earth located between the latitudes $23°26'22''$ and $66°33'39''$ north, and $23°26'22''$ and $66°33'39''$ south.

Midnight sun: It is a natural phenomenon that occurs in the summer months in places north of the Arctic Circle or south of the Antarctic Circle when the sun remains visible at the local midnight. When the midnight sun is seen in the Arctic, the Sun appears to move from left to right, but the equivalent apparent motion in Antarctica is from right to left. It occurs at latitudes from $65°44'$ to $90°$ north or south and does not stop precisely at the Arctic Circle or the Antarctic Circle due to refraction.

Monsoon: It is a seasonal change in the direction of a region's prevailing, or strongest, winds. Monsoons are most often associated with the Indian Ocean. Monsoons always blow from cold to warm areas. The summer monsoon and the winter monsoon determine the climate for most of India and Southeast Asia.

Multi-model means (MMM): It is a simple way to reduce biases in individual model outputs, and thus, it is widely used for climate change projections. The usefulness of MMM may vary from one region to the other based on the regional climate and the diagnostic variables of interest.

Northern Annular Mode (NAM): It also known as Arctic Oscillation (AO) or Northern Hemisphere Annular Mode, and is a natural form of climate variability, closely associated with the North Atlantic Oscillation (NAO), which has a similar structure over the Atlantic but when looked at from above, the shape is more annular.

Orographic terrain: Orography is the study of the topographic relief of mountains and can more broadly include hills and any part of a region's elevated terrain. Orography falls within the broader discipline of geomorphology.

Pacific decadal oscillation (PDO): This is a robust, recurring pattern of ocean-atmosphere climate variability cantered over the mid-latitude Pacific basin. The PDO is detected as warm or cool surface waters in the Pacific Ocean, north of 20°N.

Potential vorticity (PV): In fluid mechanics, potential vorticity (PV) is a quantity which is proportional to the dot product of vorticity and stratification. This quantity, following a parcel of air or water, can only be changed by diabatic or frictional processes. It is a useful concept for understanding the generation of vorticity in cyclogenesis (the birth and development of a cyclone), especially along the polar front, and in analysing flow in the ocean.

Precipitable water content (PWC): The depth of this condensed water (in millimetre) is the measure of how much water vapor is available for conversion to precipitation. This study aims to examine the potential of these PWC data to monitor precipitation systems during Indian summer monsoon.

Predictive Index® (PI®): It is a theory-based, self-report measurement of normal, adult, work-related personality that was developed and validated for use within occupational and organizational populations.

Root Mean Square Error (RMSE): It is the standard deviation of the residuals (prediction errors). Residuals are a measure of how far from the regression line data points are; RMSE is a measure of how spread out these residuals are. In other words, it tells you how concentrated the data is around the line of best fit. Root mean square error is commonly used in climatology, forecasting, and regression analysis to verify experimental results.

Sea-ice extent (SIE): Sea-ice extent is a measurement of the area of ocean where there is at least some sea ice. Usually, scientists define a threshold of minimum concentration to mark the ice edge; the most common cut off is at 15%.

Sea-ice concentration (SIC): It is a useful variable for climate scientists and nautical navigators. It is defined as the area of sea ice relative to the total at a given point in the ocean.

Sea level pressure: The atmospheric pressure at mean sea level, either directly measured or, most commonly, empirically determined from the observed station pressure.

Sea surface temperature: It is a key climate and weather measurement obtained by satellite microwave radiometers, infrared (IR) radiometers, in-situ moored and drifting buoys, and ships of opportunity. SST maps are also widely used by oceanographers, meteorologists, and climate scientists for scientific research.

Semi-annual oscillation (SAO): It is a twice-yearly northward movement (in May–June–July (MJJ) and November–December–January (NDJ)) of the circumpolar trough of sea level pressure (SLP) in the Southern Hemisphere with effects throughout the troposphere.

Single Scattering Albedo (SSA): The ratio of scattering efficiency to total extinction efficiency, is an essential parameter used to estimate the Direct Radiative Forcing (DRF) of aerosols. However, SSA is one of the large contributors to the uncertainty of DRF estimations.

Solar Proton Events (SPEs): A solar particle event or solar proton event (SPE), or prompt proton event, occurs when particles (mostly protons) emitted by the sun become accelerated either close to the sun during a flare or in interplanetary space by coronal mass ejection shocks. The events can include other nuclei such as helium ions and HZE ions. These particles cause multiple effects. They can penetrate the Earth's magnetic field and cause ionisation in the ionosphere. The effect is like auroral events, except that protons rather than electrons are involved. Energetic protons are a significant radiation hazard to spacecraft and astronauts.

Solar wind velocity: The solar wind is a stream of charged particles released from the upper atmosphere of the Sun, called the corona. At more than a few solar radii from the Sun, the solar wind reaches speeds of 250–750 km/s and is supersonic, meaning it moves faster than the speed of the fast magnetoionic wave.

Southern Annular Mode, or **SAM:** It is a climate driver that can influence rainfall and temperature in Australia. The SAM refers to the (non-seasonal) north-south movement of the strong westerly winds that blow almost continuously in the mid- to high latitudes of the Southern Hemisphere.

Southern Oscillation Index (SOI): It represents the difference in average air pressure measured at Tahiti and Darwin, Australia. More specifically, the SOI is calculated as the difference in monthly averages of standardised mean sea level pressure at each station.

Spectral refractive index: The concept of refractive index applies within the full electromagnetic spectrum, from X-rays to radio waves. It can also be applied to wave phenomena such as sound. In this case, the speed of sound is used instead of that of light, and a reference medium other than vacuum must be chosen.

TCO (Total Column Ozone): It is a measurement of the total amount of atmospheric ozone in a given column. Usually, TCO is measured in Dobson Units (DU).

Thermo-dynamical factors: The thermodynamic factor in diffusion is a parameter relating the tracer diffusion coefficient of a species in a given system to the corresponding intrinsic diffusion coefficient and may be defined for alloys as well as for oxides both pure and mixed.

Thermohaline circulation: It plays an important role in supplying heat to the polar regions. Therefore, it influences the rate of sea ice formation near the poles, which in turn affects other aspects of the climate system (such as the albedo, and thus solar heating, at high latitudes).

Traveling Atmospheric Disturbances or **TAD:** These are gusts of wind that roll through the sky, pushing along neutral atoms as they go.

Upper tropospheric atmosphere: The tropopause is the boundary in the Earth's atmosphere between the troposphere and the stratosphere. It is a thermodynamic gradient stratification layer, marking the end of the troposphere. It lies, on average, at 17 km above equatorial regions, and about 9 km over the polar regions.

Wiener process: A standard (one-dimensional) (also called Brownian motion) is a stochastic process $\{W_t\}_{t \geq 0+}$ indexed by nonnegative real numbers t with the following properties: In general, a stochastic process with stationary, independent increments is called a Lévy process.

1 Antarctic Ozone
Trends and Variability in a Changing Climate

V.K. Soni and Sanjay Bist
India Meteorological Department, Ministry of Earth
Sciences, New Delhi, India

Jagvir Singh
Ministry of Earth Sciences, New Delhi, India

CONTENTS

1.1   INTRODUCTION

The ozone-climate inter-relations and feedbacks are emerging as significant components that contribute to the understanding of Earth system science. The atmospheric ozone plays a fundamental role in both the terrestrial ecosystem and global climate. Although the study of the stratospheric ozone started in the early 20th century, observations strengthened in the 1980s after discovering the "ozone hole" over Antarctica. The stratospheric ozone layer extends between 10 and 40 km altitude, peaking at about 25 km and containing more than 90% of all atmospheric ozone. Atmospheric ozone has two significant effects on the temperature balance of the Earth. The ozone layer absorbs the most damaging part of UV radiation, which heats the stratosphere and plays a vital role in preventing UV radiation from reaching Earth's surface.

On the other hand, the ozone in the troposphere absorbs infrared radiation emitted by the Earth's surface, effectively trapping heat in the troposphere. Consequently, the climate impact of changes in ozone concentrations varies with

DOI: 10.1201/9781003203742-1

the changes in the vertical distribution of ozone in the atmosphere. Studies have shown that the tropospheric ozone has increased globally throughout the 20th century due primarily to increased human-related emissions (Ziemke et al., 2019). Stratospheric ozone depletion has also affected the climate in the Southern Hemisphere. There exist several two-way complex interactions between stratospheric ozone depletion and climate change, such as changes in stratospheric temperature and transport affect the concentration and distribution of stratospheric ozone; changes in tropospheric climate affect stratospheric circulation; changes in stratospheric ozone influence the radiative forcing of the atmosphere, and hence surface climate, as well as the chemistry of the troposphere.

Depleting stratospheric ozone is observed in the so-called Antarctic ozone hole that appears each austral spring (September–November) over Antarctica. It occasionally develops over the Arctic (Manney et al., 2011). As a result of increased ozone-depleting substances (ODSs) in the atmosphere after industrialization, stratospheric ozone started to decline in the polar regions by the late 1970s (Bodeker et al., 2005; Farman et al., 1985; WMO, 2018). The ODSs and typical seasonal atmospheric dynamics in Antarctica led to the significant ozone loss, peaks in September. The area where total column ozone drops below 220 Dobson Unit (DU) is the ozone Hole. The TCO value 220 DU is used to define ozone hole because it is lower than pre-1980 observed TCO values and also because it is almost always a middle value in a solid spatial gradient of total ozone (Newman et al., 2004). Ozone depletion is caused mainly by anthropogenic emissions of ODSs and the subsequent release of reactive halogen gases (chlorine, bromine, etc.) in the stratosphere.

The Montreal Protocol is considered one of the most influential environmental treaties. Due to regulations of the ODSs and some hydrofluorocarbons (HFCs) under the Montreal Protocol and its amendments, not only the significant increases in UV radiation at Earth's surface have been avoided, but global warming also reduced. Most ODSs are potent GHGs with global warming potentials considerably more compared to carbon dioxide and methane. Direct evidence of the strengthening of the ozone layer due to the decline of halogen species in the Antarctic stratosphere is now available from space-borne observations. Recent scientific studies show that ozone depletion has stopped or reduced significantly, but it may take years before the ozone starts to increase again as ODSs have a long lifetime. It has been estimated that the Southern Hemisphere mid-latitude ozone is expected to return to the 1980 level around mid-century (WMO, 2018). A large variability has been observed in the ozone hole extent and minimum ozone over Antarctica. The average ozone hole area in 2019 was the smallest on record since the discovery of the ozone hole in 1985, whereas, in 2020, it was one of the largest. The Antarctic ozone hole is decreasing in size. Still, the recovery can be influenced by interannual atmospheric variability and any unanticipated changes in ODSs source emissions, such as an observed unexpected increase in total global emissions of CFC-11 after 2012 despite Montreal Protocol controls. However, a substantial decline was observed in CFC-11 concentrations during 2019 and 2020.

Accurate and uninterrupted observations of atmospheric ozone on the global scale are an essential task. The atmospheric ozone is measured using in-situ and remote sensing methods. The in-situ measurement methods require direct atmospheric air sampling into the instrument and analyzing its properties and relative amount. The in-situ measurement methods of ozone involve the use of mass or optical spectroscopy or chemical reactions with the air sample. These measurement methods can be employed either from ground, aircraft, and balloon platforms. The balloon-borne ozonesonde in-situ way is most commonly used to measure the vertical distribution of ozone from the bottom up to approximately 40 km altitude. Ozonesonde interfaced with meteorological radiosonde has electrochemical concentration cell (ECC) that senses ozone as it reacts with a dilute potassium iodide solution to produce a weak electrical current proportional to the ozone concentration the sampled air.

In the remote sensing measurement method, the amount of ozone is derived indirectly by the changes in atmospheric radiation resulting from the ozone's presence. Remote sensing instruments measure spectral radiances, from which information about light absorbers such as ozone and other trace gases can be inferred using retrieval algorithms. The Dobson and Brewer spectrophotometers are used in the World Meteorological Organization (WMO) network to measure the total column ozone (TCO). The basic measurement principle of both types of spectrophotometers is the same. The TCO is estimated by comparing the intensity of solar radiation at ultraviolet wavelengths strongly and weakly absorbed by ozone. The Dobson spectrophotometer is a manually operated instrument that utilizes a variable 'optical wedge' attenuator to measure the intensity ratio of two wavelengths. The Brewer spectrophotometer is automatic mainly, which operates unattended according to programmed schedules and directly measures the intensity of solar radiation at several different ultraviolet wavelengths. The WMO's World ozone and Ultraviolet Radiation Data Centre (WOUDC), operated by Environment and Climate Change Canada, can access more details on the global ozone monitoring network and data availability. The Dobson and Brewer spectrophotometers provide reliable TCO data at various ozone monitoring stations in Antarctica, but the ground-based measurement has limited spatial coverage. The satellite-retrieved atmospheric ozone data allows extensive spatial range but with limited temporal resolution. There are several remote sensing instruments onboard satellites for measurement of TCO and vertical profile of ozone. After the pioneer Backscatter, Ultraviolet Spectrometer (BUV) onboard Nimbus-4 operated from 1970 to 1977, a range of satellite instruments have been launched to monitor TCO and vertical profile of ozone-related species. Some necessary devices are Total Ozone Mapping Spectrometer (TOMS) aboard the Nimbus-7 and Meteor-3 satellites, Ozone Monitoring Instrument (OMI) onboard the Aura satellite, TROPOspheric Monitoring Instrument (TROPOMI) onboard Sentinel-5 Precursor (S-5P) satellite, Global Ozone Monitoring Experiment-2 (GOME-2) the European MetOp satellites, Ozone Mapping and Profiler Suite (OMPS) onboard Suomi NPP (National Polar-Orbiting Partnership) satellite, Atmospheric Infrared Sounder (AIRS) onboard the Aqua satellite, and Microwave Limb Sounder (MLS) onboard NASA's Aura Earth satellite. The satellite sounder instruments also provide the atmospheric profiles of ozone. Satellites are short-lived and, therefore, the ozone data sets from different mission observations are merged to have a consistent and homogeneous global long-term data record to study the trends analysis. The satellite data used in this chapter are combined data for 1979–1992 from the TOMS instrument on the NASA/NOAA Nimbus-7 satellite, for 1993–1994 are from the TOMS instrument on the Soviet-built Meteor-3 satellite, for 1996–October 2004 are from NASA Earth Probe TOMS satellite, and for November 2004 through June 2016 from the OMI instrument (KNMI/NASA) onboard the Aura satellite and data starting July 2016 are from the OMPS instrument onboard the Suomi NPP satellite. The data are acknowledged to NASA Ozone Watch (https://ozonewatch.gsfc.nasa.gov, accessed in May 2021).

1.2 INDIAN EFFORTS IN OZONE MONITORING IN ANTARCTICA

The XIX session of the Scientific Committee on Antarctic Research (SCAR) in June 1986 and WMO Executive Committee Working Group on Antarctic Meteorology in September 1986 recommended a joint international effort for an atmospheric ozone monitoring program to understand the physical, chemical, and dynamical processes that cause the ozone depletion over polar regions. Several countries, including India, have since extended their active support to the ozone observation program.

The systematic ozone monitoring in India was started by the India Meteorological Department (IMD) during International Geophysical Year 1957–1958. Considering the importance of atmospheric ozone to weather and climate, the IMD started ozone monitoring over Antarctica during the second scientific expedition in 1982–1983, during which ozonesonde ascents were taken to obtain the vertical profile of ozone at the temporary Indian station, Dakshin Gangotri (69°S, 12°E) (Sreedharan et al., 1986) using IMD to make electrochemical ozonesonde (Sreedharan, 1968). IMD further strengthened the ozone observations at "Dakshin Gangotri" to join the international efforts for the Antarctica Ozone-Hole Investigation (Faruqui et al., 1988). Regular observations of ozone started at Dakshin Gangotri in 1987 (Koppar and Nagrath, 1991; Sreedharan et al., 1989). The ozone measurements from the Georg-Forster Station of Germany and the Novolazarevskaya station of Russia located close to the Maitri station of India gained international recognition. Ozonesonde profiles (after correction for total ozone) crossed the 20 hPa level at Dakshin Gangotri, and Maitri from 1987 to 1997 have been analyzed, and the ozone hole structure studied in detail Tiwari (1999). A significantly large ozone hole with low total ozone was observed during 1994, 1995, and 1996. The ozone hole was prolonged through November and December in 1995 due to the cold-core vortex's unusually cold stratospheric temperature persistence. Regular ozone profile measurement continued at the Dakshin Gangotri until it was abandoned in 1989. The surface and profile ozone observations started at the second station at Maitri in 1990. Fortnight ozone soundings were taken at Maitri throughout the year. Still, during the ozone hole period September–November, more frequent observations are accepted to keep a close watch on the depletion of atmospheric ozone. Several scientists have carried out the detailed study of year-to-year observations on ozonesonde data of Maitri (Peshin et al., 1996, 1997; Sreedharan et al., 1993).

The Brewer Spectrophotometer (Mark IV, No. 153) was operated at Maitri from 1999 to 2011 to measure total column ozone. Simultaneously, this instrument also provides the measurement of the total column density of NO_2 and SO_2 and the

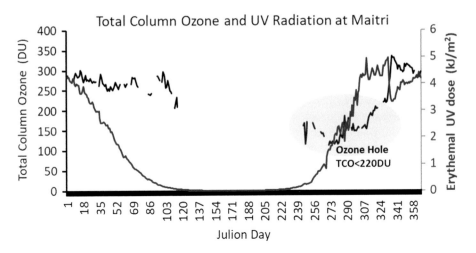

FIGURE 1.1   Total column ozone (in Dobson Unit) variation at the Indian station Maitri, Antarctica measured using the Brewer Spectrophotometer. The data gap is due to the non-availability of solar radiation due to polar nights (the sun does not rise above the horizon). The daily erythemal UV dose (kJ/m^2) is shown in blue.

maximum value of UV-B flux. The daily TCO measured at Maitri during 2000 using the Brewer Spectrophotometer is presented in Figure 1.1. Peshin (2011) found an increase in total column SO_2 during the ozone-hole event due to the low penetration of UV-B flux in the troposphere under a stratospheric ozone-depleted condition. This UV-B radiation dissociates COS in the upper troposphere and increases the production rate of SO_2. The column density of NO_2 also showed an increase after the onset of spring but was not identical with that of the SO_2 column. This increase in total column NO_2 is due to the decrease of night duration as the austral summer approaches.

Surface ozone is also measured at Maitri using the electrochemical conductivity cell method. The surface ozone reveals a minor diurnal variation of approximately 5 ppbv since the sun appears for all 24 hours during October to February, and solar radiation does not vary much from morning to night hours, resulting in minimal variation in ozone concentration. The daily variations of surface ozone have been observed to be of lesser magnitude during polar nights compared to polar days. Ali et al. (2017) reported a daily average surface ozone concentration at Larsemann Hills to vary between ~13 and ~20 ppb with an overall average value of ~16 ppb. At Maitri, it went between ~16 and ~21 ppb, with an overall average value of ~18 ppb. Surface ozone is a secondary pollutant and produced through the oxidation of precursors gases such as CO or hydrocarbons in the presence of sufficient concentrations of NO_x. As anthropogenic pollution is almost negligible at Maitri, the in-situ photochemical production of ozone may not be very significant. Depletion in the stratospheric ozone during the ozone hole period gives way to highly energetic UV radiation to reach the surface layer and initiate photolysis of oxygen and NO_x molecules in the surface boundary layer, leading to surface ozone production. The NO_x is produced from the surface snowpack. The NO_x production is directly driven by incident radiation and photolysis of nitrate deposited in the snow. Moreover, the surface ozone concentrations can also be increased by the downward transport of stratospheric ozone rich air during deep convection and stratosphere-to-troposphere exchange events. Episodes of high surface ozone in the Antarctica region associated with stratospheric intrusion have been reported at Maitri (Ganguly, 2013).

More recently, the ozone observations have been started by IMD at another Indian station, Bharati, since 2015. Bharati is close to the coast and falls in and out of the ozone hole over Antarctica due to the atmospheric dynamics of the polar vortex, resulting in extreme variability. Ozonesonde data from this station validated total ozone columns (TCOs) and vertical profiles from satellite-retrieved data. Hulswar et al. (2020, 2021) validated OMI, and MLS rescued TCO and vertical profile of ozone by comparing with the in-situ ozonesonde observations at Bharati. They found that MLS reproduced the vertical profile variation and peak heights of the ozone layer but overestimated the concentrations compared to the ozonesondes.

The ozone data collected at Indian Antarctic stations can be accessed from WMO WOUDC (https://woudc.org) and obtained from the India Meteorological Department climate data portal. The WOUDC is one of six World Data Centers that are part of the WMO's Global Atmosphere Watch program. The WOUDC is managed by the Meteorological Service of Canada, a division of Environment and Climate Change Canada.

The Montreal Protocol under the Vienna Convention for the Protection of Ozone Layer encourages global monitoring and reporting on ozone depletion. India has committed to long-term monitoring of all the components of atmospheric ozone through in-situ and remote sensing methods under the aegis of WMO. India is one of the few countries to launch satellite instruments on the INSAT-3D series geostationary platform to monitor total column ozone and vertical profile. Moreover, India has adopted and encouraged environmentally friendly and energy-efficient technologies while proactively and successfully implementing the ODSs phase-out plan.

1.3 ANTARCTIC OZONE TRENDS, VARIABILITY, AND SIGNS OF RECOVERY

The ozone depletion is showing signs of recovery. The Scientific Assessment of Ozone Depletion: 2006 (WMO, 2018) provides a formal definition of the three critical stages of ozone recovery, starting with (i) slowing of ozone decline, identified as the occurrence of a statistically significant reduction in the rate of decline in ozone due to changing equivalent effective stratospheric chlorine (EESC); (ii) beginning of ozone increases, identified as the occurrence of statistically significant increases in ozone above previous minimum values due to declining EESC; (iii) the full recovery of ozone from ODSs, identified as when ozone is no longer affected by ODSs. In the absence of changes in the sensitivity of ozone to ODSs, this is likely to occur when EESC returns to pre-1980 levels. The Montreal Protocol, its amendments and adjustments, is considered the most successful environmental agreement as the earlier decline in total ozone was successfully stopped (Pawson et al., 2008). The ODSs have a relatively long atmospheric lifetime, and their levels are expected to decline gradually. The concentrations of ODSs stopped rising in the polar stratosphere around the late 1990s due to the control under the Montreal Protocol and are slowly declining. Based on observational evidence, several studies have identified the onset of stratospheric ozone recovery over Antarctica (Kuttippurath et al., 2013; Kuttippurath and Nair, 2017; Salby et al., 2011, 2012). However, it is still unclear to what extent the Antarctic ozone hole has recovered due to significant natural interannual variability of stratospheric ozone related to dynamical perturbations of the Antarctic ozone hole, which reduces trend significance (De Laat et al., 2015; Newman et al., 2006). Solomon et al. (2016), using model calculations and comparison against observation, reported the recovery of the Antarctic ozone layer since 2000 attributed to dynamic and temperature changes along with chemistry.

Three metrics commonly used to define the Antarctic ozone hole are (i) the area of the ozone hole (adding the areas of cells falling below TCO values of 220 DU), (ii) the minimum TCO value within the ozone hole, and (iii) the Antarctic ozone mass deficit (Huck et al., 2007; Uchino et al., 1999). Interannual variability in Antarctic stratospheric dynamics influences the Antarctic stratospheric temperatures that drive significant year-to-year variation in the extent of Antarctic ozone depletion (Newman et al., 2004). Tully et al. (2019) studied linear trends in four standard metrics to describe the severity of the Antarctic ozone hole over 1979–2001 and 2001–2017, both with and without adjusting to account for inter-annual meteorological variability. They found that each metric showed a trend towards reduced ozone depletion for the period 2001–2017. They further observed that the magnitude of movements is not very much dependent on adjustment for meteorological variability but considerably reduce the scatter and, hence, the uncertainty of the trends.

1.3.1 Trend in Ozone Hole Area

The time series of ozone hole area and various other metrics are produced from merged satellite data TOMS, OMI, and OMPS beginning from 1979. The minimum ozone is obtained from total column ozone satellite measurements south of 40°S. The ozone minimum during 1979 was 212 DU and remained above 190 DU until 1981, but then the absolute column ozone minimums swiftly grew deeper: 173 DU in 1982, 154 in 1983, and 124 in 1985. The bottomless ozone hole occurred on September 30, 1994, when total column ozone minimums fell to just 73 DU. The values of absolute column ozone minimum and the most enormous ozone hole area have never been observed during the same years. The most significant ozone hole occurred in 2006, while the lowest ozone minimum was in 1994. Still, the long-term trend in both characteristics is consistent from 1979 through the late 1990s; the ozone hole rapidly grew in area and depth and has declined since 2000 (Figure 1.2). In the early 2000s, the ozone hole area roughly stabilized for annual averages. However, wide inter-annual variations in ozone hole area and depth have been observed, attributed to variations in stratospheric temperature and circulation – colder conditions in the stratosphere result in a larger ozone hole area and lower ozone. The stratospheric polar vortex plays an essential role in the stratospheric circulations and stratosphere-troposphere dynamical coupling during winter and spring in the Southern Hemisphere. Since its discovery, the smallest ozone hole area of 2019 was

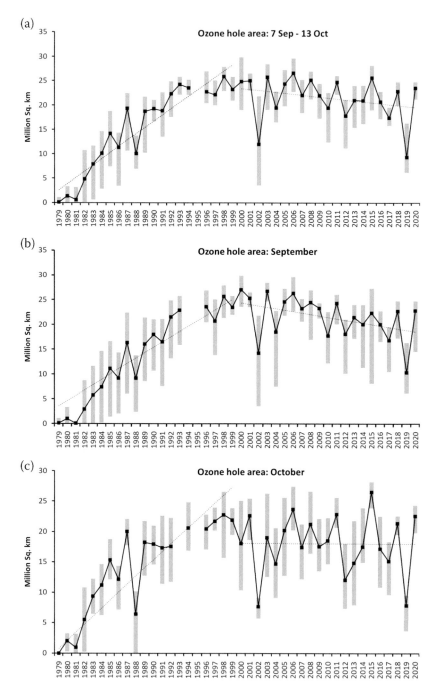

FIGURE 1.2   The monthly averaged ozone hole (defined as the region where the total ozone amount is less than 220 DU) from TOMS/OMI/OMPS satellite observations. Trends in the comments (dashed black line) for the periods from 1979 to 1999 and 2000 to 2020 are indicated. The vertical lines depict the maximum and minimum of the ozone hole area.

driven by abnormally warm stratospheric temperatures in September over Antarctica, leading to a weak polar vortex and minimizing the formation and persistence of the polar stratospheric clouds are mainly responsible for ozone destruction processes. On the other hand, the longest-lasting and one of the deepest and largest ozone hole areas were witnessed in 2020 due to a solid and stable polar vortex, which kept the temperature of the ozone layer over Antarctica consistently cold.

Updated trends derived from different merged satellite retrieved unadjusted total column ozone data sets for the period 1979–2020 have been presented in Figure 1.2. The results from the last ozone assessment have confirmed that there are evolving signs of reduction of Antarctic ozone holes in size and depth since 2000, with the most substantial changes occurring during early spring (WMO, 2018). The 2018 WMO/UNEP assessment (WMO, 2018) reported that the upper stratospheric ozone has increased by 1–3% decade^{-1} during 2000–2016. However, there is some evidence for a decrease in lower stratospheric ozone during the same period. Based on the four-step adaptive ozone trend estimation scheme to isolate the long-term zonal ozone variability related to anthropogenic forcing (Bai et al., 2017), the turning point of the ozone change was determined to occur in 2000 is later than the maximum of stratospheric ODSs. Figure 1.2 shows a significant increasing trend in the ozone hole area for September and October during 1979–1999. The ozone hole area was averaged for September 7–October 13 and September shows a significant declining trend during 2000–2020. The declining trend during October for the same period is not adequate.

1.3.2　PSC-Limited Years and PSC NAT Volume Trends

Polar stratospheric clouds (PSCs) play an essential role in the destruction of stratospheric ozone in two ways: first, they provide a surface upon which heterogeneous chemical reactions take place, which converts stable forms of halogen (leading chlorine) into reactive and ozone-destroying format (for example ClO); second, PSCs also remove nitrogen compounds that would otherwise combine with ClO to form less reactive forms of chlorine. PSCs form poleward of about 60°S latitude in the polar stratosphere at an altitude range of 10–25 km during the winter and early spring when shallow temperatures (below −78°C) favor cloud formation despite the arid conditions. Depending on their formation temperature, particle size, and chemical composition, PSCs are classified into Types I and II. Type-II PSCs, also known as nacreous or mother-of-pearl clouds, consist of ice crystals only and form when temperatures are below the ice frost point (below −83°C). Type-I PSCs contain mainly hydrated nitric acid and/or sulphuric acid droplets; they form when temperatures drop to −78°C or below and are optically much thinner. A Type I is further classified into three categories: Type-Ia (consist of large, aspherical particles, and nitric acid trihydrate (NAT), Type-Ib (consist of small, non-depolarizing spherical particles and a liquid supercooled ternary solution (STS) of sulphuric acid, nitric acid, and water), and Type-Ic (consist of metastable water-rich nitric acid in a solid phase). The classification of PSCs is still recognized but has become outdated and is no longer recommended since these "types" are known to be NAT, STS, and ice.

The PSC volume is considered an excellent proxy for Antarctic vortex dynamics and the strength of dynamical disturbances. Figure 1.3 shows the 1979–2020

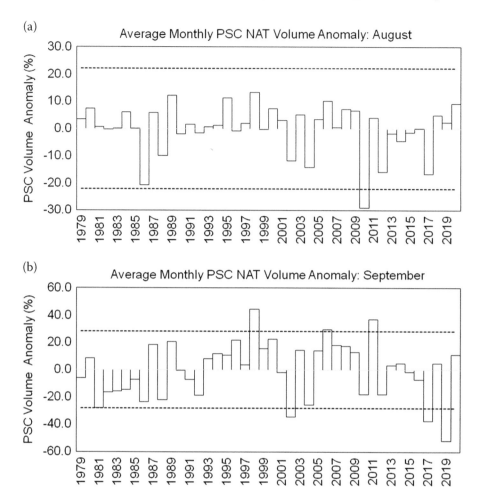

FIGURE 1.3   Annual Antarctic stratospheric potential nitric acid trihydrate (NAT) volume based on monthly mean PSC NAT volume data for 1979–2020. The data are expressed in the relative anomaly compared to the long-term mean (in percentage) for the months of (a) August and (b) September. The PSC NAT volume is calculated from NASA MERRA-2 stratospheric temperatures falling below the threshold temperature for NAT formation. The horizontal dashed lines denote the standard deviation of the NAT volume.

monthly PSC volume anomaly for August and September. A small group of years, 1986, 1988, 2002, 2004, 2010, 2012, 2017, and 2019, stand out due to significantly reduced PSC volume in August and/or September (10–20 million km^3) at or outside the standard deviation of the PSC volume. Several researchers have analyzed the observed ozone depletion and meteorological parameters and concluded that these years indeed were dynamically perturbed (De Laat and van Weele, 2011; Hoppel et al., 2005; Kanzawa and Kawaguchi, 1990; Klekociuk et al., 2014; Schoeberl et al., 1989; Shepherd et al., 2005).

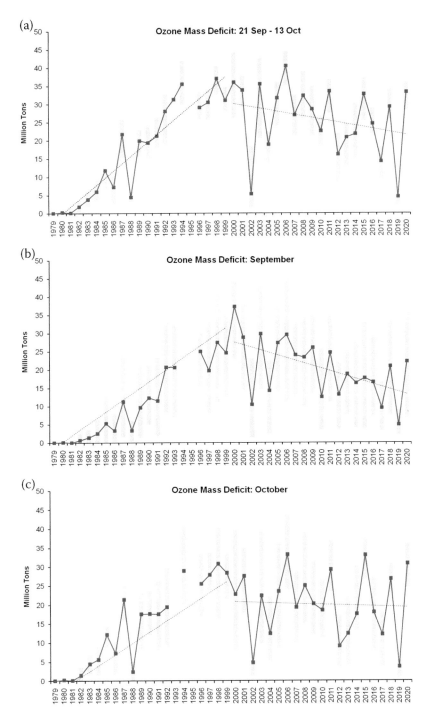

FIGURE 1.4 Antarctic ozone mass deficit within the Antarctic ozone hole expressed in millions of tons for the periods (a) September 21–October 13, (b) September, and (c) October.

1.3.3 Trend in Ozone Mass Deficit

The Ozone Mass Deficit (OMD) index determined from total ozone satellite measurements is most commonly used to quantify Antarctic ozone depletion. The OMD is defined as the difference between the actual TCO, and a reference TCO of 220 DU ($1DU = 2.69 \times 10^{16}$ molecules/cm^2) with the OMD expressed in units of ozone mass (WMO, 2018). In simple words, the OMD quantifies the abundance of ozone that would need to be added to the atmosphere to return TCO values over Antarctica to above 220 DU. This metric is more advantageous as it is independent of the vortex position.

Figure 1.4 shows the long-term record of the average OMD concerning the 220 DU threshold derived from merged satellite (TOMS/OMI/OMPS) TCO data. The long-term data reveal a significant increase in OMD in the 1980s and early 1990s and more minor, reverse changes after 2000. It is observed that the ozone mass deficit for September decreased by 0.785 ± 0.24 Mt yr^{-1} (ordinary linear regression) between 2000 and 2020. Bodeker and Kremser (2021) reported that after 2000 factors other than the decline in halogen loading of the Antarctic stratosphere are also driving the return of the Antarctic ozone layer to pre-1980 levels. It is observed that abnormal mean ozone mass deficits in 1988, 2002, and 2019 result from unusual meteorological conditions in the polar stratosphere. Unusually warmer Antarctic stratospheric temperatures restricted the heterogeneous chemical processes driving polar ozone destruction during 1988, 2002, and 2019.

1.4 SUMMARY

Under the Montreal Protocol control, the stratospheric concentrations of ozone-depleting substances have declined since their peak in the 1990s. Global ozone responds to the declining trend in ODSs and the substantial ozone depletion observed since the 1970s stopped in the late 1990s. Since then, stratospheric ozone levels have not declined further. Now general ozone increases, and a slow recovery of the ozone layer is anticipated. The apparent signs of increasing ozone are seen in the upper stratosphere and for total ozone columns above Antarctica in the spring. Sustained atmospheric ozone observations and monitoring are required to combine the evidence of ozone recovery and further advance our understanding of the complex ozone–climate interaction. Long-term monitoring activities are invaluable for assessing ozone, UV, and ODSs variations and their link with climate change.

REFERENCES

Ali, K., Trivedi, D. K., and Sahu, S. K. (2017) Surface ozone characterization at Larsemann Hills and Maitri, Antarctica. *Sci. Total Environ.*, 584–585, 1130–1137. https://doi.org/10.1016/j.scitotenv.2017.01.173.

Bai, K., Chang, N. B., Shi, R., Yu, H., and Gao, W. (2017) An intercomparison of multi-decadal observational and reanalysis datasets for global total ozone trends and variability analysis. *J. Geophys. Res. Atmos.*, 122, 7119–7139. https://doi.org/10.1002/2016JD025835.

Bodeker, G. E., and Kremser, S. (2021) Indicators of Antarctic ozone depletion: 1979 to 2019. *Atmos. Chem. Phys.*, 21, 5289–5300. https://doi.org/10.5194/acp-21-5289-2021.

Bodeker, G. E., Shiona, H., and Eskes, H. (2005) Indicators of Antarctic ozone depletion. *Atmos. Chem. Phys.*, 5, 2603–2615.

De Laat, A. T. J., van der A. R. J., and van Weele, M. (2015) Tracing the second stage of ozone recovery in the Antarctic ozone-hole with a "big data" approach to multivariate regressions. *Atmos. Chem. Phys.*, 15, 79–97. https://doi.org/10.5194/acp-15-79-2015.

De Laat, A. T. J., and van Weele, M. (2011) The 2010 Antarctic ozone hole: Observed reduction in ozone destruction by minor sudden stratospheric warmings. *Sci. Rep.*, 1, 38. https://doi.org/10.1038/srep00038,2011.

Farman, J. C., Gardiner, B. G., and Shanklin, J. D. (1985) Large losses of Total ozone in Antarctica reveal seasonal ClO x/NOx interaction. *Nature*, 315, 207–210.

Faruqui, A. A., Koppar, A. L., and Nagrath, S. C. (1988) Antarctic ozone-hole investigation programme-Preliminary result of ozonesonde ascents during January- February 1987. *Mausam*, 39(3), 313–316.

Ganguly, N. D. (2013) High surface ozone episodes at Maitri in Antarctica. *Indian J. Phys.*, 87(10), 947–951.

Hoppel, K., Nedoluha, G., Fromm, M., et al. (2005) Reduced ozone loss at the upper edge of the Antarctic ozone hole during 2001–2004. *Geophys. Res. Lett.*, 32, L20816. https://doi.org/10.1029/2005GL023968.

Huck, P. E., Tilmes, S., Bodeker, G. E., Randel, W. J., McDonald, A. J., and Nakajima, H. (2007) An improved measure of ozone depletion in the Antarctic Stratosphere. *J. Geophys. Res.*, 112, D11104. https://doi.org/10.1029/2006JD007860.

Hulswar, S., Mohite, P., Soni, V. K., and Mahajan, A. S. (2021) Differences between in-situ ozonesonde observations and satellite retrieved vertical ozone profiles across Antarctica. *Polar Sci.*, 100688. https://doi.org/10.1016/j.polar.2021.100688

Hulswar, S., Soni, V. K., Sapate, J. P., More, R. S., and Mahajan, A. S. (2020) Validation of satellite retrieved ozone profiles using in-situ ozonesonde observations over the Indian Antarctic station, Bharati. *Polar Sci.*, 25, 100547. https://doi.org/10.1016/j.polar.2020.100547

Kanzawa, H., and Kawaguchi, S. (1990) Significant stratospheric sudden warming in the Antarctic late winter and shallow ozone hole in 1988. *Geophys. Res. Lett.*, 17(1), 77–80. https://doi.org/10.1029/GL017i001p00077.

Klekociuk, A. R., Tully, M. B., Krummel, P. B., Gies, H. P., Alexander, S. P., Fraser, P. J., Henderson, S. I., Javorniczky, J., Petelina, S. V., Shanklin, J. D., Schofield, R., and Stone, K. A. (2014) The Antarctic ozone hole during 2012. *Aust. Met. Oceanog. J*, 64, 313–330. https://doi.org/10.22499/2.6404.007.

Koppar, A. L., and Nagrath, S. C. (1991) Seasonal variation in the vertical distribution of ozone over Dakshin Gangotri, Antarctica. *Mausam*, 42(3), 275–278.

Kuttippurath, J., Lefèvre, F., Pommereau, J.-P., Roscoe, H. K., Goutail, F., Pazmiño, A., and Shanklin, J. D. (2013) Antarctic ozone loss in 1979–2010: first sign of ozone recovery. *Atmos. Chem. Phys.*, 13, 1625–1635. https://doi.org/10.5194/acp-13-1625-2013.

Kuttippurath, J., and Nair, P. J. (2017) The signs of Antarctic ozone hole recovery. *Sci. Rep.-UK*, 7, 585. https://doi.org/10.1038/s41598-017-00722-7.

Manney, G. L., Santee, M. L., Rex, M., Livesey, N. J., Pitts, M. C., Veefkind, P., Nash, E. R., Wohltmann, I., Lehmann, R., Froidevaux, L., Poole, L. R., Schoeberl, M. R., Haffner, D. P., Davies, J., Dorokhov, V., Gernandt, H., Johnson, B., Kivi, R., Kyrö, E., Larsen, N., Levelt, P. F., Makshtas, A., McElroy, C. T., Nakajima, H., Parrondo, M. C., Tarasick, D. W., von der Gathen, P., Walker, K. A., and Zinoviev, N. S. (2011) Unprecedented Arctic ozone loss in 2011. *Nature*, 478, 469–475. https://doi.org/10.1038/nature10556.

Newman, P. A., Kawa, S. R., and Nash, E. R. (2004) On the size of the Antarctic ozone hole. *Geophys. Res. Lett.*. 31. https://doi.org/10.1029/2004GL020596

Newman, P. A., Nash, E. R., Kawa, S. R., Montzka, S. A., and Schauffler, S. M. (2006) When will the Antarctic ozone hole recover? *Geophys. Res. Lett.*, 33, L12814. https://doi.org/10.1029/2005GL025232.

Pawson, S., Stolarski, R. S., Douglass, A. R., Newman, P. A., Nielsen, J. E., Frith, S. M., and Gupta, M. L. (2008) Goddard Earth Observing System chemistry-climate model simulations of stratospheric ozone-temperature coupling between 1950 and 2005. *J. Geophys. Res.*, 113, D12103. https://doi.org/10.1029/2007JD009511.

Peshin, S. K. (2011) Study of SO_2 and NO_2 behaviour during the ozone-hole event at Antarctica by Brewer Spectrophotometer. *Mausam*, 62(4), 595–600.

Peshin, S. K., Rao, G. S., and Rao, P. R. (1996) ozone soundings over Maitri (Antarctica) – 1994. *Vayu Mandal*, 26, 21–23.

Peshin, S. K., Rao, P. R., and Srivastav, S. K. (1997) Antarctic ozone depletion measured by balloon sondes at Maitri – 1992. *Mausam*, 48(3), 443–446.

Salby, M. L., Titova, E. A., and Deschamps, L. (2011) Rebound of Antarctic ozone. *Geophys. Res. Lett.*, 38, L09702. https://doi.org/10.1029/2011GL047266,2011.

Salby, M. L., Titova, E. A., and Deschamps, L. (2012) Changes of the Antarctic ozone hole: Controlling mechanisms, seasonal predictability, and evolution. *J. Geophys. Res.*, 117, D10111. https://doi.org/10.1029/2011JD016285,2012.

Schoeberl, M., Stolarski, R., and Krueger, A. J. (1989) The 1988 Antarctic ozone depletion: Comparison with previous year depletions. *Geophys. Res. Lett.*, 16(5), 377–380. https://doi.org/10.1029/GL016i005p00377.

Shepherd, T. G., Plumb, R. A., and Wofsy, S. C. (2005) Preface to the special issue on the Antarctic stratospheric sudden warming and split ozone hole of 2002. *J. Atmos. Sci.*, 62(3), 565–566. https://doi.org/10.1175/JAS-9999.1.

Solomon, S., Ivy, D. J., Kinnison, D., Mills, M. J., Neely, R. R., and Schmidt, A. (2016) Emergence of healing in the Antarctic ozone layer, *Science*, 353, 269–274. https://doi.org/10.1126/science.aae0061.

Sreedharan, C. R. (1968) An Indian electrochemical ozonesonde. *Jour. Phys. E. Sci. Inst.* 2(1), 195–197.

Sreedharan, C. R., Chopra, A. N., and Sharma, A. K. (1986) Vertical ozone distribution over Dakshin Gangotri, Antarctica. *Indian J. Radio Space Phys.*, 15, 159–162.

Sreedharan, C. R., Gulhane, P. M., and Kataria, S. S. (1993) ozone soundings over Antarctica. *Curr. Sci.*, 64(9), 634–636.

Sreedharan, C. R., Rao, G. S., and Gulhane, P. M. (1989) ozone measurements from Dakshin Gangotri, Antarctica, 1988–89. *Indian J. Radio Space Phys.*, 18, 188–193.

Tiwari, V. S. (1999) Measurement of ozone at Maitri, Antarctica. *Mausam*, 50(2), 203–210.

Tully, M. B., Krummel, P. B., and Klekociuk, A. R. (2019) Trends in Antarctic ozone hole metrics 2001–17. *J. South. Hemisph. Earth Syst. Sci.*, 69. https://doi.org/10.1071/es19020.

Uchino, O., Bojkov, R., Balis, D. S., Akagi, K., Hayashi, M., and Kajihara, R. (1999) Essential characteristics of the Antarctic-spring ozone decline: update to 1998. *Geophys. Res. Lett.*, 26, 1377–1380.

WMO (2018) Scientific Assessment of ozone Depletion: 2018, Global ozone Research and Monitoring Project – Report No. 58, 588 pp., Geneva, Switzerland. ISBN: 978-1-7329317-1-8.

Ziemke, J. R., Oman, L. D., Strode, S. A., Douglass, A. R., Olsen, M. A., McPeters, R. D., Bhartia, P. K., Froidevaux, L., Labow, G. J., Witte, J. C., Thompson, A. M., Haffner, D. P., Kramarova, N. A., Frith, S. M., Huang, L.-K., Jaross, G. R., Seftor, C. J., Deland, M. T., and Taylor, S. L. (2019) Trends in global tropospheric ozone inferred from a composite record of TOMS/OMI/MLS/OMPS satellite measurements and the MERRA-2 GMI simulation. *Atmos. Chem. Phys.*, 19, 3257–3269. https://doi.org/10.5194/acp-19-3257-2019.

2 Stratospheric Dynamics and Its Role in Climate over Antarctica

Karanam Kishore Kumar and N. Koushik
Space Physics Laboratory, Vikram Sarabhai Space Centre,
Thiruvananthapuram, India

CONTENTS

2.1 INTRODUCTION

Though the Earth's atmosphere is divided into several layers: troposphere (0–18 km over the low latitudes; 0–10 km over the mid-latitude; 0–7 km over the high latitudes), stratosphere (18–50 km over the low latitudes; 10–50 km over the mid-latitude; 7–50/60 km over the high latitudes), mesosphere (50–90/100 km over the low latitudes; 50–90/100 km over the mid-latitude; 50/55–85/100 km over the high latitudes), and thermosphere (100–600 km across the globe); a seamless interaction exists among these layers. The physical and chemical processes in one layer affect the mean state of the other layers, and several investigations illustrate this vertical coupling process in the Earth's atmosphere. However, the studies about the vertical coupling processes in the atmosphere predominantly focused on upward coupling, i.e. influence of the tropospheric processes on the middle and upper atmosphere (Eckermann and Vincent, 1994; Eckermann et al., 1997; Lieberman, 1999; Isoda et al., 2004; Kumar et al., 2007; Kumari et al., 2021). However, the downward coupling processes, i.e. middle and upper atmosphere influence on tropospheric weather and climate, is yet to be explored in detail (Kodera et al., 1990; Ramaswamy et al., 1992; Rind and Lacis, 1993; Kolstad et al., 2010; Karpechko and Manzini, 2012; Shaw and Perlwitz, 2013; Kidston et al., 2015; Son et al., 2017; Gray et al., 2018; Charlton-Perez et al., 2018; Afargan-Gerstman et al., 2020).

It is now known that the troposphere and vice versa influence changes in the stratosphere. For example, the stratospheric composition is modulated by the

DOI: 10.1201/9781003203742-2

transport from the troposphere. The deep convective systems in the tropics inject a large amount of water vapour into the lower stratosphere, which has a profound impact on the stratospheric composition (Holton et al., 1995; Sherwood and Dessler, 2000; Levine et al., 2007; Wang et al., 2009; Uma et al., 2014). The composition changes in the stratosphere, in turn, affect the troposphere through radiative forcing. Apart from composition, the dynamical processes such as quasi-biennial oscillation (QBO) in the stratosphere influence the tropospheric weather (Collimore et al., 2003; Liess and Geller, 2012; Nie and Sobel, 2015; Gray et al., 2018; Klotzbach et al., 2019). Thus, there exist strong troposphere-stratosphere interactions, which are essential from the climatological standpoint. It is imperative to identify all the physical processes through which the stratospheric state affects the tropospheric variability. Primarily, changes in the chemical composition of the stratosphere are expected to have a profound impact on the troposphere and surface climate. There is increasing evidence that a better representation of the stratosphere in numerical models results in improved predictability of the state of the climate. Though there is ample evidence on the influence of the stratospheric process on climate, investigations are minimal to arrive at any general conclusion.

Among low-, mid-, high-latitudes, the influence of stratospheric dynamics on the troposphere and surface climate is relatively significant over the high latitudes, especially over the Southern Hemisphere. By now, there is ample evidence for the stratospheric impact on the Antarctica climate (Thompson and Solomon, 2002; Cai, 2006; Son et al., 2008; Lenton et al., 2009; Son et al., 2010; Sigmond and Fyfe, 2010; Thompson et al., 2011; Smith et al., 2012; Wang et al., 2019; Kwon et al., 2020). It is always thought that the coupling of the lower and middle atmosphere is one way from the troposphere to the stratosphere over Antarctica. However, recent investigations show that dynamic, chemical, and radiative processes in the stratosphere immensely influence the troposphere and surface climate (Kidston et al., 2015; Lim et al., 2019; Domeisen and Butler, 2020). This particular aspect of downward coupling over the Antarctic came into prominence after discovering the ozone hole in 1985. Farman et al. (1985) were the first to report the unexpected large-scale decrease of ozone over Antarctica using ground-based total ozone measurements at the Argentine Islands (65°S 64°W) and Halley Bay (76°S 27°W). This discovery has changed the perspective of the role of the stratosphere in the Earth's climate system. Now it is known that the ozone hole is an annual phenomenon occurring in the stratosphere over Antarctica during September and October, and it is slowly recovering to its original level. Several studies focus on the reasons for the observed decrease in ozone over the Antarctic stratosphere, and it is found that chlorine is the main dominant species in depleting the ozone (Solomon et al., 1986; Hofmann and Solomon, 1989; Manzer, 1990; Anderson et al., 1991; Brune et al., 1991). It was also found that these chlorine atoms come from chlorofluorocarbons (CFCs), which were extensively employed in refrigeration. The CFCs undergo photodissociation by UV radiation, which breaks a chlorine atom from a CFC molecule. This detached chlorine atom reacts with ozone and splits it into an oxygen molecule and a chlorine monoxide (ClO) molecule.

Further, atomic oxygen reacts with ClO and releases the chlorine atom from ClO, reacting with ozone. However, these chemical reactions need a surface to take place. The Polar Stratospheric Clouds (PSCs) provide such a surface for these chemical species to react (Crutzen and Arnold, 1986; Drdla et al., 1993, 2002; Nakajima et al., 2016; Tritscher et al., 2021). These clouds convert stable chlorine to radicals that react with ozone. Thus, PSCs play a vital role in stratospheric ozone depletion over Antarctica, and this is still an active area of research in the realms of Antarctic climate investigations.

PSCs mainly form during winter in the Antarctic stratosphere, due to frigid temperatures occurring in this part of the globe. The cold temperature in the Antarctic stratosphere is linked with the strong polar vortex, which acts as a barrier for mixing mid-latitude and polar air masses in the stratosphere during the polar nights. The cold air trapped inside the polar vortex is conducive to the formation of PSCs. Though the polar stratosphere is extremely dry, the severe freezing conditions aid in constructing the PSCs. On the other hand, the polar vortex over the Arctic is relatively weak compared to over the Antarctic. Thus, developing icy conditions, which is a prerequisite for forming PSCs, are relatively rare. It is one of the reasons why the ozone hole does not appear over the Arctic. Since discovering the ozone hole, there has been tremendous interest in exploring the PSCs and their associated dynamics. Both ground- and space-based observations of PSCs contributed immensely in understanding the ozone loss processes over Antarctica (Fromm et al., 1997; Goodman et al., 1997; Maturilli et al., 2005; Lowe and MacKenzie, 2008; Lambert et al., 2012; Nakajima et al., 2016; Höpfner et al., 2018; Tritscher et al., 2021). Some of the observations and modelling studies provided evidence for the role of atmospheric gravity waves in the formation of PSCs (Fueglistaler et al., 2003; Pagan et al., 2004; Eckermann et al., 2009; McDonald et al., 2009; Noel et al., 2009; Noel and Pitts, 2012; Orr et al., 2015; Tritscher et al., 2021). The mesoscale fluctuations in the stratospheric temperature field induced by the gravity waves play a vital role in forming PSCs, facilitating the ozone depletion in the Antarctic stratosphere.

The observed depletion of ozone in the Antarctic stratosphere had far-reaching consequences beyond the increasing the amount of ultraviolet radiation, especially on the climate (Russell et al., 2006; Lenton et al., 2009; Sigmond and Fyfe, 2010; Thompson et al., 2011; Previdi and Polvani, 2014; Li et al., 2016; Son et al., 2018). For example, coolling of the atmosphere above Antarctica associated with ozone loss increased the thermal gradients between the Southern Hemispheric pole and tropics, which modulated the weather and climate over Antarctica. As the ozone hole has started to recover, reversal or slowing down of Southern Hemisphere circulation trends have been observed (Banerjee et al., 2020). Long-term observations reported that the stratospheric polar vortex also controls tropospheric variability, especially on Northern or Southern Annular Modes (NAM, SAM). The annular modes are fundamental indicators of high latitude weather and climate. The SAM represents the zonal mean pressure difference between the latitudes of 40S and 65S. The trends in the SAM index show the intensification of the stratospheric polar vortex related to ozone depletion in the stratosphere, as discussed by Thompson and Solomon (2002).

Ample evidence is emerging now on the relation between the SAM and the surface temperature trend pattern over Antarctica (Bandoro et al., 2014). It implies that ozone depletion in the Antarctic stratosphere modulates the surface temperature through the modulation of SAM. Many climate models could capture the ozone depletion in the Antarctic stratosphere and their effect on the tropospheric and surface climate. However, it is observed that the prescribed ozone in the model underestimates the ozone depletion and associated impact compared to the interactive stratospheric chemistry. Li et al. (2016) showed that the interactive chemistry-based model is relatively better in simulating the climate change features over Antarctica. These authors reported that an interactive chemistry-based model could capture climate change in the atmosphere and surface and the ocean and sea ice over Antarctica. Cagnazzo et al. (2013) investigated the role of stratospheric dynamics in the long-term changes of the Southern Hemisphere. They found that stratospheric dynamics have a profound effect on oceanic carbon uptake. These authors, using two sets of coupled atmosphere–ocean-sea-ice model simulations (one with a high top in which stratospheric dynamics is well described and another with a low top in which stratospheric dynamics are not defined explicitly), reported that though both high- and low-top models reproduce the changes in Southern Hemispheric circulation due to ozone depletion, the stratospheric cooling in the high-top model is 1.5 times larger than that found in low-top models.

Further, these results suggested that the difference in the annual southern ocean carbon uptake estimation by the high-top model is ~20–25% less than the low-top model. A poleward shift of tropospheric mid-latitude jet during the Austral summer has been reported by Son et al. (2010). There were also attempts to explore the pathways through which stratospheric variations are propagated to the troposphere and affecting its structure and dynamics in the Southern Hemisphere (Son et al., 2008; McLandress et al. 2010; Orr et al., 2012). Thus, there is increasing evidence for stratospheric dynamics on the Antarctic climate, primarily triggered by ozone depletion. It is now recognized that the climate over Antarctica and the surrounding southern ocean have a profound effect on Earth's climate and thus needs urgent attention from the atmospheric research community. The present chapter uses reanalysis data sets to discuss the long-term changes in the stratospheric dynamics and their signatures in the tropospheric circulations and surface climate. Section 2 provides data and methods, Section three provides results and discussion, and a summary is given in Section 4.

2.2   DATA AND METHODOLOGY

Reanalysis data sets provide high-quality information on various atmospheric, oceanic, and land surface parameters over a sufficiently long time, typically spanning several decades. Different agencies across the world, such as National Center for Atmospheric Research (NCAR), National Center for Environmental Prediction (NCEP), National Aeronautics and Space Administration (NASA), Japan Meteorological Agency (JMA), European Centre for Medium-Range Weather Forecasts (ECMWF), and National Oceanic and Atmospheric Administration (NOAA) are involved in the development of reanalysis data sets, which are ultimately aimed at creating a comprehensive and accurate long-term record of data about weather and climate research.

In reanalysis products, measurements from various observational platforms such as radiosondes, aircraft, satellites, weather radars, ships, buoys, and surface observatories are combined with an a priori estimate from a global numerical model to create an 'analysis'. Dynamical and physical processes in the Earth system are mathematically represented as equations in the global numerical model. The technique used by observations from diverse platforms with different spatial and temporal coverage, sampling errors, and varying observational densities is called 'data assimilation'. It is achieved by combining the numerical modelling forecasts with the available observations to reduce the overall errors in the analysis to the minimum. Thus from millions of observations from various platforms, the climate analysis generates a realistic estimate of the state of the atmosphere at a specific time.

Both the numerical prediction models and the data assimilation schemes are updated frequently to improve the accuracy of daily weather predictions. It results in artificial changes in the climate records produced by the analyses. Reanalysis products overcome this difficulty by using a single global numerical prediction model and data assimilation scheme. Reanalysis datasets serve as one of the most reliable datasets for the study of stratospheric processes by providing valuable global, high temporal resolution information on meteorological variables in stratospheric heights. Various global reanalysis products have been found to reproduce stratospheric dynamical features such as the polar vortex (Martineau and Son, 2010; Martineau et al., 2016), the Quasi-Biennial Oscillation (Kawatani et al., 2016), and the Stratopause Semi-annual Oscillation (Kawatani et al., 2020).

This chapter uses the primary data set of the Modern-Era Retrospective analysis for Research and Applications outputs, version 2 (MERRA-2). NASA's Global Modeling and Assimilation Office released MERRA-2 with significant improvements from its predecessor, MERRA. Goddard Earth Observing System (GOES) 5.12.4 assimilation scheme is employed in MERRA-2, and it uses three-dimensional variational assimilation with incremental analysis update (IAU; Bloom et al., 1996). MERRA-2 includes several observations previously not assimilated into MERRA data sets. Initial evaluation of MERRA-2 products can be found in Bosilovich et al. (2015). Here we use the assimilated state (ASM) fields from MERRA-2 rather than the analyzed (ANA) fields that do not include the IAU. The spatial resolution for MERRA-2 is $0.5° \times 0.625°$ (latitude $\times$ longitude), and it has 42 pressure levels from the surface to 0.1 hPa. MERRA-2 data is available at a temporal resolution of 3 hours. More details on the MERRA-2 reanalysis project can be found in Gelaro et al. (2017). MERRA-2 data sets can be downloaded from https://gmao.gsfc.nasa.gov/reanalysis/MERRA-2/.

Compared to MERRA, the updated MERRA-2 uses the vertically resolved ozone measurements from AURA Microwave Limb Sounder (MLS) for assimilation. As a result of this, the performance of MERRA-2 is found to be particularly well throughout much of the Stratosphere (Davis et al., 2017), especially since observations became available. MERRA-2 data sets have been found to produce a reliable representation of Southern Hemispheric stratospheric dynamics (Friedrich et al., 2017), near-surface wind speeds (Carvalho, 2019) and circulation changes associated with ozone recovery in the Southern Hemisphere (Orr et al., 2021).

2.3   RESULTS AND DISCUSSION

Among various geographical locations, the climate over polar regions is of paramount interest, as discussed in Section 1. The present chapter focuses on the role of stratospheric processes over the Antarctic climate. Figure 2.1 shows the height-latitude structure of zonal and meridional wind climatology during January and July. The MERRA-2 reanalysis data sets during 1980–2020 are used to construct these climatologies. The zonal winds during the Austral summer month of January depicted in Figure 2.1(a) readily reveal the presence of solid westward winds in the 30–50 km with peak magnitudes of the order of -60 ms^{-1} over the low latitudes. The winds are very benign over the high latitudes during January. During the Austral summer (December–January–February), the temperature gradient between the tropics and high latitude is not very prominent.

The polar stratosphere is warmer than the tropical stratosphere; thus, the winds are westwards in the stratosphere. Enhanced eastward winds of the order of 30–40 ms^{-1} positioned over mid-latitude, and the sub-tropics can be noted in the 10–12 km in Figure 2.1(a). These winds are associated with the subtropical jet (STJ), which forms at the edges of Hadley Cell, due to the latitudinal thermal contrast in the troposphere.

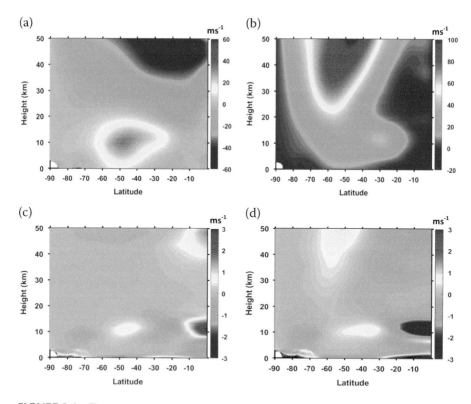

FIGURE 2.1   Forty-one year mean latitude-height cross-sections of the zonal winds during (a) January and (b) July in the Southern Hemisphere obtained using MERRA-2 reanalysis data set during 1980–2020. (c) and (d) are also the same as (a) and (b), respectively, but for the southerly winds.

During the Austral winter (June–July–August), when the Southern Hemispheric polar region is cut off from the sunlight, a strong temperature gradient develops between polar and tropics in the stratosphere, which results in substantial eastward winds, as shown in Figure 2.1(b). Apart from the absence of solar heating, the radiative cooling in the polar stratosphere due to the infrared emissions to the space by ozone, carbon dioxide, and water vapor further enhances the latitudinal temperature gradient. Figure 2.1(b) depicts the meridional cross-section of zonal winds during July in the Southern Hemisphere. A solid eastward wind of the order of 100 ms^{-1} can be noted at the stratopause level (~50 km). These winds are associated with the polar vortex, the region of sizeable atmospheric vorticity over the high latitudes. The polar vortex extends from the tropopause to stratopause region in the vertical direction. As mentioned earlier, the sizeable latitudinal temperature gradient present during the Austral winter in the stratosphere results in a strong pressure gradient between the polar and tropical stratosphere. This pressure gradient moves the air towards the south from the equator, which in turn results in the eastward motion due to the action of Coriolis force. Thus, strong eastward winds associated with the polar vortex dominate the wind regime in the polar stratosphere, which has important implications in the polar stratospheric dynamics. The strong zonal winds depicted in Figure 2.1(b) isolate the polar stratosphere from the rest, resulting in a very cold polar stratosphere. Now, it is well known that the icy conditions in the Southern Hemispheric stratosphere lead to the formation of Polar Stratospheric Clouds (PSCs), which aid in ozone loss processes through heterogeneous chemical reactions involving chlorine. One more critical aspect of polar vortex–associated winds is that they are relatively strong in the Southern Hemisphere than their northern counterpart. It is attributed to the weak planetary wave activity in the Southern Hemisphere (due to somewhat less zonal asymmetries in land-ocean contrast) than in the Northern Hemisphere. The signature of STJ in the troposphere can also be noted in July. However, the wintertime STJ is more robust than that observed during the summer (note that the ranges of color bars depicted in Figures 2.1(a) and (b) are different). Thus, Figure 2.1 provides the meridional cross-section of zonal winds in the Southern Hemisphere, depicting essential features such as the wintertime polar vortex.

Figures 2.1(c) and (d) show the latitudinal cross-section of southerly winds during January and July, respectively. Unlike zonal winds, southerly winds are very weak and do not exhibits any large scale features. However, near the equatorial tropopause, a relatively strong northward wind solid breeze can be noted during January. These winds are associated with the large-scale Hadley Cell, which arises due to the solar radiation's differential heating of the Earth's surface. Due to the excess solar radiation received at the Earth's surface over the tropics. A stable layer opposes the vertical motions and diverges towards the poles. The air mass moving poleward in both hemispheres deflects towards the east in either hemisphere due to the Coriolis force and a strong subtropical jet forms at the edges of the Hadley Cell, as shown in Figures 2.1(a) and (b). At the same time, the rising tropical air loses most of the water vapor through precipitation, and the poleward-moving air is thus dry and radiatively cools. In the sub-tropics, the air becomes dense, starts descending, and creates high-pressure zones at the surface. The air from this high-pressure region flows towards the equator, relatively at low pressure due to the rising air. It completes the Hadley Circulation with a rising branch at the equator and descending branch at the sub-tropics. These branches are connected by poleward flow at the tropopause and equatorward flow at the surface. They are responsible for wet climate over the tropics and dry weather over the sub-tropics. Figure 2.1(c) shows the northward flow at the tropopause and southward flow at the surface near the equator depicting the Hadley Circulation during the Austral summer. Similarly, Figure 2.1(d) shows the Hadley Circulation during the Austral winter with the southward flow at the tropopause and northward flow at the surface near the equator. Thus, Figure 2.1 depicts the zonal and meridional winds in the Southern Hemispheric troposphere and stratosphere.

Figures 2.2(a) and (b) show the meridional cross-section of the climatology of the vertical distribution of ozone mixing ratio in the troposphere and stratosphere in January and July, respectively. Figure 2.2(a) shows that the ozone mixing ratio maximizes at the stratosphere's 30–35 km altitude. The large mixing ratios extend from the equator to mid-latitudes during the Austral summer. Most of the ozone is confined to the stratosphere (90% in the stratosphere and 10% in the troposphere). There are several ways in which ozone is specified; for example, in Figures 2.2(a) and 2.2(b), it is determined in parts per million by volume (ppmv), which represent number of ozone molecules present in a million air molecules at a given height.

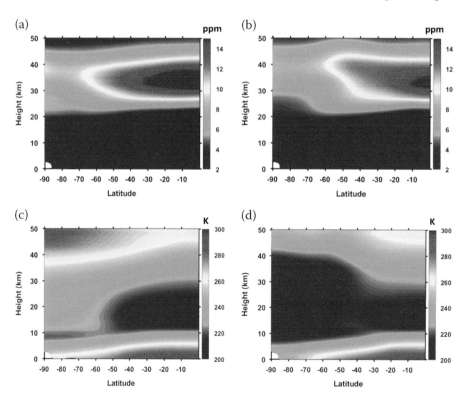

FIGURE 2.2   Same as Figure 2.1 but for ozone [(a) January and (b) July] and temperature [(c) January and (d) July].

More popular ozone units are ozone partial pressure, number density, and total ozone in Dobson Units. Among these, the total ozone column, which represents the sum of all the ozone in the atmosphere directly above a given location, is the most commonly used representation of ozone.

The total ozone column is reported in Dobson Units (DU), and the typical values range between 200 and 500 DU. This unit measures the thickness of the ozone layer if all the ozone present in the atmosphere is brought near the Earth surface. For example, 400 DU represent the ozone layer thickness of 4 mm at the surface. In the present chapter, vertical profiles of ozone are described in ppmv. The vertical profile of ozone is represented in terms of partial pressure (pressure contributor by the ozone to the total pressure at a given altitude) and number density (number of ozone molecules per cubic centimeter) peak at 20–25 km, whereas ozone mixing ratio shown in Figures 2.2(a) and (b) peak at 30–35 km. Though the ozone mixing ratio shows relatively large magnitudes over the equator compared to polar latitudes, the polar and high latitudes show a maximum total ozone content. It is now known that the ozone production is maximum over tropics region due to the availability of relative more solar UV radiation over the tropics. However, a large-scale circulation known as Brewer-Dobson circulation in the stratosphere transport the ozone from the tropic to the mid- and high latitudes. Due to this circulation, ozone accumulates at mid- and high latitudes, thus increasing the thickness of the layer and hence the total ozone content. The Brewer-Dobson circulation has a rising limb in the tropics, poleward transport, and sinking motion in the high latitudes. Earlier studies reported a significant hemispheric difference in the intensity and the width of the Brewer-Dobson circulation. Thus, Figures 2.2(a) and (b) depict the vertical distribution of ozone in the Southern Hemispheric latitude.

Figures 2.2(c) and (d) show the meridional cross-section of temperature climatology in the Southern Hemispheric troposphere and stratosphere during January and July, respectively. During the Austral summer, it can be seen that the temperature in the upper stratosphere over the high latitude is relatively higher than that over the tropics. Thus, the temperature gradient is from high latitude to the equator. The prolonged daytime hours during the summer over high latitudes are responsible for the observed temperature distribution in the upper stratosphere. This heating is attributed to the ozone absorption of solar UV radiation. This observed temperature gradient further drives the westward winds depicted in Figure 2.1(a) at this altitude region. The coldest temperature in the troposphere and stratosphere is found at a tropical tropopause region ~ 17 km altitude, which is the entry point to tropical region trace gases such as water vapor and methane and particulate matter. It is now known that the cold tropical tropopause temperature is responsible for the dryness of the stratosphere. While entering into the stratosphere, the water vapor in the tropical troposphere will be frozen and precipitate out, thus leaving the stratospheric air dry. As discussed earlier, the air thus entering into the stratosphere will be transported towards the poles by the Brewer-Dobson Circulation. It is how pollutants/trace gases in the tropics are transported to the polar stratosphere. In July, the temperature gradient reverses, as shown in Figure 2.2(d), driving the solid eastward winds (Figure 2.1(b)). These winds, representing the polar vortex, play an important role in isolating the polar stratosphere from its mid-latitude counterpart, thus cooling it further.

As discussed earlier, one can notice the relatively frigid temperature in the polar stratosphere, conducive to forming PSC. The low planetary wave activity and the solid polar vortex are critical aspects of the Southern Hemispheric high latitude compared to their Northern Hemispheric counterpart. It is now known that the westward propagating planetary waves triggered by the longitudinal contrast in land-ocean distribution and orography can impart westward momentum at upper stratospheric altitudes and thus weaken the polar vortex, as discussed earlier. It is a widespread phenomenon over the Northern Hemispheric high latitudes during boreal winters. The westward propagating planetary waves interact with the polar vortex and break or displace it from its original position, leading to a widespread phenomenon known as sudden stratospheric warming (SSW). During this event the stratospheric temperature increases dramatically by 40–70°C from the climatological values, and the winds reverse from eastward to westward in the mid-stratosphere. However, it is scarce to observe the SSW events over the Southern Hemispheric high latitudes due to the solid polar vortex. Thus, the high-temperature gradient and low planetary wave activity in the Southern Hemispheric stratosphere keeps the polar vortex strong during the Austral winter. Figures 2.2(c) and (d) thus provide climatology of the vertical temperature structure as a function of latitude in the Southern Hemisphere.

The zonal mean MERRA-2 reanalysis data set is averaged between 90°S and 60°S (known as the polar cap region) during 1980–2020 to discuss the annual cycle of ozone, temperature, and zonal winds as a function of height. The discussion of meridional winds is excluded as there are no large-scale features of importance as far as the present chapter is concerned. Figure 2.3(a) depicts the annual cycle of ozone in the troposphere and stratosphere. The striking feature of this figure is the stark seasonal contrast of the ozone mixing ratio with the Austral summer maximum and winter minimum in the stratosphere. The maximum ozone mixing ratio is observed in 12–14 ppmv between November and December. It is known that the ozone is produced by the solar UV radiation in the stratosphere. During winters, the polar region is cut off from the solar radiation and the ozone transport from the tropics to the polar region by the Brewer-Dobson Circulation aids in maintaining the ozone levels, as shown in Figure 2.3(a).

At any given location, the ozone levels depend on production, loss, and the advection processes. During the Austral summer, the production process dominates, while during the winter, the advection dominates. The ozone peak is also at lower heights during the summer compared to the winter. As discussed earlier, the ozone layer is spread in a broad region of 20–50 km, resulting in a relatively sizeable total ozone compared to other latitudes. The summertime ozone layer is relatively narrow compared to other seasons. However, the region of large ozone concentration is fairly broad during summer, as shown in the Figure 2.3. During autumn and spring, there are no notable changes in the climatological ozone mixing ratio. As discussed in Section 1, the ozone hole over Antarctica is predominantly noted during September, and there is no sign of it in the climatology shown in Figure 2.3(a). This figure covers both the ozone loss and recovery period, this may be the reason for not

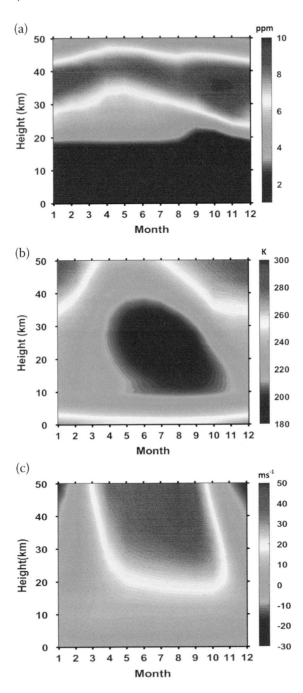

FIGURE 2.3 Annual cycle of monthly mean climatology of (a) ozone, (b) temperature, and (c) zonal winds as a function of height. The data set is averaged in the 60–90°S latitudes to construct the climatology for the polar cap region.

observing the ozone hole signature Apart from this, the data is averaged over 90–60°S latitudes, which might have masked this feature. Figure 2.3(a) thus depicts the annual cycle of the ozone mixing ratio over the polar cap region.

Figure 2.3(b) shows the annual cycle of the vertical structure of temperature in the troposphere and stratosphere over the polar cap region. As discussed previously, the coldest temperatures are found during the Austral winters in the 20–30 km altitude region and relatively warmer temperatures in the 40–50 km during summer, as shown in this figure. The thermal structure of the polar stratosphere plays a key role not only in driving the circulation through temperature gradients but also in triggering chemical reactions. The temperature in the stratosphere, especially during the summer, is controlled by ozone through the absorption of solar UV radiation. Figure 2.3(a) shows that the ozone is peaking in the 30–40 km region during the Austral summer, but the maximum heating is noted in the 40–50 km altitude region, as shown in Figure 2.3(b). It is due to the availability of solar UV radiation for absorption and subsequent heating. Relatively more radiation is available at the 40–50 km region than at the 30–40 km, and thus maximum heating is noted in the upper stratosphere. During the Austral winter, the absence of sunlight and the radiative cooling due to ozone, water vapor, and methane are responsible for observed cold temperatures, as discussed earlier. Though ozone, water vapor, and methane are greenhouse gases warming the lower troposphere by trapping the terrestrial radiation, it cools the surroundings through infrared radiation emission in the stratosphere. Thus, the warm and cold temperatures during summer and winter, respectively, shape the structure and dynamics of the polar stratosphere.

Figure 2.3(c) shows the annual cycle of zonal winds as a function of altitude over the polar cap. This figure shows that the significant eastward winds of 50–60 ms^{-1} dominate the entire stratosphere during the Austral winter. The maximum eastward winds of the order of 100 ms^{-1} are noted near 50–55°S latitudes, as shown in Figure 2.1(b) during July. However, the annual cycle depicted in Figure 2.3(c) is constructed using mean winds within the polar caps and thus has a relatively smaller magnitude. These winds surrounding the polar cap region play a key role in isolating this region from the mid-latitude and thus further cooling the winter stratosphere. In the 40–50 km region, winds are westward during the Austral summer. The transition of winds from westward during summer to eastward during winter is per the reversal temperature gradient during respective seasons. The westward winds result from the northward motion and eastward winds from the southward motions in the Southern Hemisphere due to the Coriolis forces. The stratospheric ozone is responsible for the heating during the Austral summer, generating a thermal gradient between the polar and mid-latitude region, which drives the circulation. Thus, all the three parameters depicted in Figure 2.3 are co-varying, and any changes in any one of them will be reflected in the other two parameters. The ozone depletion in the Antarctic stratosphere discussed in Section 1 thus affected the thermal and dynamical structure of the stratosphere, which in turn affected the lower atmospheric climate. It forms the basis of the present discussion.

Forty-one years of MERRA-2 data set is analyzed for studying the long-term evolution of ozone, temperature, and zonal winds in the polar cap region. Figure 2.4(a) shows the monthly mean time series of ozone mixing ratio in the

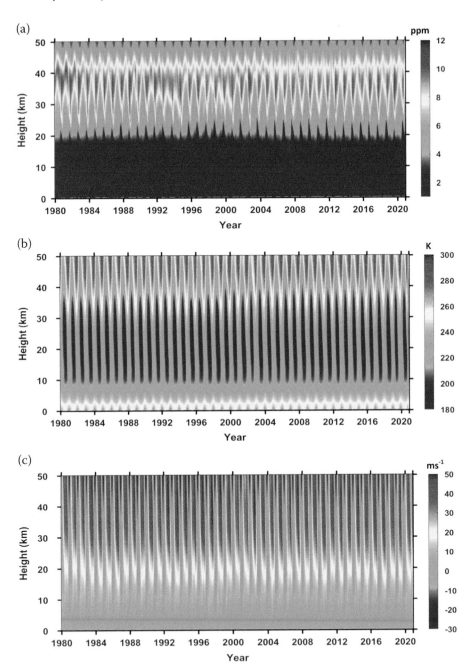

FIGURE 2.4 Time-height sections of (a) ozone, (b) temperature, and (c) zonal winds during 1980–2020.

troposphere and stratosphere during 1980–2020. A regular seasonal pattern in the ozone mixing ratio is evident from this figure throughout the study period, and the ozone layer extends from 20 to 50 km. Similarly, the long-term time series of temperature and zonal winds are shown in Figures 2.4(b) and (c), respectively. Interestingly, the upper stratospheric temperature is relatively warmer than the near-surface temperature over the polar cap. The zonal winds are dominated by the eastward winds associated with the polar vortex. It is difficult to discern any long-term changes in the geophysical parameters shown in this figure as they exhibit pronounced seasonal variability. One has to deseasonalize these parameters to investigate the long-term trends.

To depersonalize, the mean annual cycle of all three parameters is constructed, and the same is subtracted from the individual years. This procedure is widely employed in trend analysis to obtain the deseasonalized perturbations of geophysical parameters. Figure 2.5(a) shows the deseasonalized perturbations of ozone at 20 hPa (~28 km) altitude for September. The ozone hole over Antarctica is predominately observed during September and, thus, this month is chosen for illustrating the deseasonalized perturbations. The perturbations of the order of ± 2 ppmv can be observed from this figure. It is evident from this figure that ozone decreased drastically starting from the year 1980 until the year 2002 and then showed an increasing trend. There are four very prominently seen ozone anomaly in Figure 2.5(a) corresponding to 1988, 1994, 2002, and 2019.

Three peaks (1988, 2002, and 2012) out of these 4 years have corresponding peaks in temperature, as shown in Figure 2.5(b). From previous reports, it is known that SSW events occurred during 2002 and 2019, and thus there was a drastic increase in the temperature during September, as shown in Figure 2.5(b). The other two peaks observed during 1988 and 1994 are yet to be investigated. Especially, a prominent peak observed during 1994 seems to be very interesting and worth further investigation. Overall, the time series of ozone shows a decreasing trend for the first 20 years and, after that, shows an increasing trend. The time series of deseasonalized temperature perturbations are shown in Figure 2.5(b), corresponding to 20 hPa. From this time series, it is evident that the temperature perturbations show an increasing trend with time at this pressure level. Though the ozone depicted in Figure 2.5(a) shows decreasing trends during the initial years, the temperature shows an increasing trend throughout the observational period at this altitude. This feature will be discussed further later in this chapter. In temperature perturbations, two prominent peaks can be noticed during 2002 (~15 K) and 2019 (~25 K).

Both of these peaks correspond to SSW events, as mentioned earlier. The time series of deseasonalized zonal wind perturbations are depicted in Figure 2.5(c), which readily shows an increasing trend representing the strengthening of eastward winds. The striking feature of this figure is the prominent anomaly peak (~−7 ms^{-1}) observed during the year 2002. This anomaly, again, corresponds to the significant SSW observed during this year. During major SSW events, the stratospheric temperature rises and the zonal winds reverse from eastward to westward. However, during minor SSW events, zonal winds do not change, and thus significant wind anomalies are not observed during the year 2019, during which a minor SSW event

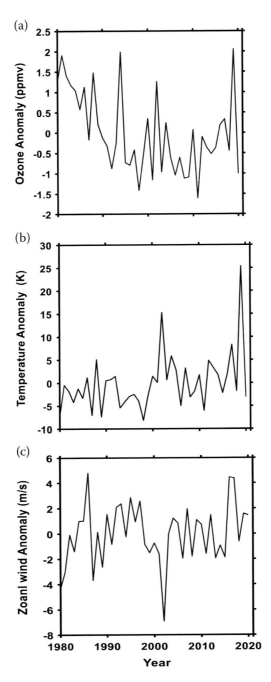

FIGURE 2.5 Time series of deseasonalized perturbations of (a) ozone, (b) temperature, and (c) zonal winds during September at 20 hPa pressure level (~28 km) in the polar cap region. Deseasonalized perturbations are obtained by subtracting the mean annual cycle from individual yearly cycles.

is observed. Therefore, from Figure 2.5, it can be noted that at the 28 km altitude, the ozone is showing a decreasing trend until 2002 and then an increasing trend representing the recovery of ozone, whereas temperature and zonal winds show a growing trend.

The ozone, temperature, and zonal wind perturbations at all the altitudes in the troposphere and stratosphere are subjected to trend analysis for each month to obtain the month-height section of trends. Figure 2.6(a) shows the month-height section of trends per decade in the ozone variability obtained using 41 years of time series. This figure provides the wide variability of ozone during both the declining as well as the recovery phase. Most of the trends depicted here are significant at a 90% level, and the white patches represent no/negligible trends. A dominant decreasing trend of the order of 0.5 ppmv/decade is observed in most altitudes (25–45 km) during all the months. However, a definite decreasing trend is observed in the 35–45 km altitude during autumn and winter, whereas during spring and summer, a definite decreasing trend is noted in the 25–30 km altitude. It may be attributed to the maximum ozone concentration observed in the stratospheric altitude during these seasons, as shown in Figure 2.3(a). Though the decreasing trends dominate the stratosphere, there are a few altitude regions where an increasing trend is observed. During the autumn and winter seasons, a rising trend is observed at the 20 km region. However, these trends are very weak in the order of 0.1 ppmv/decade. As discussed in Section 1, ozone depletion in the Antarctic stratosphere is widely reported in climate sciences. The observed ozone depletion, which came to light in the early 1980s, is attributed to the release of CFCs into the atmosphere.

The photodissociation of CFCs by solar UV radiation releases chlorine atoms, which react with the ozone molecule and give rise to oxygen and ClO molecules. It is noted that ClO reacts with atomic oxygen and frees the chlorine atom, which further destructs the ozone molecule and thus triggers a cyclic reaction. It is also pointed out that these reactions need some sort of surface, which is provided by the PSC, as discussed earlier. The leading role of a PSC is to make non-reactant chlorine reactive ones. The stratosphere needs to be at a very low temperature (below 195 K) to form PSCs. Though the winter stratosphere is the coldest over the polar cap, as shown in Figure 2.3(b), the absence of solar radiation during this time, which is a prerequisite for ozone-depleting reactions to take place, delay the ozone loss process and during the late winter and early spring months (August, September, and October) when solar radiation is returned over the polar cap region and the temperature is also optimum for the formation of PSC, the ozone depletion reactions peak and thus the biggest ozone hole is observed during September and October. During the summer months, the stratospheric temperature is relatively high, and the occurrence of PSCs will be relatively low, thus slowing down the ozone-depleting reactions.

Figure 2.7 depicts a typical Cloud-Aerosol Lidar and Infrared Pathfinder Satellite Observations (CALIPSO) measurements of PSCs on 17 July 2008. This figure shows a height-latitude/longitude section of total attenuated backscatter at 532 nm. A detailed description of PSC observations by CALIPSO is given by Tritscher et al. (2021). One can note the signature of PSCs in the height region of 15–25 km over the Southern Hemispheric high latitudes. Using CALIPSO measurements,

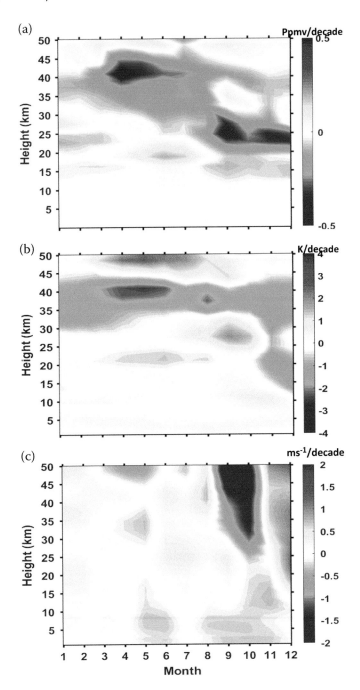

FIGURE 2.6 Month-height sections of trends in (a) ozone, (b) temperature, and (c) zonal winds. Trends in most parts of the stratosphere are significant at the 90% level. The white color indicates no trends or insignificant trends.

FIGURE 2.7   Height-latitude/longitude section of total attenuated backscatter measured by space-based Lidar operating at 532 nm onboard CALIPSO on 17 July 2008 at 03:27 UTC. The signature of PSCs is noted over the Southern Hemispheric high latitudes.

numerous studies are focusing on the physical and chemical characterization of these clouds. As mentioned earlier, these clouds need a frigid temperature to form.

In some cases where the temperature is not cold enough, propagating atmospheric waves such as gravy waves can modulate the background temperature, and the PSCs can form at the trough region of the waves. Figure 2.8(a) shows CALIPSO observations of PSCs over the Antarctic area on 17 July 2008 (same as Figure 2.7 but different times). It is exciting to note the descending nature of PSCs, revealing a wavelike structure, thus providing evidence for the role of gravity waves in the formation of PSC. Further, using the observations collected during the 34th, 35th, and 36th Indian Scientific Expeditions to Antarctica, the gravity wave characteristics are investigated over the Indian research station, Bharati (69.4°S, 76.2°E) using high-resolution GPS-sonde observations. Figures 2.8(b) and (c) show the perturbations in zonal and meridional winds during 29–30 January 2016. From this figure, it is fascinating to note the signature of inertia gravity waves. It is pointed out that these types of wave signatures are frequently observed over the Antarctic stratosphere. Though the observational period of CALIPSO and GPS-sondes are not the same, these illustrations are used to demonstrate the role of gravity waves in the formation of PSCs.

Similarly, the height-month section of trends in the temperature is shown in Figure 2.6(b). A cooling trend can be noted in this figure during all the months with varying degrees of cooling rates in the 30–40 km altitude region with a maximum cooling rate of the order of −2 K/decade during autumn and early winter season. It is the dominant feature of this figure. The observed cooling trend can be directly

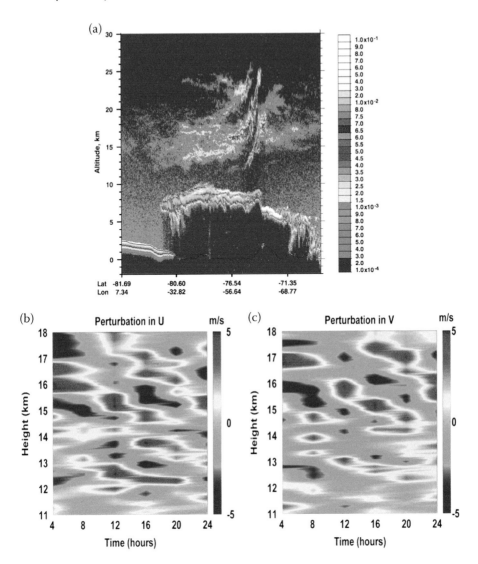

FIGURE 2.8 (a) Same as Figure 2.7 but for CALIPSO observations on 17 July 2008 at 20:02 UTC. Time-height section of mean removed perturbations in (b) zonal and (c) meridional winds measured using high vertical resolution GPS-sondes over the Indian Antarctic station, Bharati (69.4°S, 76.2°E) during 29–30 January 2016. The temporal resolution of the observations is 4 hours.

attributed to the decreasing trend in the ozone shown in Figure 2.6(a). However, the altitude structure of ozone and temperature trends is not similar.

Moreover, interestingly a warming trend of the order of ~1.5 K/decade is observed during September and October in the 25–30 km altitude region where ozone exhibits a decreasing trend. During summer, the ozone absorption of solar UV radiation is the dominant mode of stratospheric heating and thus, decreasing trend in ozone results in cooling the stratosphere. However, it is to be noted that BDC also induces dynamical heating in the polar stratosphere. Thus, both ozone depletion and BDC strength changes can alter the polar stratosphere's temperature trend. The observed heating trend in the 20–30 km region during September and October can be attributed to BDC strengthening. Earlier studies by Johanson and Fu (2007) also reported the warming trend during early spring, despite the decreasing ozone trend. From Figure 2.6(b), one more patch of the heating trend of the order of 2 K/decade can be noticed in the upper stratosphere (45–50 km) during the autumn and winter seasons. This feature requires further analysis to identify the potential reasons behind this warming in the upper stratosphere. However, the ozone variability and BDC strength shape the temperature trends in the mid- and lower stratosphere. The height-month section of trends in zonal winds is shown in Figure 2.6(c) in the troposphere and stratosphere. As mentioned earlier, ozone-induced stratospheric temperature changes modify the wind system over the polar cap. The striking feature in this figure is the relatively large decreasing trends (westward) in the zonal winds in the 30–50 km region during the spring season followed by strong increasing trends (eastward) during the summer. It is interesting to note the contrasting trends during consecutive seasons representing weakening and strengthening of the winds, respectively. During the winter, a weak positive trend is observed in the entire stratosphere, meaning the polar vortex's strengthening. As discussed earlier, the zonal winds are primarily driven by the thermal contrast between tropical and polar latitudes. The thermal difference observed during the spring and winter, as shown in Figure 2.6(b), may be the potential reasons for these observed trends in the zonal winds. One more exciting feature from Figure 2.6(c) is the weak westward trends throughout the troposphere during all months, which may significantly affect the Antarctic climate. Thus, Figure 2.6 brings out the observed trends using 41 years of reanalysis data set and show the variability in ozone, temperature, and zonal winds in the Antarctic stratosphere.

Figure 2.5(a) shows that the ozone levels are recovering from 2002. Thus, it will be interesting to carry out trend analysis separately during 2002–2020. Figure 2.9(a) shows the height-month section annual cycle of trends in ozone estimated using data sets during 2002–2020. It is interesting to note the positive trend in the ozone of 0.5 ppmv/decade in most parts of the stratosphere except in the upper stratosphere, where it shows weak negative trends. Apart from this, a weak decreasing trend is observed in the ozone during the late spring and early summer. Overall, it is clear from Figures 2.6(a) and 2.9(a) that the ozone recovery is steadily happening in the Antarctic stratosphere and, for this, one should thank the Montreal Protocol.

Similarly, trend analysis is carried out during the ozone recovery phase for the temperature, as shown in Figure 2.9(b). The striking feature of this figure is the cooling trend in most parts of the stratosphere, except for September in the lower stratosphere, where it shows a weak warming trend. A relatively sizeable cooling trend is observed during October. It is interesting to note that the Antarctic stratosphere is cooling down even during the ozone recovery phase. It may be due to other reasons, such as increasing greenhouse gases in the Antarctic stratosphere, which radiatively cools the stratosphere. As of now, it isn't easy to separate the observed trends in temperature from ozone recovery and increase in greenhouse gas

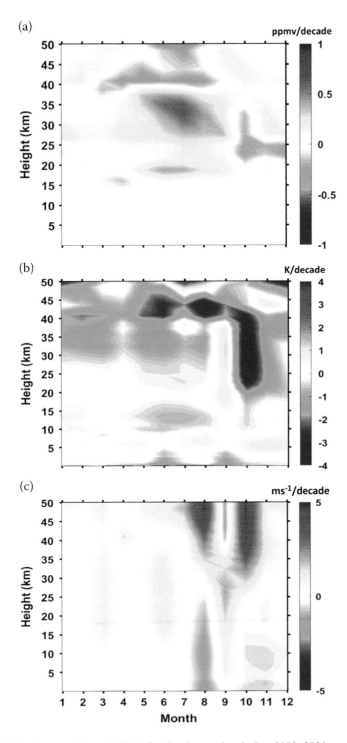

FIGURE 2.9 Same as Figure 2.6 but for the time series during 2002–2020.

concentrations. Moreover, as discussed earlier, the Antarctic stratosphere temperature is also controlled through dynamical heating by BDC.

Similarly, Figure 2.9(c) shows the trends in the zonal winds during the ozone recovery phase. This figure shows that the eastward winds are strengthening in the upper stratosphere during the late winter and spring season, indicating the polar vortex's strengthening during this period. The strengthening of the polar vortex has profound implications in the tropospheric climate. Thus, Figure 2.9 brings out the present status of trends during the recovery phase of ozone in the Antarctic stratosphere.

Now the critical question is how the stratospheric variability discussed above influences the Antarctic climate? Though other essential processes affect the Antarctic climate, such as greenhouse warming, the present chapter focuses on the impact of stratospheric processes on the Antarctic climate. As discussed earlier, one of the dominant variability observed in the Antarctic stratosphere during the study period is ozone depletion, which affected the stratosphere's thermal and dynamical structure. The direct impact of ozone depletion in the Antarctic stratosphere is through radiative forcing at the tropopause level. It is noted that the radiative forces exerted by the ozone depletion, as shown in Figure 2.6(a), is comparable with that of other trace gases in the troposphere. However, the ozone recovery in recent years is restoring the radiative balance before forming the ozone hole. The enhanced solar radiation received at the surface due to the depletion of stratospheric ozone is thought to be adding to the warming at the surface in Antarctica. A study by Li et al. (2016) suggested that the ozone depletion in the Antarctic stratosphere affects the troposphere, texture, the southern ocean, and sea ice in the region. It is known that the sea ice over Antarctica shows a slightly increasing trend compared to their arctic counterpart, which offers a decreasing trend. There is no consensus among many research communities on the observed positive trend in the sea ice extent over Antarctica. Though there is greenhouse warming, still a positive direction is seen in the sea ice extent. There may be some potential stratospheric processes compensating the greenhouse warming. However, the pathway through which it happens is yet to be identified. Figure 2.10 shows the long-term time series of sea ice extent during January and July over Antarctica. From this figure, it is evident that the sea ice extent is increasing during both summer and winter months until 2015, a drastic decrease is noted. Recent studies reported that the sea ice extent, after showing an upward trend since 1980, the sea ice extent declined dramatically during the spring season of 2016 and showed record low values during December 2016. It is a quite remarkable observation as the factors that kept sea ice extent increasing all these years and the factors responsible for the drastic decrease in sea ice extent observed in recent years is yet to be explored. It is one of the crucial challenges posed to climate scientists as the drastic decrease in sea ice extent will have profound climate implications through the sea level rising across the globe. The direct radiative forcing of stratospheric ozone depletion and its recovery may hold the key in explaining these observations depicted in Figure 2.10.

Figure 2.6(c) shows that the circumpolar eastward winds are increasing during Austral summer. Earlier studies suggested that these increasing trends in eastward winds enhance the stratosphere-troposphere coupling processes and thus can change

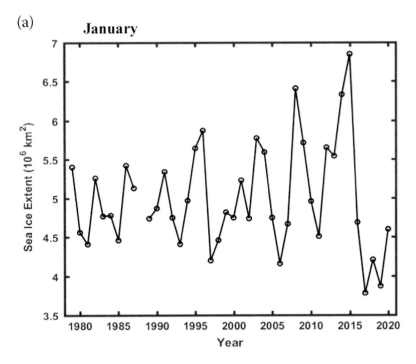

(a)

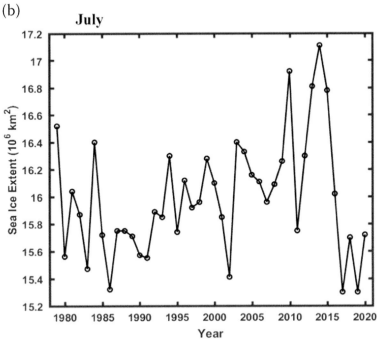

(b)

FIGURE 2.10 Time series of sea ice extent during (a) January and (b) July during 1979–2020.

the surface eastward winds. One can note the strengthening of stratospheric east-ward winds in terms of positive trends in Figure 2.6(c) during the summer months. It is interesting to note the positive trends throughout the troposphere, especially in January. Though the trends are relatively weak in the troposphere, this magnitude will be significant in terms of momentum imparted by tropospheric winds. The wind stress over the southern ocean has a considerable influence on the sea's surface temperature. This pathway of stratospheric winds modifying the sea surface temperature through stratosphere-troposphere coupling is a key to assess the stra-tospheric impact on the Antarctic climate. At the same time, a strong negative trend is also seen in the zonal winds in Figure 2.6(c) in the upper stratosphere re-presenting the weakening of the vortex. Corresponding to this weakening of polar vortex, there are weak negative trends in the troposphere. These changes in the wind pattern in the troposphere will also influence the surface climate. As discussed in Section 1, one of the primary stratosphere and troposphere coupling modes in the Antarctic in the Southern Annual Mode (SAM). The SAM index certainly reflects the state of the stratospheric polar vortex and thus is an essential metric for as-sessing the stratospheric influence on the troposphere. Recent studies are reporting that the trends in the SAM index point towards the strengthening of the polar vortex, as seen in Figure 2.6(c) during the Austral summer. Earlier studies revealed that the stratospheric polar vortex over the high southern latitudes play an essential role in downward coupling with the troposphere and thus impact the climate (Thompson and Solomon, 2002). From Figure 2.6, it is evident there is significant variability in zonal winds in the stratosphere associated with the polar vortex. Again there is a known co-variability between SAM and the surface temperature trend pattern, as reported by Bandoro et al. (2014). Thus, these results suggest that any changes in the polar vortex strength can affect the Antarctic surface temperature. The results discussed in this chapter indicate that cooling the Antarctic stratosphere triggered by ozone depletion increased the thermal contrast between tropics and polar regions, affecting the weather and climate over Antarctica. It is the most straightforward pathway. However, an integrated approach to monitor surface, troposphere, and stratosphere over Antarctica coupled with numerical modeling is the need of the hour to arrive at any general conclusion.

2.4   SUMMARY AND CONCLUDING REMARKS

The present chapter discussed stratospheric structure and dynamics in regulating the tropospheric and surface climate over Antarctica. Forty-one years of MERRA-2 data sets during 1980–2020 are used to construct the mean climatology of the Antarctic stratosphere, with various aspects such as meridional cross-sections and annual cycles of climate variables such as ozone, temperature and zonal winds are discussed. Further, the data sets are averaged in the 60–90°S latitude to study the time evolution of these climate variables in the polar cap region. Using deseaso-nalized perturbations in these three climate variables, a height-month section of trends are estimated. Overall, the ozone shows a negative trend throughout the stratosphere, except at a few altitude regions during certain months. It was inter-esting to note that the decreasing trends in ozone are observed relatively at higher altitudes during the autumn and winter, compared to spring and summer. It is consistent with the ozone peak observed during these months. During summer, the temperature trends show a layer of cooling in the 30–45 km altitude regions and a relatively thin layer of cooling trends in the 35–40 km altitudes. Overall cooling directions are prominently noted in the stratosphere. However, a patch of warming trends in the lower stratosphere around 25–30 km during September. It is pointed out that the heating of the Antarctic stratosphere is not only driven by the ozone absorption of solar UV radiation but lo due to the dynamic heating induced by BDC. The imbalance between radiative and dynamical heating may result in the observed trends.

Further, the trends in the zonal winds show relatively large positive trends representing the strengthening of the vortex during the summer months and weakening during the spring. A detailed discussion is made on the co-variability of trends in these three climate variables. From earlier studies, it is noted that strengthening and weakening the polar vortex has profound effects on the tropospheric climate through modification of the SAM index.

Further, typical PSC observations of CALIPSO and the role of gravity waves in their formation are briefly discussed. Data sets limited to 2002–2020 are analysed to investigate the trends during the ozone recovery phase, which show steady ozone recovery and slowing down of cooling in the stratosphere except during October, where the trend shows relatively large cooling. Further, the zonal winds emphatically show the strengthening of the vortex during the late winter and spring seasons. Thus, the present chapter brought out a comprehensive analysis of the Antarctic stratosphere during the ozone depletion and recovery phase. The pathways through which the stratospheric processes interact with tropospheric and surface climate are discussed in this chapter.

The prime candidates for the stratosphere-troposphere coupling over the Antarctic region are direct radiative forcing of ozone depletion on the troposphere, polar vortex-induced changes in the tropospheric wind system, and Southern Hemispheric high-latitude circulation due to ozone depletion induced thermal contrast between tropical and polar latitudes. However, quantitatively, further insights into these coupling processes are needed to elucidate the role of the stratosphere in impacting the Antarctic climate.

ACKNOWLEDGEMENTS

The authors gratefully acknowledge MERRA-2 reanalysis data sets, a product of the Global Modeling and Assimilation Office at NASA GSFC, supported by NASA's Modeling, Analysis, and Prediction (MAP) program. The sea ice extent dataset is obtained from NSIDC (U.S. National Snow and Ice Data Center) website. The authors also acknowledge the CALIPSO team for making the observations available in the public domain. N. Koushik gratefully acknowledges the financial support provided by the Indian Space Research Organization for his research work. The authors also thankfully acknowledge the support given by Dr. Neloy Khare, Ministry of Earth Sciences, in completing this chapter.

REFERENCES

Afargan-Gerstman, H., Polkova, I., Papritz, L., Ruggieri, P., King, M. P., Athanasiadis, P. J., Baehr, J., and Domeisen, D. I. V. (2020). Stratospheric influence on North Atlantic marine cold air outbreaks following sudden stratospheric warming events. *Weather Clim. Dyn.* 1, 541–553. https://doi.org/10.5194/wcd-1-541-2020

Anderson, J. G., Toohey, D. W., and Brune, W. H. (1991). Free radicals within the Antarctic vortex: The role of CFCs in Antarctic ozone loss. *Science*. 251, 39–46. https://doi.org/10.1126/science.251.4989.39

Bandoro, J., Solomon, S., Donohoe, A., Thompson, D. W. J., and Santer, B. D. (2014). Influences of the Antarctic ozone hole on southern hemispheric summer climate change. *J. Clim.* 27, 6245–6264. https://doi.org/10.1175/JCLI-D-13-00698.1

Banerjee, A., Fyfe, J. C., Polvani, L. M., Waugh, D., and Chang, K. L. (2020). A pause in Southern Hemisphere circulation trends due to the Montreal protocol. *Nature*. 579, 544–548. https://doi.org/10.1038/s41586-020-2120-4

Bloom, S. C., Takacs, L. L., da Silva, A. M., and Ledvina, D. (1996). Data assimilation using incremental analysis updates. *Mon. Weather Rev.* 124, 1256–1271. https://doi.org/10.1175/1520-0493(1996)124<1256:DAUIAU>2.0.CO;2

Bosilovich, M., Akella, S., Coy, L., Cullather, R., Draper, C., Gelaro, R., Kovach, R., Liu, Q., Molod, A., Norris, P., Wargan, K., Chao, W., Reichle, R., Takacs, L., Vikhliaev, Y., Bloom, S., Collow, A., Firth, S., Labow, G., Partyka, G., Pawson, S., Reale, O., Schubert, S. D., and Suarez, M. (2015). MERRA-2: Initial Evaluation of the Climate, Series on Global Modeling and Data Assimilation, Greenbelt, MD, USA, NASA/TM–2015-104606, Vol. 43, NASA.

Brune, W. H., Anderson, J. G., Toohey, D. W., Fahey, D. W., Kawa, S. R., Jones, R. L., Mckenna, D. S., and Poole, L. R. (1991). The potential for ozone depletion in the Arctic Polar Stratosphere. *Science*. 252, 1260–1266. https://doi.org/10.1126/science.252.5010.1260

Cagnazzo, C., Manzini, E., Fogli, P.G. et al. (2013). Role of stratospheric dynamics in the ozone–carbon connection in the Southern Hemisphere. *Clim.Dyn.* 41, 3039–3054. https://doi.org/10.1007/s00382-013-1745-5

Cai, W. (2006). Antarctic ozone depletion causes an intensification of the Southern Ocean super-gyre circulation. *Geophys. Res. Lett.* 33, L03712. https://doi.org/10.1029/2005GL024911

Carvalho, D. (2019). An assessment of NASA's GMAO MERRA-2 reanalysis surface winds. *J. Clim.* 32(23), 8261–8281. https://doi.org/10.1175/jcli-d-19-0199.1.

Charlton-Perez, A. J., Ferranti, L., and Lee, R. W. (2018). The influence of the stratospheric state on North Atlantic weather regimes. *Q. J. R. Meteorol. Soc.* 144, 1140–1151. https://doi.org/10.1002/qj.3280

Collimore, C. C., Martin, D. W., Hitchman, M. H., Huesmann, A., and Waliser, D. E. (2003). On the relationship between the QBO and tropical deep convection. *J. Clim.* 16, 2552–2568. https://doi.org/10.1175/1520-0442(2003)016<2552:OTRBTQ>2.0.CO;2

Crutzen, P. J., and Arnold, F. (1986). Nitric acid cloud formation in the cold Antarctic Stratosphere: A major cause for the springtime' ozone hole.' *Nature*. 324, 651–655. https://doi.org/10.1038/324651a0

Davis, S. M., Hegglin, M. I., Fujiwara, M., Dragani, R., Harada, Y., Kobayashi, C., Long, C., Manney, G. L., Nash, E. R., Potter, G. L., Tegtmeier, S., Wang, T., Wargan, K., and Wright, J. S. (2017). Assessment of upper tropospheric and stratospheric water vapor and ozone in reanalyses as part of S-RIP. *Atmos. Chem. Phys.* 17(20), 12743–12778.

Domeisen, D. I. V., and Butler, A. H. (2020). Stratospheric drivers of extreme events at the Earth's surface. *Commun. Earth Environ.* 1, 1–8. https://doi.org/10.1038/s43247-020-00060-z

Drdla, K., Turco, R. P., and Elliott, S. (1993). Heterogeneous chemistry on Antarctic polar stratospheric clouds: A microphysical estimate of the extent of chemical processing. *J. Geophys. Res. Atmos.* 98, 8965–8981. https://doi.org/10.1029/93JD00164

Eckermann, S., and Vincent, R. (1994). First observations of intraseasonal oscillations in the equatorial mesosphere and lower thermosphere. *Geophys. Res. Lett.* 21(4), 265–268. https://doi.org/10.1029/93gl02835.

Eckermann, S. D., Hoffmann, L., Höpfner, M., Wu, D. L., and Alexander, M. J. (2009). Antarctic NAT PSC belt of June 2003: Observational validation of the mountain wave seeding hypothesis. *Geophys. Res. Lett.* 36, L02807. https://doi.org/10.1029/2008GL036629

Farman, J. C., Gardiner, B. G., and Shanklin, J. D. (1985). Large losses of total ozone in Antarctica reveal seasonal ClOx/NOx interaction. *Nature.* 315, 207–210. https://doi.org/10.1038/315207a0

Friedrich, L. S., McDonald, A. J., Bodeker, G. E., Cooper, K. E., Lewis, J., and Paterson, A. J. (2017). A comparison of Loon balloon observations and stratospheric reanalysis products. *Atmos. Chem. Phys.* 17(2), 855–866. https://doi.org/10.5194/acp-17-855-2017.

Fromm, M. D., Lumpe, J. D., Bevilacqua, R. M., Shettle, E. P., Hornstein, J., Massie, S. T., and Fricke, K. H. (1997). Observations of Antarctic polar stratospheric clouds by POAM II: 1994-1996. *J. Geophys. Res. Atmos.* 102, 23659–23672. https://doi.org/10.1029/97JD00794

Fueglistaler, S., Buss, S., Luo, B. P., Wernli, H., Flentje, H., Hostetler, C. A., Poole, L. R., Carslaw, K. S., and Peter, T. (2003). Detailed modelling of mountain wave PSCs. *Atmos. Chem. Phys.* 3, 697–712. https://doi.org/10.5194/acp-3-697-2003

Gelaro, R., McCarty, W., Suárez, M. J., Todling, R., Molod, A., Takacs, L., Randles, C. A., Darmenov, A., Bosilovich, M. G., Reichle, R., Wargan, K., Coy, L., Cullather, R., Draper, C., Akella, S., Buchard, V., Conaty, A., da Silva, A. M., Gu, W., Kim, G.-K., Koster, R., Lucchesi, R., Merkova, D., Nielsen, J. E., Partyka, G., Pawson, S., Putman, W., Rienecker, M., Schubert, S. D., Sienkiewicz, M., and Zhao, B. (2017). The Modern-Era Retrospective Analysis for Research and Applications, Version 2 (MERRA-2). *J. Clim.* 30, 5419–5454. https://doi.org/10.1175/JCLI-D-16-0758.1

Goodman, J., Verma, S., Pueschel, R. F., Hamill, P., Ferry, G. V., and Webster, D. (1997). New evidence of size and composition of polar stratospheric cloud particles. *Geophys. Res. Lett.* 24, 615–618. https://doi.org/10.1029/97GL00256

Gray, L. J., Anstey, J. A., Kawatani, Y., Lu, H., Osprey, S., and Schenzinger, V. (2018). Surface impacts of the quasi-biennial oscillation. *Atmos. Chem. Phys.* 18, 8227–8247. https://doi.org/10.5194/acp-18-8227-2018

Hofmann, D. J., and Solomon, S. (1989). Ozone destruction through heterogeneous chemistry following the eruption of El Chichón. *J. Geophys. Res.* 94, 5029–5040. https://doi.org/10.1029/JD094iD04p05029

Holton, J. R., Haynes, P. H., McIntyre, M. E., Douglass, A. R., Rood, R. B., and Pfister, L. (1995). Stratosphere-troposphere exchange. *Rev. Geophys.* 33, 403–439. https://doi.org/10.1029/95RG02097

Höpfner, M., Deshler, T., Pitts, M., Poole, L., Spang, R., Stiller, G., and von Clarmann, T. (2018). The MIPAS/Envisat climatology (2002–2012) of polar stratospheric cloud volume density profiles. *Atmos. Meas. Tech.* 11, 5901–5923. https://doi.org/10.5194/amt-11-5901-2018

Isoda, F., Tsuda, T., Nakamura, T., Vincent, R. A., Reid, I. M., Achmad, E., Sadewo, A., and Nuryanto, A. (2004). Intraseasonal oscillations of the zonal wind near the mesopause were observed with medium-frequency and meteor radars in the tropics. *J. Geophys. Res. Atmos.* 109, D21108. https://doi.org/10.1029/2003JD003378

Johanson, C. M., and. Fu, Q. (2007). Antarctic atmospheric temperature trend patterns from satellite observations. *Geophys. Res.Lett.* 34, L12703. https://doi.org/10.1029/2006GL029108

Karpechko, A. Y., and Manzini, E. (2012). Stratospheric influence on tropospheric climate change in the northern hemisphere. *J. Geophys. Res. Atmos.* 117, D05133. https://doi.org/10.1029/2011JD017036

Kawatani, Y., Hamilton, K., Miyazaki, K., Fujiwara, M., and Anstey, J. A. (2016). Representation of the tropical stratospheric zonal wind in global atmospheric re-analyses. *Atmos. Chem. Phys.* 16, 6681–6699. https://doi.org/10.5194/acp-16-6681-2016

Kawatani, Y., Hirooka, T., Hamilton, K., Smith, A. K., and Fujiwara, M. (2020). Representation of the equatorial stratopause semiannual oscillation in global atmospheric reanalyses. *Atmos. Chem. Phys.* 20, 9115–9133. https://doi.org/10.5194/acp-20-9115-2020

Kidston, J., Scaife, A. A., Hardiman, S. C., Mitchell, D. M., Butchart, N., Baldwin, M. P., and Gray, L. J. (2015). Stratospheric influence on tropospheric jet streams, storm tracks and surface weather. *Nat. Geosci.* 8, 433–440. https://doi.org/10.1038/ngeo2424

Klotzbach, P., Abhik, S., Hendon, H. H., Bell, M., Lucas, C., G. Marshall, A., and Oliver, E.C.J. (2019). On the emerging relationship between the stratospheric Quasi-Biennial oscillation and the Madden-Julian oscillation. *Sci. Rep.* 9, 2981. https://doi.org/10.1038/s41598-019-40034-6

Kodera, K., Yamazaki, K., Chiba, M., and Shibata, K. (1990). Downward propagation of upper stratospheric mean zonal wind perturbation to the Troposphere. *Geophys. Res. Lett.* 17, 1263–1266. https://doi.org/10.1029/GL017i009p01263

Kolstad, E. W., Breiteig, T., and Scaife, A. A. (2010). The association between stratospheric weak polar vortex events and cold air outbreaks in the Northern Hemisphere. *Q. J. R. Meteorol. Soc.* 136, 886–893. https://doi.org/10.1002/qj.620

Kumar, K. K., Antonita, T. M., Ramkumar, G., Deepa, V., Gurubaran, S., and Rajaram, R. (2007). On the tropospheric origin of Mesosphere Lower Thermosphere region intraseasonal wind variability. *J. Geophys. Res.* 112, D07109. https://doi.org/10.1029/2006JD007962

Kumari, K., Wu, H., Long, A., Lu, X., and Oberheide, J. (2021). Mechanism studies of Madden-Julian oscillation coupling into the mesosphere/lower thermosphere tides using SABER, MERRA-2, and SD-WACCMX. *J. Geophys. Res. Atmos.* 126. https://doi.org/10.1029/2021JD034595

Kwon, H., Choi, H., Kim, B.-M., Kim, S.-W., and Kim, S. J. (2020). The recent weakening of the southern stratospheric polar vortex and its impact on the surface climate over Antarctica. *Environ. Res. Lett.* 15, 094072. https://doi.org/10.1088/1748-9326/ab9d3d

Lambert, A., Santee, M. L., Wu, D. L., and Chae, J. H. (2012). A-train CALIOP and MLS observations of early winter Antarctic polar stratospheric clouds and nitric acid in 2008. *Atmos. Chem. Phys.* 12, 2899–2931. https://doi.org/10.5194/acp-12-2899-2012

Lenton, A., Codron, F., Bopp, L., Metzl, N., Cadule, P., Tagliabue, A., and Le Sommer, J. (2009). Stratospheric ozone depletion reduces ocean carbon uptake and enhances ocean acidification. *Geophys. Res. Lett.* 36, L12606. https://doi.org/10.1029/2009GL038227

Levine, J. G., Braesicke, P., Harris, N. R. P., Savage, N. H., and Pyle, J. A. (2007). Pathways and timescales for troposphere-to-stratosphere transport via the tropical tropopause layer and their relevance for very short-lived substances. *J. Geophys. Res.* 112, D04308. https://doi.org/10.1029/2005JD006940

Li, F. et al. (2016). Impacts of interactive stratospheric chemistry on the Antarctic and Southern Ocean climate change in the Goddard Earth Observing System, version 5 (GEOS-5). *J. Clim.* 29, 3199–3218. https://doi.org/10.1175/JCLI-D-15-0572.1

Lieberman, R. S. (1999). Eliassen–Palm fluxes of the 2-day wave. *Am. Meteorol. Soc.* 56(16), 2846–2861.10.1175/1520-0469(1999)056<2846:EPFOTD>2.0.CO;2.

Liess, S., and Geller, M. A. (2012). On the relationship between QBO and distribution of tropical deep convection. *J. Geophys. Res. Atmos.* 117, 2552–2568. https://doi.org/10.1029/2011JD016317

Lim, E.-P., Hendon, H. H., Boschat, G., Hudson, D., Thompson, D. W. J., Dowdy, A. J., and Arblaster, J. M. (2019). Australian hot and dry extremes induced by weakenings of the stratospheric polar vortex. *Nat. Geosci.* 12, 896–901. https://doi.org/10.1038/s41561-019-0456-x

Lowe, D., and MacKenzie, A. R. (2008). Polar stratospheric cloud microphysics and chemistry. *J. Atmos. Solar-Terrestrial Phys.* 70, 13–40. https://doi.org/10.1016/j.jastp.2007.09.011

Manzer, L.E. (1990). The CFC-Ozone issue: Progress on the development of alternatives to CFCs. *Science.* 249, 31–35. https://doi.org/10.1126/science.249.4964.31

Martineau, P., Son, S.-W., and Taguchi, M. (2016). Dynamical consistency of reanalysis datasets in the extratropical stratosphere. *J. Clim.* 29, 3057–3074. https://doi.org/10.1175/JCLI-D-15-0469.1

Martineau, P., and Son, S. W. (2010). Quality of reanalysis data during stratospheric vortex weakening and intensification events. *Geophys. Res. Lett.* 37, 1–15. https://doi.org/10.1029/2010GL045237

Maturilli, M., Neuber, R., Massoli, P., Cairo, F., Adriani, A., Moriconi, M. L., and Di Donfrancesco, G. (2005). Differences in Arctic and Antarctic PSC occurrence as observed by Lidar in Ny-Ålesund (79° N, 12° E) and McMurdo (78° S, 167° E). *Atmos. Chem. Phys.* 5, 2081–2090. https://doi.org/10.5194/acp-5-2081-2005

McDonald, A. J., George, S. E., and Woollands, R. M. (2009). Can gravity waves significantly impact PSC occurrence in the Antarctic? *Atmos. Chem. Phys.* 9, 8825–8840. https://doi.org/10.5194/acp-9-8825-2009

McLandress, C., Jonsson, A. I., Plummer, D. A., Reader, M. C., Scinocca, J. F., and Shepherd, T. G. (2010). Separating the dynamical effects of climate change and ozone depletion. Part I: Southern Hemisphere stratosphere. *J Clim.* 23, 5002–5020.

Nakajima, H., Wohltmann, I., Wegner, T., Takeda, M., Pitts, M.C., Poole, L. R., Lehmann, R., Santee, M. L., and Rex, M. (2016). Polar stratospheric cloud evolution and chlorine activation measured by CALIPSO and MLS and modelled by ATLAS. *Atmos. Chem. Phys.* 16, 3311–3325. https://doi.org/10.5194/acp-16-3311-2016

Nie, J., and Sobel, A. H. (2015). Responses of tropical deep convection to the QBO: Cloud-resolving simulations. *J. Atmos. Sci.* 72, 3625–3638. https://doi.org/10.1175/JAS-D-15-0035.1

Noel, V., Hertzog, A., and Chepfer, H. (2009). CALIPSO observations of wave-induced PSCs with near-unity optical depth over Antarctica in 2006–2007. *J. Geophys. Res.* 114, D05202. https://doi.org/10.1029/2008JD010604

Noel, V., and Pitts, M. (2012). Gravity wave events from mesoscale simulations, compared to polar stratospheric clouds observed from spaceborne Lidar over the Antarctic Peninsula. *J. Geophys. Res. Atmos.* 117. https://doi.org/10.1029/2011JD017318

Orr, A., Bracegirdle, T. J., Hosking, J. S., Jung, T., Haigh, J. D., Phillips, T., and Feng, W. (2012). Possible dynamical mechanisms for Southern Hemisphere climate change due to the ozone hole. *J. Atmos. Sci.* 69, 2917–2932. https://doi.org/10.1175/JAS-D-11-0210.1

Orr, A., Hosking, J. S., Hoffmann, L., Keeble, J., Dean, S. M., Roscoe, H. K., Abraham, N. L., Vosper, S., and Braesicke, P. (2015). Inclusion of mountain-wave-induced cooling for the formation of PSCs over the Antarctic Peninsula in a chemistry-climate model. *Atmos. Chem. Phys.* 15, 1071–1086. https://doi.org/10.5194/acp-15-1071-2015

Pagan, K. L., Tabazadeh, A., Drdla, K., Hervig, M. E., Eckermann, S. D., Browell, E. V., Legg, M. J., and Foschi, P. G. (2004). Observational evidence against the mountain-wave generation of ice nuclei as a prerequisite for the formation of three solid nitric acid polar stratospheric clouds observed in the Arctic in early December 1999. *J. Geophys. Res. Atmos.* 109. https://doi.org/10.1029/2003JD003846

Previdi, M., and Polvani, L. M. (2014). Climate system response to stratospheric ozone depletion and recovery. *Q. J. R. Meteorol. Soc.* 140, 2401–2419. https://doi.org/10.1002/qj.2330

Ramaswamy, V., Schwarzkopf, M. D., and Shine, K. P. (1992). Radiative forcing of climate from halocarbon-induced global stratospheric ozone loss. *Nature*, 355, 810–812. https://doi.org/10.1038/355810a0

Rind, D., and Lacis, A. (1993). The role of the stratosphere in climate change. *Surv. Geophys.* 14, 133–165. https://doi.org/10.1007/BF02179221

Russell, J. L., Dixon, K. W., Gnanadesikan, A., Stouffer, R. J., and Toggweiler, J. R. (2006). The Southern Hemisphere westerlies in a warming world: propping open the door to the deep ocean. *J. Clim.* 19, 6382–6390. https://doi.org/10.1175/JCLI3984.1

Shaw, T. A., and Perlwitz, J. (2013). The life cycle of Northern Hemisphere downward wave coupling between the stratosphere and troposphere. *J. Clim.* 26, 1745–1763. https://doi.org/10.1175/JCLI-D-12-00251.1

Sherwood, S. C., and Dessler, A. E. (2000). On the control of stratospheric humidity. *Geophys. Res. Lett.* 27, 2513–2516. https://doi.org/10.1029/2000GL011438

Sigmond, M., and Fyfe, J. C. (2010). Has the ozone hole contributed to increased Antarctic sea ice extent? *Geophys. Res. Lett.* 37, L18502. https://doi.org/10.1029/2010GL044301

Smith, K. L., Polvani, L. M., and Marsh, D. R. (2012). Mitigation of 21st-century Antarctic sea ice loss by stratospheric ozone recovery. *Geophys. Res. Lett.* 39, 2012GL053325. https://doi.org/10.1029/2012GL053325

Solomon, S., Garcia, R. R., Sherwood Rowland, F., and Wuebbles, D. J. (1986). On the depletion of Antarctic ozone. *Nature*. 321, 755–758.

Son, S.-W., Gerber, E. P., Perlwitz, J., Polvani, L. M., Gillett, N. P., Seo, K.-H., Eyring, V., Shepherd, T. G., Waugh, D., Akiyoshi, H., Austin, J., Baumgaertner, A., Bekki, S., Braesicke, P., Brühl, C., Butchart, N., Chipperfield, M. P., Cugnet, D., Dameris, M., Dhomse, S., Frith, S., Garny, H., Garcia, R., Hardiman, S. C., Jöckel, P., Lamarque, J. F., Mancini, E., Marchand, M., Michou, M., Nakamura, T., Morgenstern, O., Pitari, G., Plummer, D. A., Pyle, J., Rozanov, E., Scinocca, J. F., Shibata, K., Smale, D., Teyssèdre, H., Tian, W., and Yamashita, Y. (2010). Impact of stratospheric ozone on Southern Hemisphere circulation change: A multimodel assessment. *J. Geophys. Res.* 115, D00M07. https://doi.org/10.1029/2010JD014271

Son, S.-W., Han, B.-R., Garfinkel, C. I., Kim, S.-Y., Park, R., Abraham, N. L., Akiyoshi, H., Archibald, A. T., Butchart, N., Chipperfield, M. P., Dameris, M., Deushi, M., Dhomse, S. S., Hardiman, S. C., Jöckel, P., Kinnison, D., Michou, M., Morgenstern, O., O'Connor, F. M., Oman, L. D., Plummer, D. A., Pozzer, A., Revell, L. E., Rozanov, E., Stenke, A., Stone, K., Tilmes, S., Yamashita, Y., and Zeng, G. (2018). Tropospheric jet response to Antarctic ozone depletion: An update with Chemistry-Climate Model Initiative (CCMI) models. *Environ. Res. Lett.* 13, 054024. https://doi.org/10.1088/1748-9326/aabf21

Son, S.-W., Lim, Y., Yoo, C., Hendon, H. H., and Kim, J. (2017). Stratospheric Control of the Madden–Julian Oscillation. *J. Clim.* 30, 1909–1922. https://doi.org/10.1175/JCLI-D-16-0620.1

Son, S.-W., Polvani, L. M., Waugh, D. W., Akiyoshi, H., Garcia, R., Kinnison, D., Pawson, S., Rozanov, E., Shepherd, T. G., and Shibata, K. (2008). The impact of stratospheric ozone recovery on the Southern Hemisphere westerly jet. *Science*. 320, 1486–1489. https://doi.org/10.1126/science.1155939

Thompson, D. W. J., and Solomon, S. (2002). Interpretation of recent Southern Hemisphere climate change. *Science*. 296, 895–899. https://doi.org/10.1126/science.1069270

Thompson, D. W. J., Solomon, S., Kushner, P. J., England, M. H., Grise, K. M., and Karoly, D. J. (2011). Signatures of the Antarctic ozone hole in Southern Hemisphere surface climate change. *Nat. Geosci.* 4, 741–749. https://doi.org/10.1038/ngeo1296

Tritscher, I., Pitts, M. C., Poole, L. R., Alexander, S. P., Cairo, F., Chipperfield, M. P., Grooß, J., Höpfner, M., Lambert, A., Luo, B., Molleker, S., Orr, A., Salawitch, R., Snels, M., Spang, R., Woiwode, W., and Peter, T. (2021). Polar stratospheric clouds: Satellite observations, processes, and role in ozone depletion. *Rev. Geophys.* 59. https://doi.org/10.1029/2020RG000702

Uma, K. N., Das, S. K., and Das, S. S. (2014). A climatological perspective of water vapour at the UTLS region over different global monsoon regions: observations inferred from the Aura-MLS and reanalysis data. *Clim. Dyn.* 43, 407–420. https://doi.org/10.1007/s00382-014-2085-9

Wang, G., Hendon, H. H., Arblaster, J. M., Lim, E. P., Abhik, S., and van Rensch, P. (2019). Compounding tropical and stratospheric forcing of the record low Antarctic sea-ice in 2016. *Nat. Commun.* 10, 1–9. https://doi.org/10.1038/s41467-018-07689-7

Wang, P. K., Setvák, M., Lyons, W., Schmid, W., and Lin, H.-M. (2009). Further evidence of deep convective vertical transport of water vapour through the tropopause. *Atmos. Res.* 94, 400–408. https://doi.org/10.1016/j.atmosres.2009.06.018

3 Antarctic Weather and Climate Patterns

S Sunitha Devi
India Meteorological Department, Ministry of Earth
Sciences, New Delhi, India

R.S. Maheskumar
Ministry of Earth Sciences, New Delhi, India

CONTENTS

DOI: 10.1201/9781003203742-3

3.1   INTRODUCTION

Weather-wise, Antarctica is the coldest, windiest, and driest continent on planet Earth. It is mainly due to a combination of geographical, physical, and thermo-dynamic factors involving rather steep orographic terrain, less solar energy received all through the year, and high albedo owing to persistent snow cover. Though these factors are applicable more or less the same to both the North and South Poles of the Earth, what makes Antarctica, the central region of the South Pole, even colder than its northern counterpart, the Arctic, is partly due to the massive and thick ice sheet that covers about 98% of the continent. What makes Antarctica the windiest place on Earth's surface is its peculiar topography. It has the highest average elevation of any continent (4,892 meters, or 8,200 feet) surrounded by water (which holds its temperature longer than land). The steep temperature gradient thus created causes katabatic solid winds. As a result of these meteorological factors, the average temperature experienced over the coastal belt of Antarctica is around −10°C (14°F) and the temperature in its inland region is around −55°C (−67°F).

Due to the above-mentioned topographic feature and the resultant katabatic winds, which inhibit moisture incursion and cloud formation, Antarctica receives too little rainfall – it averages about 5 cm per year over the interior parts of the continent, usually in the form of snow. Hence, it is climatologically classified as a desert. Also, in terms of the areal coverage, Antarctica can be considered the largest desert on Earth (14.2 million square km), more significant than the world's three large deserts: the Gobi, Arabian, and Sahara put together.

We will discuss the significant weather and climatic features of this continent and attempt to unravel the fundamental causes behind the above-depicted pecu-liarities exhibited by the region. This compilation is based on the review of several peer-reviewed publications and scientific facts available from open sources.

3.1.1   GEOGRAPHY OF ANTARCTICA

Antarctica is in the south polar region of the Earth, encircled within the latitudinal belt of 60°S, and is surrounded by the Southern Ocean. Its surface is primarily covered with ice, which accounts for 90% of its freshwater ice. Geographically, the continent can be divided into three broad regions: East Antarctica, West Antarctica, and the Antarctic Peninsula. The Trans-Antarctic Mountains act as a natural boundary between the eastern and western continents (Figure 3.1). East Antarctica has a more or less circular and symmetric form. In contrast, West Antarctica has a spiral shape, dominated by the Antarctic Peninsula and two great embayments containing the Ross and Weddell Seas (King and Turner, 1997).

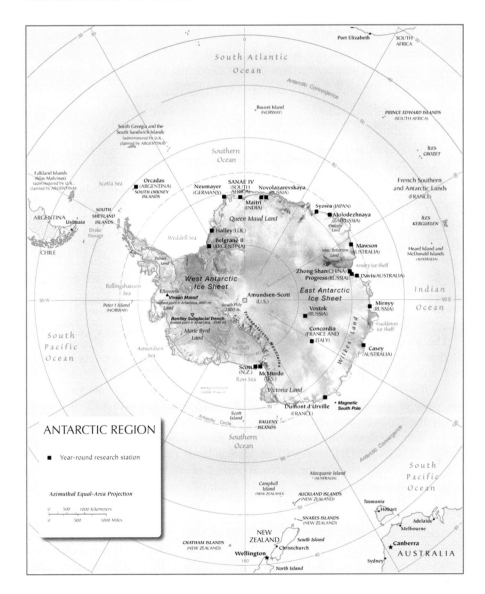

FIGURE 3.1   Map of Antarctica (Courtesy: https://www.nationsonline.org/).

3.1.2   SEASONS IN ANTARCTICA

Antarctica has only two seasons: summer and winter. Summer is from October to February. During this time, the sun is almost always in the sky. Days rapidly get longer there in summer until, eventually, the sun doesn't set at all. This phenomenon is called the midnight sun. The remaining are the winter months, devoid of much sunlight and prolonged nights.

3.2   WEATHER PATTERNS

The weather over Antarctica is quite harsh and variable and consequently threat posing. Significant weather features consist of mainly katabatic winds and meso-cyclones, apart from a few synoptic-scale storms forming over the Southern Ocean.

3.2.1   EXTREMES

All kinds of extremes characterize the weather over Antarctica, be it temperatures, winds, or high variability in local conditions and extreme precipitation events (McCormick, 2002).

3.2.1.1   Temperature

As has already been stated, what makes Antarctica the coldest place on Earth is that most of the continent is a high, flat plateau of ice covered in snow throughout the year. The high albedo of snow which reflects a large percentage of solar radiation, together with the high altitude (an average of 14,000 feet above sea level) and latitude (60°S–90°S, where the sun's rays reach at a smaller elevation angle, causing it to spread over a larger area, resulting in less heating) makes the interior places of Antarctica the coldest regions on Earth.

The lowest ever recorded temperature on the Earth's surface was at Vostok Station (a station maintained by Russia) at −89.2°C (−128.6°F) on July 21, 1983, and at the South Pole −82.8°C (−117.0°F), on June 23, 1982.

Until 2011, the warmest temperature ever recorded at the South Pole, recorded at the peak of the Southern Hemispheric summer had been −12.3°C (9.9°F) on December 25, 2011, was still well below freezing. However, of late, there had been still warmer temperature records. The highest temperature ever recorded on Antarctica was 20.75°C (69.3°F) at Comandante Ferraz Antarctic Station on February 9, 2020, beating the previous record of 18.3°C (64.9°F) at Esperanza Base, on the northern tip of the Antarctic Peninsula, on February 6, 2020.

3.2.1.2   Wind

The peculiar orographic feature of the continent makes the large-scale wind system katabatic. Since the highlands radiate heat and get cooled, the air in contact with these highlands also gets cooled and thus becomes denser than the air at the same elevation and therefore begins to flow downhill. These winds are often quite intense and blow out from the polar plateau of the interior part of the continent, down the steep vertical edges along the coast. It is at the steep edge of Antarctica that the strong katabatic winds form.

The highest wind speed recorded in Antarctica was 327 km/h (199 mph) at Dumont d'Urville Station in July 1972. Because the South Pole is well inland and on a flat plateau area, the katabatic winds in this region are relatively mild. The highest re-corded wind at the South Pole was only 92.6 km/h (58 mph) on September 27, 2011.

3.2.1.3   Precipitation

Due to the prevalence of katabatic winds, interior parts of Antarctica are dominated by dry air and subsidence motion. Hence, these regions will not have much cloud development. In contrast, the availability of moisture and low-pressure systems make the coastal belt often cloudy. Thus, we may notice the presence of clouds over the Antarctic Peninsula region.

The precipitation (hydro meteors reaching the Earth's surface) generally occurs in snow or ice crystals, though rain is observed occasionally near the coast. Due to the nearly omnipresent gale-force winds, an accurate measurement of snowfall is difficult as the snow crystals would be blown off and miss the measuring equipment. However, the estimated average accumulation of snow over the continent is about 150 mm of water equivalent per year. Over the elevated plateau region, the annual value is less than 50 mm. Near the coast, it exceeds 200 mm in general, though the heaviest being over 1,000 mm for an area near the Bellingshausen Sea.

Snowfall in Antarctica consists of frequent clear-sky precipitation and heavier falls, also referred to as the 'extreme precipitation events (EPEs)' from intrusions of maritime air masses associated with amplified planetary waves. EPEs consisting of the most significant 10% of daily totals are shown to contribute more than 40% of the total annual precipitation across much of the continent, with some areas receiving over 60% of the total from these events. The most significant contribution of extreme precipitation events to the annual total is in the coastal areas and especially on the ice shelves, with the Amery Ice Shelf receiving 50% of its annual precipitation in less than the 10 days of heaviest rainfall. For the continent as a whole, 70% of the variance of the yearly rainfall is explained by variability in precipitation from extreme precipitation events, with this figure rising to over 90% in some areas (Turner et al., 2019).

Maritime climate prevails over the western side of the continent where depressions form to the west give light to moderate precipitation for several days. Hence, the role of EPEs is more minor in this region. In contrast, the eastern side has a relatively dry climatological southerly flow. However, occasionally strong westerlies can bring air masses to the region with little moisture because of the Foehn effect. Incursions of moist air from the north result in significant moisture convergence so that EPEs is much more important over this region. One region where EPEs have a critical role in controlling the total precipitation is Victoria Land, especially the region on the western side of the Ross Sea and inland of the Terra Nova Bay region. This dry area has experienced a significant decrease in snow accumulation since the 1950s (Thomas et al., 2017). Katabatic solid winds flow down to the coast (Bromwich, 1989), impeding the penetration of many maritime air masses into the interior, thus increasing the influence of the heavy snowfall events that occasionally arrive. Over the very highest parts of the Antarctic plateau in the Dome A, Ridge B, and toward Dome F areas, intrusions of maritime air are much rarer than in the coastal region, so the annual precipitation amount is dictated more by frequent clear-sky precipitation episodes (Sato et al., 1981). Therefore, the role of EPEs is less critical at the highest elevations, with more than 40 of the highest precipitation days required to give 50% of the annual total. The importance of orography in influencing the nature of the precipitation is evident in the central part of West Antarctica along 90°W. This area has higher orography compared to the regions to the east and west. Still, it is a region of frequent intrusions of maritime air from the Southern Ocean, which are forced up to higher elevations, giving more incredible cloud and precipitation (Nicolas and Bromwich, 2011) and reducing the importance of EPEs, with more than 50 days of the heaviest rainfall required to give 50% of the annual total. Cyclones, whether local or remotely located, are essential in providing EPEs in all parts of the Antarctic, except the highest parts of the central plateau.

The variability in the number of EPEs throughout the year at many locations is therefore strongly influenced by the semi-annual oscillation, which is the cycle that controls the southward (northward) movement and deepening (weakening) of depressions in summer and winter (Van den Broeke, 2000). The seasonal fields showing the importance of EPE in dictating the total precipitation are all very similar to the annual data. However, when there is a less cyclonic activity in summer, EPEs have a slightly smaller influence over the ocean and coastal areas.

3.2.2 Major Weather Systems

3.2.2.1 Synoptic-Scale Weather Systems and Fronts

Synoptic-scale weather systems, comprising the extra-tropical cyclones (depressions) and anticyclones (highs), are the central atmospheric systems in the Antarctic coastal region and over the Southern Ocean. They typically have a horizontal length scale in the range of 1,000–6,000 km and a lifetime of between 1 day and a week. Synoptic scale disturbances lie between the mesoscale (less than 1,000 km diameter) atmospheric phenomena, like cloud clusters, mesocyclones, squall lines, and the planetary-scale long waves with a few thousand wavelength kilometers. Due to the sparsity of observational networks, satellite imageries are the primary tool in locating the synoptic-scale, low-pressure systems and associated frontal cloud bands and tracking them over the Southern Ocean.

Case studies and climatological investigations have shown that depressions form mainly on horizontal temperature gradients (baroclinic zones) in the troposphere and grow through baroclinic instability. In the Southern Hemisphere, the significant meridional temperature change in the troposphere occurs at the polar front separating the temperate mid-latitude air to the north and the cold polar air masses to the south (King and Turner, 1997).

The following discussion is mainly adopted from a Scholarly Paper (Keaveney, 2004), which analyzes in depth the definitive requirements, constraints, and methodology for generating accurate weather forecasts for aircraft operations catering to the research facilities established over the continent.

3.2.2.2 Katabatic Winds

When a surface cools radiatively, the air near the surface becomes colder and denser than the aloft, thereby achieving negative buoyancy. When this process occurs over sloping terrain, the cold, thick air close to the surface accelerates down the slope in response to the buoyancy force. This downslope flow is known as a katabatic wind. This type of flow is expected in the mid-latitudes as well as in Antarctica. Mid-latitude katabatic winds only occur at night and are considered a small-scale feature of the boundary layer (King and Turner, 1997); however, in Antarctica, the katabatic winds are unique as it controls the entire low-level circulation, all through the year, due to the perennial cold surface, stable boundary layer with a nearly persistent inversion layer, and the overall dome-shaped topography. The overall general circulation pattern consists of winds moving from the polar plateau towards the coasts, turning left under the effect of the Coriolis force, and merging with the coastal polar easterlies (King and Turner, 1997). Wind speeds are correlated to the degree of terrain slope: the steeper the hill, the stronger the winds. Their flow direction is also determined by topography, as the interior winds tend to follow the topographic features, while most coastal winds have varying approaches. This is a function of the degree to which synoptic and mesoscale features influence the wind regime. Though the interior is relatively blocked from the impacts of such weather systems, the coast is not. This results in an interior flow characterized by persistent, directionally constant winds (purely katabatic winds) and a coastal flow that combines katabatic and synoptic/mesoscale forcing. In general, the katabatic winds are more intense during the winter when the surface radiative cooling and stable boundary layer are at their peak.

Antarctica's katabatic flow is well known for producing extreme, sustained winds. Specific points along East Antarctica's coast are especially famous for their high winds, which routinely reach hurricane force. Cape Denison, on the coast of Adélie Land, has a mean annual wind speed of 22 m s^{-1} [50 mph], and its maximum winds have approached 89 m s^{-1} [200 mph] (GLACIER, 1998). Douglas Mawson's expedition, based at Cape Denison, recorded gale-force winds on all but 1 of 203 consecutive winter days during 1912–1913 (King and Turner, 1997). While places like Cape Denison are extreme, the remainder of coastal Antarctica has mean winds of only 5–10 m s^{-1} (King and Turner, 1997). The reason behind the constrained locations of intense coastal winds is local topography.

While researching surface winds over East Antarctica, Parish (1982) discovered an exciting drainage pattern as the winds approached the coast. The streamlines of the flow converged in certain areas, one of which was upstream of Cape Denison. He concluded that the positioning of this confluence zone would provide Cape Denison with a "nearly inexhaustible supply of cold air", which flows through a narrow to-pographic gap at the coast, allowing for the formation of the persistent, strong ka-tabatic winds observed there. He also found another confluence zone upstream of Terra Nova Bay, where similarly intense, prolonged katabatic winds were also monitored. Bromwich and Kurtz (1984) investigated the confluence zone theory for Terra Nova Bay's katabatic winds. The winds at this location are essential for two reasons: first, they help form and maintain a large polynya [open water in the ice] throughout the winter, and second, the winds are linked to mesocyclone development.

Utilizing an updated and more precise topographic map of Antarctica, Parish and Bromwich (1987) calculated the near-surface wind pattern over the whole con-tinent. Their results further verified the confluence zones found by Parish (1982), which had also been identified by several others, specifically a zone near the Byrd Glacier [southwest portion of the Ross Ice Sheet] and a zone near the Siple Coast [southeast portion of Ross Ice Sheet] (Figure 3.2).

The winds from the confluence zone near the Siple Coast will, with no synoptic or mesoscale pressure gradient present, both turn to the left under the Coriolis force

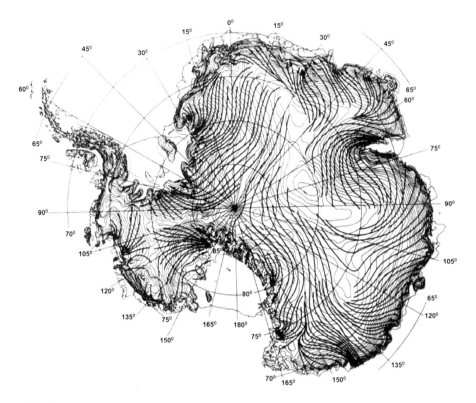

FIGURE 3.2 Simulated streamlines of katabatic wind drainage (Reproduced from Parish and Bromwich 1987).

and be retarded by surface friction as they flow across the Ross Ice Sheet (King and Turner, 1997). This will force them up against the Transantarctic Mountains, but they will not cross the mountain ranges. As a result, the cold air will continue to pile up, deepening toward the mountains, creating a pressure gradient perpendicular to the barrier (King and Turner, 1997). This will form a northward-flowing barrier wind with the mountain on its left. If a synoptic-scale cyclone is in the Amundsen Sea, its pressure gradient will act to balance the Coriolis force and overcome friction. This will allow the katabatic winds from the Siple Coast to flow over 1,000 km across the Ross Ice Shelf, reaching the Ross Sea near Ross Island, without significant deceleration or change in direction (King and Turner, 1997). This phenomenon is known as a katabatic surge.

Barrier winds will also form if either a synoptic or mesoscale cyclone is present over the central Ross Ice Sheet or the southern Ross Sea. In this case, the easterly winds of the hurricane will be forced against the Transantarctic Mountains (O'Connor et al., 1994) along with the katabatic winds.

3.2.2.3 Mesocyclones
Mesocyclones are relatively short-lived, sub-synoptic-scale, low-pressure systems that occur in polar regions. They generally exist for less than 24 hours and have a

horizontal extent of less than 1,000 km. Though they can produce hazardous weather conditions over the oceans and along the coast of Antarctica, due to the sparse surface observational network, their existence was unknown until the advent of satellites. (King and Turner, 1997). Furthermore, research into this phenomenon of Antarctic meteorology was not started until high-resolution satellite imagery was readily available for this region in the early 1980s (King and Turner, 1997). For this reason, mesocyclones are not understood to the level that has been achieved for Antarctica's synoptic features; however, recent research is helping to close this gap and clarify the role of mesocyclones in Antarctica's complex weather regime.

Typical Antarctic mesocyclones are cold air vortices with diameters less than 500 km and lifetimes less than 12 hours. As observed from satellite images, their cloud patterns can be classified into spiraliform and comma-shaped (King and Turner, 1997). Spiraliform mesocyclones have a circular, symmetric pattern with several cloud bands encircling the low center. In contrast, the comma-shaped mesocyclones have a comma-shaped appearance, with the head of the cloud near the center of the common and the tail stretching toward higher pressure (King and Turner, 1997). The formation of both types is based on the extent of synoptic support present. Spiraliform cyclones form in areas with no synoptic forcing, whereas comma-shaped lows need strong synoptic flow (Turner and Pendlebury, 2002). Mesocyclones develop throughout the year along the coast of Antarctica, over its ice shelves, at the northern edges of sea ice, and over the ice-free Southern Ocean (Carrasco et al., 2003; Simmonds et al., 2003).

On average, half the mesoscale vortices are comma-shaped. Carrasco and Bromwich (1994) found a greater prevalence of more than 70% for comma-shaped mesocyclones in these areas [minus the Siple Coast] during 1988. These results indicate the importance of solid synoptic support to mesocyclone activity in the Ross Sea/Ross Ice Shelf region. The horizontal extents of the vortices are all below 300 km, and only a tiny percentage of the total is considered deep [defined by Carrasco et al. (2003) as containing middle/white cloud signatures on a grayscale satellite image]. Their lack of vertical growth (often less than 700 hPa level) indicates the prevalence of stable conditions. The stability of the Ross Sea/Ross Ice Shelf region is a by-product of the katabatic winds that flow through this area.

Carrasco et al. (2003) also compared monthly distributions during 1991 and found the most fantastic mesoscale activity occurred in the summer. In addition, they observed that the annual average weekly mesoscale cyclone formation equalled two to three in Terra Nova and one to two in Byrd Glacier. This matches results found by Bromwich (1991) and Carrasco and Bromwich (1994).

To assess 1991's activity against other years, Carrasco et al. (2003) compared the synoptic situation during that year to climatology. Since mesoscale cyclogenesis is a function of synoptic variability, this type of comparison would identify how far 1991's activity varied from climatology. They found that the Ross Sea/Ross Ice Shelf region's sea level pressure anomaly was nearly zero, indicating that mesocyclone activity for 1991 was close to climatology. Therefore, the mesoscale cyclone activity observed in this region is not a unique characteristic but a regular one.

The process of mesocyclone genesis is not entirely understood at present. However, certain factors appear to be important: baroclinic instability, barotropic instability, vortex stretching, surface fluxes of heat and moisture, and low-level convergence (Turner and Pendlebury, 2002; Gallée, 1995). The top three areas of mesoscale cyclone activity highlighted by Carrasco et al. (2003) [Terra Nova Bay, Byrd Glacier, and the Siple Coast] can provide most of these factors through their respective confluence zones and the resultant katabatic wind regimes they create.

Baroclinic instability in Antarctica results from "moderate-to-strong low-level thermal gradients" (King and Turner, 1997). Katabatic winds in the three areas produce low-level baroclinicity by transporting cold air from East Antarctica to the warmer air over the Ross Sea and Ross Ice Shelf, establishing a baroclinic zone that can initiate mesocyclone formation (Carrasco and Bromwich, 1993, 1994; King and Turner, 1997; Carrasco et al., 2003; Heinemann and Klein, 2003). Gallée (1995) conducted a numerical simulation over the southwestern Ross Sea [Terra Nova Bay area] to "determine if pure katabatic winds [were] able to force mesocyclonic activity during the open water season (February-March)". He found that the katabatic winds formed boundary layer fronts [baroclinic zones] that played an essential role in forming and deepening southwestern Ross Sea mesocyclones. Bromwich (1991) used AWS data and satellite images to study mesoscale cyclogenesis at Terra Nova Bay and Byrd Glacier. He noted that both regions were active mesocyclone areas and intense discharge of cold East Antarctic boundary-layer air. He found that baroclinicity and cyclonic vorticity in the boundary layer was linked to the areas' high mesoscale cyclone activity. Although the confluence zone over the Siple Coast differs dynamically from those near Terra Nova Bay and Byrd Glacier, Bromwich and Liu (1996) found that a baroclinic zone could form "where the northern edge of the cold East Antarctic katabatic flow meets the edge of the warmer katabatic flow from West Antarctica".

Barotropic instability as a mechanism for mesoscale cyclone formation was found in the Terra Nova Bay region. Numerical studies by J. Carrasco found mesocyclones formed south of the low-level katabatic jet near Terra Nova Bay (Carrasco et al., 2003). This region, south of the katabatic winds, is associated with cyclonic shear and therefore is a source of barotropic instability. This simulation acts as the initial trigger for mesocyclone activity (Carrasco et al., 2003).

The stretching of vertical air columns helps produce cyclonic vorticity and enhances mesoscale cyclone activity. In the low levels, katabatic wind descent from the higher East Antarctic plateau to the lower Ross Sea/Ross Ice Shelf region helps induce vertical stretching (Gallée, 1995; Heinemann and Klein, 2003). However, the vertical extent of these airflows is pretty tiny [only several hundred meters], so synoptic support is required to produce significant stretching over a more extensive layer (Heinemann and Klein, 2003). In particular, approval in the form of synoptic-scale advection of cyclonic vorticity and/or synoptic-scale warm air advection is required (Gallée, 1995). Low tropospheric troughs provide a means for significant cyclonic vorticity advection (Carrasco and Bromwich, 1993; Turner and Pendlebury, 2002; Heinemann and Klein, 2003). Low-level warm air advection produced by the synoptic circulation has been shown to play an essential role in mesocyclone formation (Carrasco and Bromwich, 1993; Carrasco et al., 2003). The warm air advection forms/enhances the baroclinic zone. Bromwich (1991) discovered that while synoptic forcing influenced formation, it was not required for cyclogenesis; however, upper-level support was vital for subsequent development of the cyclones.

Surface fluxes of heat and moisture are also necessary. Gallée (1995) found that the sensible heat input from the ocean enhanced the temperature contrast in the boundary layer front created by the katabatic winds. Moisture fluxes are needed to form large-scale, thick clouds and precipitation, usually in snow. When the ocean is ice covered, no significant mesocyclone activity or precipitation occurs (Gallée 1995). In this situation, other processes are required to produce mesocyclones, primarily synoptic-scale circulation/support.

Finally, low-level convergence is produced at all three katabatic wind confluence zones. This convergence provides cyclonic shear necessary for mesocyclone initiation.

Mesocyclones can produce large amounts of snow, gale-force winds, and severe reductions in invisibility. Mesocyclones that form around Terra Nova Bay generally move to the northeast or east-southeast. Byrd Glacier cyclones have a northeast tendency, while the Siple Coast storms move to the northwest, along the Transantarctic Mountains. These winds, in general, could be strong from the surface up to at least 2 km. Low-pressure systems forming over the surrounding seas can often influence and dominate the prevailing wind pattern over coastal areas.

The term *mesoscale cyclone* covers a wide range of weather systems from insignificant minor vortices with only weak surface circulation and no distinct cloud signature to the very active disturbances known as polar lows, which have been known to be associated with winds of hurricane force and heavy snowfall (Rasmussen and Turner, 2003). The incidence of mesoscale cyclone activity increases dramatically with latitude in both hemispheres, typically reaching a maximum in the vicinity of the polar front (Rasmussen and Turner, 2003). Mesoscale cyclones are usually classified as "polar lows" associated with maximum surface winds exceeding 15 m/s, diameter less than 1,000 km, and are located poleward of the main polar front (Rasmussen and Turner, 2003).

3.2.2.4   Other Weather Phenomena Observed

Windy conditions might pick up loose snow and carry it along. When the snow is still below eye level, it is called drifting snow, but it is called blowing snow when raised above eye level. Blowing snow conditions will affect the visibility quite severely. Wind speeds over 30 km/h can lead to drifting snow, while wind speeds over 60 km/h are more likely to produce blowing snow.

a Blizzards

Blizzards occur when wind speeds are gale force or more vital (i.e. > 60 km/h) for at least an hour, the temperature is less than 0°C, and visibility is reduced to 100 m or less. These conditions are dangerous and disruptive for outdoor activities. Sometimes blizzards persist for days.

b Whiteout

A whiteout is an optical phenomenon, periodically happens in Antarctica, in which uniform light conditions effectively make it impossible to distinguish shadows, landmarks, or the horizon. This can occur when the snow cover is unbroken and the sky is overcast. A whiteout is a severe hazard as it causes a loss of perspective and direction.

3.3   CLIMATOLOGY

So far, we have discussed the weather and climatic extremes prevailing over the continent. Due to remoteness and weather extremities, meteorological observations are not available at an extensive scale to arrive at a well-defined climatology for the region. The extreme conditions and remote location of the Antarctic continent inhibited the systematic study of its climate for a long. With the establishment of a few weather stations following the International Geophysical Year 1957, routine records of local weather conditions at these stations became available. However, nearly all these stations were established along the Antarctic coast; the few in the plateau's interior were widely separated and provided spatially limited information. Observations from polar-orbiting satellites have partly overcome the scarcity of stations, but the usefulness of satellite observations is fixed over the polar regions. Physical and radiative similarities between clouds and the snow-covered surfaces hinder the determination of critical elements of the energy balance over Antarctica from satellites alone (Yamanouchi et al., 1987). Of late, the realization among the scientific world that the polar regions with their ice cover are more sensitive to a warming globe than any other place on Earth has led to the establishment of more weather stations as well as scientific expeditions, thereby aiding in increasing the frequency and density of weather observations from the region.

Here we will discuss the long-term average meteorological features and, during the course, also reiterate a few of the first ever recorded extreme weather parameters like temperatures (highest and lowest), wind, etc. Also, the climatic patterns are discussed for certain meteorological variables like temperature, surface pressure, wind, precipitation, and relative humidity.

3.3.1   Climatic Characteristics of the South Pole

The South Pole is at an altitude of 9,300 feet. The temperature remains too low because of the altitude and location in the middle of the polar plateau. There is only one sunrise and one sunset each year at the South Pole: the sun rises on September 21 and sets on March 21. During the summer months, the sun slowly spirals up to its maximum height and then back down to the horizon again over 6 months, never setting until March 21, while the sun then remains below the horizon for the next 6 months. It is during the long nights during winter months that the South Pole experiences extreme cold.

Highest Recorded Temperature	+9.9°F (−12.3°C) December 25, 2011
Lowest Recorded Temperature	−117.0°F (−82.8°C) June 23, 1982
Average Annual Temperature	−57.1°F (−49.5°C)
Peak Wind	50 kts (58 mph) on September 27, 2011
Average Wind	10.7 kts (12.3 mph)
Maximum Pressure	719.0 hPa on August 25, 1996
Minimum Pressure	641.7 hPa on July 25, 1985
Average Pressure	681.3 hPa

3.3.2 Mean Climatic Patterns of the Continent

Turner et al. (1999) compared global analyses from the National Centers for Environmental Prediction, the European Centre for Medium-Range Weather Forecasts, the Australian Bureau of Meteorology, and the United Kingdom Met Office. They found an enormous discrepancy between the four analyses that occurred over Antarctica. Using data from the Antarctic First Regional Observing Study of the Troposphere (FROST) project, they verified that the analyses of the synoptic situation over the Southern Ocean and coastal regions are of reasonably high quality. Still, there were difficulties creating comments over the interior of Antarctica. The role of sparse observational data in this problem is twofold: first, the lack of data results in preliminary analyses, especially for more minor weather features; second, without sufficient data, it is difficult to determine which of the four studies is the most accurate.

The following description uses the National Centers for Environmental Prediction–National Center for Atmospheric Research (NCEP–NCAR) reanalysis data set, which assimilates and reprocesses in-situ meteorological data and satellite data to produce a comprehensive global dataset of meteorological parameters at 2.5° latitude/longitude resolution (Kalnay et al., 1996). The NCEP–NCAR data set extends from 1948 to the present, but parameters for the high latitudes are more reliable since the incorporation of satellite-based observations in 1979 (Bromwich and Fogt, 2004; Bromwich et al., 2007).

3.3.2.1 Surface Temperature

The average annual temperature ranges from about −10°C on the Antarctic coast to −60°C at the highest parts of the interior. January is the warmest month in Antarctica, during which average temperatures reach as high as 0 degrees in the Antarctic Peninsula. However, the average temperatures range from −10°C to −60°C, even during the summer months, as we go from coastal to the far interior part of the continent. Near the coast, the temperature can exceed +10°C at times in summer and fall to below −40°C in winter. Over the high inland, it can rise to about −30°C in summer but fall below −80°C in winter. Figure 3.3 shows the monthly temperature distribution.

3.3.2.2 Surface Pressure

The polar front (or circumpolar trough) encircles the Antarctic continent between 60°S and 70°S, with its actual location varying with both longitude and the time of year – it is generally farthest south during the transition months and moves north during summer and winter (Streten and Pike, 1980; Simmonds and Jones, 1998). Hence, there is usually a belt of low pressure, the circumpolar trough, surrounding Antarctica, containing multiple common centers. The continent is dominated by high pressure, but meaningful analysis of surface pressure data is complicated because of the elevated topographic features of much of Antarctica.

Radiative cooling over the Antarctic ice sheet produces freezing, dense air that flows away from elevated areas and is replaced by subsiding air from above. The resulting katabatic winds accelerate downhill, enhanced by the confluence of glacial valleys. Katabatic winds blow with great consistency over large areas. At the coast, they lose their driving force and soon dissipate offshore.

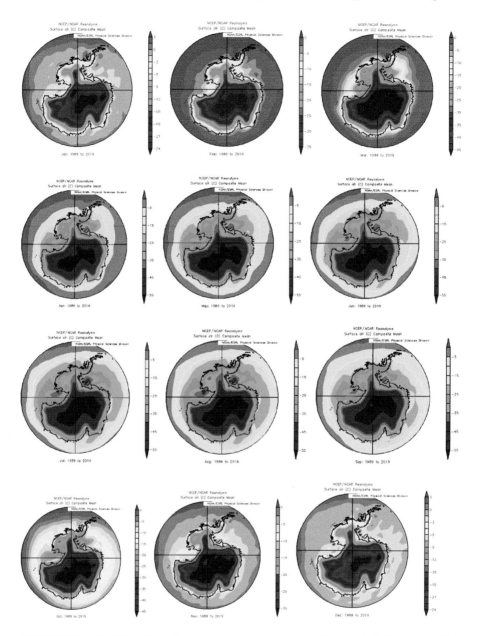

FIGURE 3.3 Monthly climatology of temperature at a surface level during the period 1989–2019. These graphs are generated online at www.cdc.noaa.gov maintained by NOAA/ESRL Physical Sciences Division.

Low-pressure systems near the Antarctic coast can interact with katabatic winds to increase their strength. The resulting wind speeds can exceed 100 km/h for days at a time. Wind gusts well over 200 km/h have been measured.

3.3.2.3 Winds

The continent's high plateau anchors a robust polar vortex of westerly winds and is the source of a near-constant katabatic airflow from a frequent inversion layer near the ice sheet surface. Along its coastline, a narrow zone of easterly flow occurs south of the girdling westerlies. Variations in this westerly flow and the relative air pressure between the vortex interior and the encircling regions are the primary oscillation of the far southern climate system, the so-called Southern Annular Mode (SAM) (Marshall, 2003). Figure 3.4 depicts the monthly average surface wind pattern.

3.3.2.4 Precipitation

Since Antarctica receives so little rainfall – the interior averages about 2 inches per year, usually as snow – it is recognized as a desert (Figure 3.5).

Surface Mass Balance – A. Monaghan and D. Bromwich Snowfall accumulation is the mass input to the Antarctic ice sheets and is the net result of precipitation, sublimation/vapor deposition, drifting snow processes, and melt (Bromwich, 1988). Rainfall is the Antarctic surface mass balance dominant at regional and larger scales (Genthon, 2004). To date, atmospheric models have been the most widely used means of assessing the temporal variability of the Antarctic snowfall for periods more extended than a decade (e.g. Bromwich et al., 2004). The most recent modeling studies indicate there has been no trend in snowfall since about 1980 (Van den Broeke and van Lipzig, 2005; Monaghan et al., 2006). Precipitation fields from the NCEP/DOE Reanalysis II (NN2) were employed to assess Antarctic snowfall for 2007. The snowfall in NN2 has been found to have an anomalously upward trend from 1979 onward compared to other model-based records and measurements from snow-stake farms and ice core records (Bromwich et al., 2004); however, the inter-annual variability of the snowfall is in very good agreement with other models (Monaghan et al., 2006). Therefore, if detrended, the NN2 record roughly approximates the "flat" trends that more accurate models predict. The 2007 snowfall can be compared to the average for 1979–2007 to give a proxy of whether the snowfall for the year was above or below the mean. Figure 5.24 shows the 2007 detrended annual precipitation anomalies from the 1979 to 2007 mean. In general, the anomalies over the continent were negative, consistent with above-average 850-hpa yearly geopotential height anomalies (which indicate lower-than-normal synoptic activity) over nearly the entire continent (Figure 5.20a).

Conversely, precipitation anomalies over the Antarctic Peninsula and near the front of the Ross Ice Shelf and Marie Byrd Land (~135°W) were positive, consistent with lower-than-normal geopotential height anomalies nearby. Overall, the pattern was very similar to 2006. The 2007 detrended monthly precipitation anomalies from the 1979 to 2007 mean, averaged over the grounded Antarctic ice sheets, were also derived (not shown). None of the anomalies exceeded two standard deviations, but the irregularities in February, November, and the annual anomaly varied from the mean by more than one standard deviation. Continent-wide precipitation was below

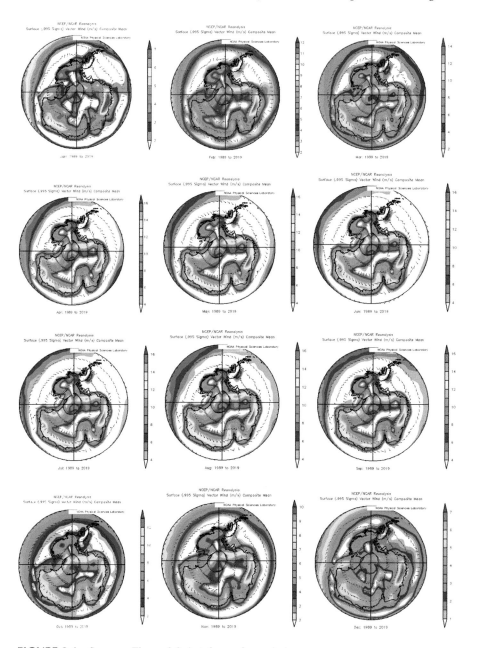

FIGURE 3.4   Same as Figure 3.3, but for surface winds.

average during winter and spring 2007. This may have been due to below-normal synoptic activity during these months, as indicated by positive 850-hPa geopotential height anomalies across the continent and near the coastal margins (Figure 5.22c), which projected strongly onto the annual mean. Overall, the 2007 yearly precipitation was about 7% below average. This downward fluctuation had a positive contribution

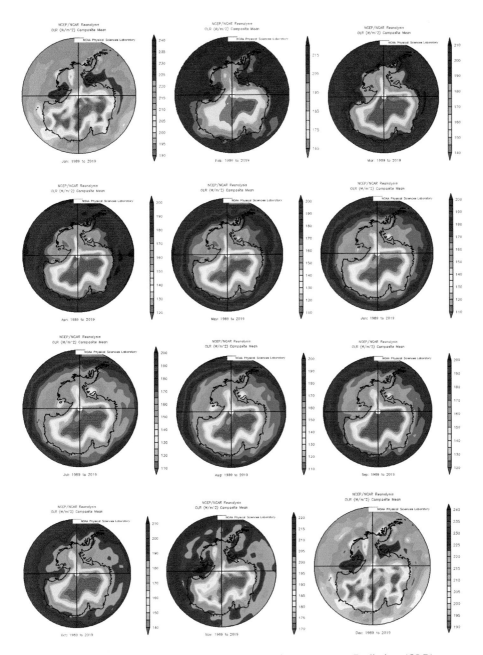

FIGURE 3.5 Same as Figure 3.3, but for the Outgoing Longwave Radiation (OLR).

to the sea level of approximately 0.35 mm. The 2006 annual precipitation anomaly at the continental scale was similarly below average, by 6% (Arguez et al., 2007). However, the yearly similar precipitation causality between the 2 years appears to have been due to different reasons, as the 850-hPa annual and seasonal anomalies were quite other. For example, precipitation was above average during April/May 2007, while it was below average during the same period in 2006; weaker SAM conditions during austral autumn 2007 led to enhanced penetration of maritime air at that time.

3.3.2.5   Relative Humidity

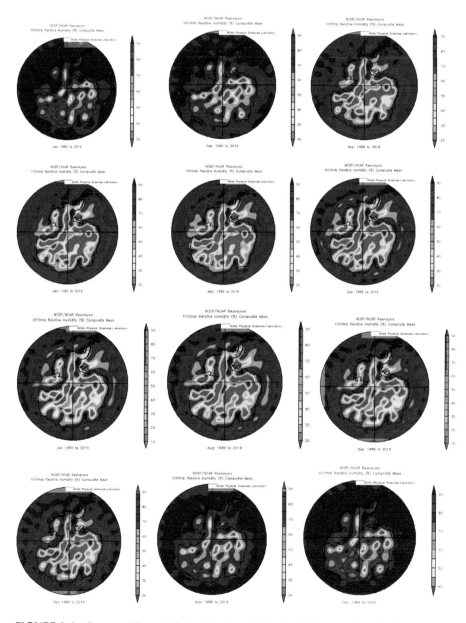

FIGURE 3.6   Same as Figure 3.3, but for the relative humidity at surface level.

3.3.3 Climate Variability and Change

The Southern Annular Mode (SAM) is the primary mode of climate variability at high southern latitudes. The atmospheric circulation changes associated with variability in its phase include a deeper (weaker) ASL when the SAM index is positive (negative; Hosking et al., 2013). A deeper (softer) ASL, therefore, gives stronger (weaker) northwesterly flow toward the western side of the Antarctic Peninsula and more (fewer) Extreme Precipitation Events (EPEs) in this area. Similarly, a more profound (weaker) ASL gives more robust (more inadequate) flow off the Ross Ice Shelf and fewer (more) EPEs on the ice shelf. This pattern of correlation and anticorrelation is found in spring, fall, and winter but is not apparent in summer when there is a minimum in cyclonic activity around the Antarctic. The most extensive and coherent seasonal correlation pattern across the continent is in fall when the precipitation from EPEs over much of East Antarctica is anticorrelated with the SAM. Even though El Niño–Southern Oscillation/SAM effects are partly coupled (Fogt et al., 2011), overall, the eastern Ross Ice Shelf and western Marie Byrd Land receive most (least) EPEs when the SAM is negative (positive) and the tropical Pacific is in the El Niño (La Niña) phase. Other parts of the continent also have more than 90% of the interannual variability of the total annual precipitation explained by variability in the number of EPEs, but this results primarily from local variability in the occurrence of depressions, especially those developing within the circumpolar trough, rather than the major modes of climate variability. There are several areas across the continent where EPEs explain minor precipitation variability. These include a small area inland of the coast of East Antarctica near 68°S, 135°E and an inland region of West Antarctica from the Antarctic Peninsula to 120°W. These are both areas where the climatological easterly flow is forced up an orographic slope, increasing the smaller amounts of precipitation and decreasing the relative contribution from EPEs. There is a large area on the plateau along 100°E from the South Pole to 75°S where EPEs account for a smaller fraction of the interannual precipitation variability than the rest of the continent. This is an area of very high elevation where most precipitation comes from the smaller daily amounts, and maritime intrusions are particularly rare. The non-EPE rain most strongly influences the interannual variability of the precipitation. The trends in the annual total of precipitation from EPEs over 1979–2016 are small across most of the continent with only two large areas of statistically significant ($p < 0.05$) change in southern Dronning Maud Land and inland of the coast across 100–120°E. These are both areas where there have been significant trends in the annual precipitation total due to greater ridging and amplification of planetary waves over East Antarctica, resulting in more onshore (offshore) flow close to 50°E (100°E). One area with large temperature and circulation changes over 1979–2016 is the Antarctic Peninsula (Turner et al., 2005). This region experienced some of the most significant temperature increases observed in the Southern Hemisphere during the second half of the 20th century. Still, since the late 1990s, there has been a regional cooling (Turner et al., 2016). This was reflected in an increase in EPEs on the western side of the Antarctic Peninsula until the end of the 20th century, followed by a subsequent decrease. During the 20th century, positive trends in snow accumulation on the Antarctic Peninsula and eastern WAIS have contrasted with negative trends in snow accumulation in the western WAIS and Victoria Land (Thomas et al., 2017; Wang et al., 2019), consistent with a deepening ASL. However, there has been no significant change in the precipitation from EPEs over the period considered here.

Studies of Antarctic climate indicate that climate variability is not homogeneous over the entire continent (Turner et al., 2005; Rignot et al., 2008; O'Donnell et al., 2011).

3.3.3.1 It Was Once Warm in Antarctica

Despite how frigid the Antarctic is now, it was once as warm as the sun-soaked beaches of California. Studies at Yale suggest that some 40–50 million years ago, during the Eocene epoch, high atmospheric levels of CO_2 created greenhouse-like conditions on Earth. Antarctica's weather at that time averaged 14°C (57°F), with a high of 17°C (63°F), states that would quickly reduce the current Antarctic's titanic icebergs and mountainous glaciers to familiar ocean swell.

New reconstructions of past atmospheric pressures over Antarctica show much of the continent experienced unusually high pressures during portions of the early and middle 20th century. According to the new research, those high pressures were associated with warm weather that may have significantly impacted the South Pole's race during the Southern Hemisphere's summer of 1911–1912.

A team of Norwegian explorers led by Roald Amundsen was the first to reach the South Pole on December 14, 1911. A competing British party led by Captain Robert Falcon Scott reached the pole on January 17, 1912. Amundsen's team returned safely home, but Scott's team perished on the ice in March of 1912.

In a new study in *Geophysical Research Letters*, a journal of the American Geophysical Union, researchers used data from weather stations in Antarctica and elsewhere in the Southern Hemisphere to reconstruct atmospheric pressures for Antarctica from 1905 to the present. This new research shows that natural variability played an essential role in Antarctica's climate changes over the past century. The study also finds that while the ozone hole is primarily responsible for recent low pressures observed over Antarctica in summer, other factors like sea surface temperatures are needed to explain these trends fully.

An accompanying study published in the October issue of the *Bulletin of the American Meteorological Society* used these pressure reconstructions and measurements from early Antarctic explorers to see how the unusual weather of 1911–1912 fit into the larger context Antarctic climate variability. This new research shows that the Amundsen and Scott expeditions experienced exceptionally high pressures often associated with unseasonably warm summer temperatures on their expedition routes. According to the study's authors, the warm temperatures were a boon to Amundsen's crew but hindered the Scott party's progress in two instances, which may have contributed to their deaths.

"In context, the pressures and temperatures were exceptional," said Ryan Fogt, an atmospheric scientist at Ohio University in Athens, Ohio, and lead author of both new studies. "The high pressure and the warmer conditions were measurements we haven't seen much of since."

The research also found the higher pressures experienced by Amundsen and Scott extended across much of the continent many other times during the first half of the 20th century. These higher pressures were likely due to natural climate variability because they occurred before the ozone hole formed, according to Fogt.

He said that studying Antarctica's past climate can help researchers better understand what will happen to the continent's ice sheets in a warming world.

"I think it's a new element to the South Pole expeditions that hasn't to my knowledge been considered," Fogt said of the new research. "Certainly, there were large differences in leadership styles that led to the outcomes, and that played a major role, but researchers haven't, to my knowledge, looked at the summer conditions during the race itself. This year was quite exceptional."

3.3.3.2   Reconstructing Past Climate

Antarctica has experienced lower-than-average summer surface pressures over the past three decades, mainly because the ozone hole has cooled the stratosphere above the continent. Lower pressures bring stronger westerly winds around Antarctica, which keep most of the continent colder than average. Fogt has been interested in understanding Antarctica's past climate to put the current low-pressure trends into a larger context.

Scientists have only had consistent temperature and pressure records for Antarctica since the 1950s, when weather stations were permanently installed on the southernmost continent. In the new research, Fogt and his colleagues used Southern Hemisphere mid-latitude pressure observations to reconstruct Antarctica's atmospheric pressure back to 1905. The researchers compared their reconstruction to weather data recorded by members of the Amundsen and Scott expeditions.

Previous studies of the Amundsen and Scott expeditions found the weather during the summer of 1911–1912 was warm, but no one had yet looked in detail at how unusual those values were, Fogt said.

Fogt's new research found December of 1911 was hot over the Ross Ice Shelf and near the South Pole. While on the Antarctic Plateau, Amundsen's crew experienced temperatures that peaked above −16°C (3°F) – more than 10°C (18°F) above average. The warmest temperature ever recorded on the Antarctic Plateau was −7°C (19°F) in December 1989.

"It was this impressive warm signature that we haven't seen in South Pole observations that early in December, and only a few times in a nearby automatic weather station," Fogt said.

3.3.3.3   Cold Weather Spells Tragedy

Scott's crew also experienced warm temperatures in December – about 5°C (9°F) above average – but they were further behind Amundsen. A wet snowstorm struck Scott and his crew during this time, slowing their progress across the Ross Ice Shelf on their way to the Antarctic Plateau and delaying them for several days, Fogt said. After reaching the South Pole in mid-December, Amundsen's crew returned to their base on the Ross Ice Shelf on January 26, 1912, and left the continent shortly. Scott's crew reached the pole in mid-January and made their way back to base in January and February 1912. Temperatures were well above average again during Scott's return trek in early February. Still, temperatures dropped in late February and March when the explorers crossed back over the Ross Ice Shelf. Fogt's new research suggests the exceptionally warm conditions in early February likely made the March cold spell feel more extreme, which could have affected the Scott party's morale, their perception of the cold, and their sensitivity to it, he said. "You adapt to cold conditions in Antarctica, and they start to feel amazingly warm after a while," Fogt said. "But when it warms up and then gets cold again, wow, you feel it."

3.3.3.4   Indian Antarctic Program

Indians set their first foot on Antarctic ice about 38 years ago. The Indian Antarctic Programme was initiated in 1981 with a selected team of 21 members under the leadership of Dr. S Z Qasim, secretary of Department of Environment and former director of National Institute of Oceanography (NIO), to conduct scientific research in the frozen continent. On December 6, 1981, the expedition started from Goa onboard Marine Vessel Polar Circle, a chartered ship from Norway. The troupe landed in Antarctica on January 9, 1982, and returned to Goa on February 21, 1982, thus marking the end of their 77-day expedition. Research stations set up by India in Antarctica: until now, under the Antarctic Treaty's environmental protocol (1959), India has set three research stations.

i Dakshin Gangotri

The first Indian scientific research base station was established in Antarctica as part of the third Antarctic program. Located at 2,500 kilometers from the South Pole, it was established in 1983–1984. This was the first time an Indian team spent a winter in Antarctica to carry out scientific work. In 1989, it was excavated and was being used again as a supply base and transit camp (Figure 3.7).

ii Maitri

In 1988, an ice-free, rocky area on the Schirmacher Oasis was selected to build Maitri's second research station (Figure 3.8).

   Maitri is the gateway for Indian scientists to venture into interior Antarctic mountains and hosts summer and winter research teams every year.

iii Bharati

Located beside Larsmann Hill, about 3,000 km east of Maitri, Bharati was established in 2015. It is situated between Thala Fjord and Quilty Bay, east of Stornes Peninsula in Antarctica (Figure 3.9).

   Bharati made India an elite member of the club of nine nations that have multiple stations in the region. In line with the Antarctic Treaty System, Bharati can be completely disassembled and removed without leaving even a brick behind.

FIGURE 3.7   Dakshin Gangotri.

FIGURE 3.8   Maitri is the gateway for Indian scientists to venture into interior Antarctic mountains and has been hosting summer and winter research teams every year.

3.3.3.5   National Center for Polar and Ocean Research

The National Center for Antarctic and Ocean Research is a research and development body. It functions under the Ministry of Earth Sciences, Government of India. The same controls the Indian Antarctic program. Logistical support to the various activities of the Indian Antarctic program is provided by the relevant branches of the Indian armed forces. The launching point of Indian expeditions has varied from Goa in India to Cape Town in South Africa.

FIGURE 3.9 Bharati/National Center for Antarctic and Ocean Research.

3.3.3.6 Observations from Maitri

The India Meteorological Department (IMD) has initiated the surface meteor-ological observations at the Maitri station since 1988. Observations of temperature (maximum and minimum), surface pressure, winds were measured. The data re-ported here are for a period from 1990 to 2015.

3.3.3.7 Surface Temperature

Figures 3.10–3.12 depict the annual variation of maximum and minimum tem-peratures during January and July, representing the summer and winter months, respectively. On average, the maximum temperatures were around 4°C during summer and around −13°C during winter. At the same time, the minimum tem-peratures varied between −3°C and −20°C, during summer and winter, respectively.

The mean monthly variation of maximum and minimum temperatures from 1990 to 2015 over the Maitri station is plotted in the figure. It is seen that the tem-peratures started falling from January to August and, after that, start increasing rapidly, following the sun's position over the Antarctic.

3.3.3.8 Wind Speed

Figure 3.13 shows the variation of mean wind speed over different months for the period 1990–2015. The wind speed increases from summer to winter and starts decreasing afterwards. Due to these higher wind speeds during the winter season, blizzards are standard features over this region.

3.4 SUMMARY AND CONCLUSIONS

There are many potential applications for using synoptic classification to help un-derstand how local climate parameters relate to climate variability in Antarctica.

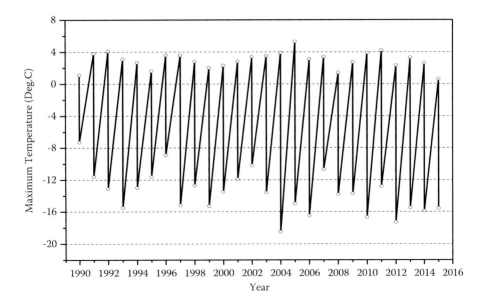

FIGURE 3.10   Temperature variation from 1990 to 2018.

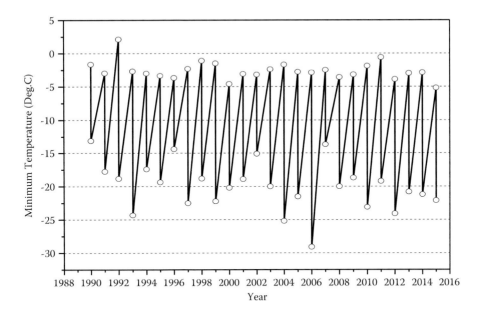

FIGURE 3.11   Temperature variation from 1988 to 2016.

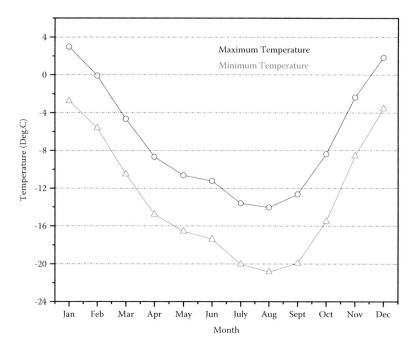

FIGURE 3.12   Monthly temperature variation.

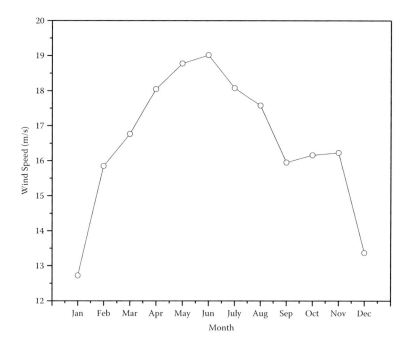

FIGURE 3.13   Variation of mean wind speed over this region.

Future work with the Ross Sea synoptic types will focus on investigating the relationship to meteorological parameters at other sites and relating the synoptic types to other climate parameters (e.g. sea ice and snow chemistry).

Although many uncertainties exist regarding the mechanisms of climate pattern changes in each polar region, a few factors appear to play dominant roles in the circulation response to external forcing. Paramount among these is the contrast in orography in the Arctic and Antarctic, leading to very different patterns of Rossby wave propagation (including the stability of tropical teleconnections). Meridional mountain ranges and the distribution of land and oceans in the Northern Hemisphere generate strong standing wave patterns, while the Southern Ocean and nearly zonally symmetric Antarctic continent allow for generally more substantial zonal flow and annular patterns. Further, stratospheric ozone depletion is playing a more crucial role in the Antarctic circulation (i.e. Thompson and Solomon, 2002; Keeley et al., 2007) than in the Northern Hemisphere and could be amplifying the changes expected from greenhouse gases in some seasons (Turner et al., 2007).

ACKNOWLEDGEMENTS

The authors are thankful to NCEP and NCAR for the reanalysis data sets. Authors are also grateful to Secretary, Ministry of Earth Sciences (MoES). Authors acknowledge other datasets collected from Indian Meteorological Department (IMD).

REFERENCES

Arguez, A. and 158 others (2007). State of the Climate in 2006, SpecialSupplement, *Bull. Am. Met. Soc.*, 88, 135.

Bromwich, D. H. (1989). Satellite Analyses of Antarctic Katabatic Wind Behavior. *Bull. Amer. Meteor. Soc.*, 70, 738–749.

Bromwich, D.H. (1988). Snowfall in High Southern Latitudes. *Rev. Geophys.*, 26(1), 149–168.

Bromwich, D. H. (1991). Mesoscale Cyclogenesis over the Southwestern Ross Sea Linked to Strong Katabatic Winds. *Mon. Wea. Rev.*, 119, 1736–1752.

Bromwich, D.H. (1998). Antarctic Meteorology and Climatology. J.C. King and J. Turner. 1997. Cambridge: Cambridge University Press, xi 409 p, illustrated, *Polar Record*, 34(190), 256–257. https://doi.org/10.1017/S0032247400025754

Bromwich, D. H., and Kurtz, D. D. (1984). Katabatic Wind Forcing of the Terra Nova Bay Polynya. *J. Geophys. Res.*, 89, 3561–3572.

Bromwich, D. H., and Liu, Z. (1996). An Observational Study of the Katabatic Wind Confluence Zone near Siple Coast West Antarctica. *Mon. Wea. Rev.*, 124, 462–477.

Bromwich, D. H., Monaghan, A. J., Powers, J. G., Cassano, J. J., Wei, H.-L., Kuo, Y.-H., and Pellegrini, A. (2003). Antarctic Mesoscale Prediction System (AMPS): A Case Study from the 2000-01 Field Season. *Mon. Wea. Rev.*, 131, 412–434.

Bromwich, D. H., & Fogt, R. L. (2004). Strong Trends in the Skill of the ERA-40 and NCEP–NCAR Reanalyses in the High and Midlatitudes of the Southern Hemisphere, 1958–2001. *J. Climate*, 17, 4603–4619. https://doi.org/10.1175/3241.1.

Bromwich, D. H., Guo, Z., Bai, L., and Chen, Q.-S. (2004). Modeled Antarctic precipitation. Part I: Spatial and temporal variability. *J. Climate*, 17, 427–447.

Bromwich, D. H., Fogt, R. L., Hodges, K. I., and Walsh, J. E. (2007). A Tropospheric Assessment of the ERA-40, NCEP, and JRA-25 Global Reanalyses in the Polar Regions. *J. Geophys. Res.*, 112, D10111. https://doi.org/10.1029/2006jd007859.

Carrasco, J. F., and Bromwich, D. H. (1993). Mesoscale Cyclogenesis Dynamics over the Southwestern Ross Sea, Antarctica. *J. Geophys. Res.*, 98, 12 973–12 995.

Carrasco, J. F., and Bromwich, D. H. (1994). Climatological Aspects of Mesoscale Cyclogenesis over the Ross Sea and Ross Ice Shelf regions of Antarctica. *Mon. Wea. Rev.*, 122, 2405–2425.

Carrasco, J. F., Bromwich, D. H., and Monaghan, A. J. (2003). Distribution and Characteristics of Mesoscale Cyclones in the Antarctic: Ross Sea Eastward to the Weddell Sea. *Mon. Wea. Rev.*, 131, 289–301.

Fogt, R.L. (2003). Trip Report: McMurdo Station, Antarctica, December 18 02-07 Jan 03. http://polarmet.mps.ohio-state.edu/PolarMet/index.html.

Fogt, R. L., Bromwich, D. H., and Hines, K. M. (2011). Understanding the SAM Influence on the South Pacific ENSO Teleconnection. *Clim. Dyn.*, 36, 1555–1576.

Gallée, H. (1995). Simulation of the Mesocyclonic Activity in the Ross Sea, Antarctica. *Mon. Wea. Rev.*, 123, 2051–2069.

Genthon, C. (2004). Space-time Antarctic surface mass-balance variability from climate models. Ann. Glaciol., 39, 271–275.

GLACIER, (1998). Winds over Antarctica! http://www.glacier.rice.edu/weather/3_antwindcirc.html.

Heinemann, G., and Klein, T. (2003). Simulations of Topographically Forced Mesocyclones in the Weddell Sea and the Ross Sea Region of Antarctica. *Mon. Wea. Rev.*, 131, 302–316.

Hosking, J. S., Orr, A., Marshall, G. J., Turner, J., and Phillips, T. (2013). The Influence of the Amundsen-Bellingshausen Seas Low on the Climate of West Antarctica and Its Representation in Coupled Climate Model Simulations, *J. Clim.*, 26(17), 6633– 6648.

Kalnay, Eugenia Kanamitsu, M. Kistler, R. Collins, William Deaven, D. L. S. Gandin Iredell, M. Saha, Satyajit White, G. Woollen, J. Zhu, Yuejian Chelliah, Muthuvel Ebisuzaki, W. Higgins, Wayne Janowiak, John C. K. Ropelewski, C. Wang, Julian and Leetmaa, A. (1996). The NMC/NCAR 40-year reanalysis project. *Bull Am Meteorol Soc. Bulletin of the American Meteorological Society*, 77. https://doi.org/10.1175/1520-0477(1996).

Keaveney, S. R., (2004). Antarctic Operational Weather Forecasting, Maryland University College Park Department of Meteorology (CI04-386).

Keeley, S. P. E., Gillett, N. P., Thompson, D. W. J., Solomon, S., and Forster, P. M. (2007). Is Antarctic Climate Most Sensitive to Ozone Depletion in the Middle or Lower Stratosphere? *Geophys. Res. Lett.*, 34.

King, J. C., and Turner, J. (1997). *Antarctic Meteorology and Climatology*. Cambridge University Press, 409 pp.

Marshall, G. J. (2003). Trends in the Southern Annular Mode from Observations and Reanalyses. *J. Climate*, 16(24), 4134–4143. https://doi.org/10.1175/1520-0442(2003) 016<4134:titsam>2.0.co;2.

McCormick, B., Ed. (2002). *Field Manual for the United States Antarctic Program*. http://www.polar.org/usapserv/Field%20Manual.pdf.

Monaghan, A. J., Bromwich, D. H., Wei, H.-L., Cayette, A. M., Powers, J. G., Kuo, Y.-H., and Lazzara, M. A. (2003). Performance of Weather Forecast Models in the Rescue of Dr Ronald Shemenski from the South Pole in April 2001. *Wea. Forecasting*, 18, 142–160.

Monaghan, A. J., Bromwich, D. H., Fogt, R. L., Wang, S.-H., Mayewski, P. A., Dixon, D. A., Ekaykin, A., Frezzotti, M., Goodwin, I., Isaksson, E., Kaspari, S. D., Morgan, V. I., Oerter, H., Van Ommen, T. D., Van der Veen, C. J., and Wen, J. (2006). Insignificant Change in Antarctic Snowfall Since the International Geophysical Year. *Science*, 313, 827–831.

Nicolas, J. P., and Bromwich, D. (2011). Climate of West Antarctica and Influence of Marine Air Intrusions. *J. Climate*, 24(1), 49–67. https://doi.org/10.1175/2010jcli3522.1.

O'Connor, W. P., Bromwich, D. H., and Carrasco, J. F. (1994). Cyclonically Forced Barrier Winds along the Transantarctic Mountains near Ross Island. *Mon. Wea. Rev.*, 122, 137–150.

O'Donnell, A.J., Boer, M.M., McCaw, W.L., and Grierson, P.F. (2011). Climatic Anomalies Drive Wildfire Occurrence and Extent in Semi-Arid Shrublands and Woodlands of Southwest Australia. *Ecosphere*, 2(11).

Parish, T. R. (1982). Surface Airflow over East Antarctica. *Mon. Wea. Rev.*, 110, 84–90.

Parish, T. R., and Bromwich, D. H. (1987). The Surface Windfield over the Antarctic Ice Sheet. *Nature*, 328, 51–54.

Rasmussen, E.A. , and Turner, J. (2003).*Polar Lows: Mesoscale Weather Systems in the Polar Regions*. Cambridge University Press, Cambridge.

Rignot, E. et al., (2008). Recent Antarctic Ice Mass Loss From Radar Interferometry and Regional Climate Modelling. *Nat. Geosci.*, 1, 106–110.

Sato, N., Kikuchi, K., Barnard, S. C., and Hogan, A. W. (1981). Some Characteristic Properties of Ice Crystal Precipitation in the Summer Season at South Pole Station, Antarctica. *J. Meteor. Soc. Japan*, 59, 772–780.

Simmonds, I., Keay, K., and Lim, E.-P. (2003). Synoptic Activity in the Seas around Antarctica. *Mon. Wea. Rev.*, 131, 272–288.

Simmonds, I., and Jones, D. A. (1998). The Mean Structure and Temporal Variability of the Semiannual Oscillation in the Southern Extratropics. *Int. J. Climatol.*, 18, 473–504. https://doi.org/10.1002/(SICI)1097-0088(199804)18:5<473::AID-JOC266>3.0.CO;2-0.

Streten, N. A., and Pike, D. J. (1980). Characteristics of the Broadscale Antarctic Sea Ice Extent and the Associated Atmospheric Circulation 1972–1977. *Arch. Met. Geophy Biokl.*, 29, 279–299.

Thomas, E. R., van Wessem, J. M., Roberts, J., Isaksson, E., Schlosser, E., Fudge, T. J., Vallelonga, P., Medley, B., Lenaerts, J., Bertler, N., van den Broeke, M. R., Dixon, D. A., Frezzotti, M., Stenni, B., Curran, M., and Ekaykin, A. A. (2017). Regional Antarctic snow accumulation over the past 1000 years. *Climate of the Past*, 13, 1491–1513. https://doi.org/10.5194/cp-13-1491-2017.

Thompson, D. W., and Solomon, S. (2002). Interpretation of Recent Southern Hemisphere Climate Change. *Science*, 296(5569), 895–899.

Turner, J., Leonard, S., Marshall, G. J., Pook, M., Cowled, L., Jardine, R., Pendlebury, S., and Adams, N. (1999). An Assessment of Operational Antarctic Analyses Based on Data from the FROST Project. *Wea. Forecasting*, 14, 817–834.

Turner, J., and S. Pendlebury, Eds. (2002). *The International Antarctic Weather Forecasting Handbook*. http://www.comnap.aq/comnap/comnap.nsf/P/Pages/About.Publications/?Open#weather.

Turner, J., Colwell, S. R., Marshall, G. J., Lachlan-Cope, T. A., Carleton, A. M., Jones, P. D., Lagun, V., Reid, P. A., and Iagovkina, S. (2005). Antarctic Climate Change During The Last 50 Years. *Int. J. Climatol.*, 25, 279–294.

Turner, J., Overland, J. E., and Walsh, J. E. (2007). An Arctic and antarctic perspective on recent climate change. *Int. J. Climatol.*, 27, 277–293.

Turner, J., Hosking, J. S., Bracegirdle, T. J., Phillips, T., and Marshall, G. J. (2016). Variability and Trends in the Southern Hemisphere High Latitude, Quasi-Stationary Planetary Waves, *Int. J. Climatol.*, 25(3). https://doi.org/1002/joc.4848.

Turner, J., Phillips, T., Thamban, M., Rahaman, W., Marshall, G. J., Wille, J. D., et al. (2019). The Dominant Role of Extreme Precipitation Events in Antarctic Snowfall Variability. *Geophys. Res. Lett.*, 46, 3502–3511. https://doi.org/10.1029/2018GL081517.

Yamanouchi, T., Hirasawa, N., Hayashi, M., Takahashi, S., and Kaneto, S. (2003). Meteorological characteristics of Antarctic inland station, Dome Fuji. Mem. Natl Inst. Polar Res., Spec. Issue, 57, 94–104, National Institute of Polar Research.

Van den Broeke, M. R. (2000). On the Interpretation of Antarctic Temperature Trends. *J. Climate*, 13(21), 3885–3889.

van den Broeke, M. R., and van Lipzig, N. P. M. (2005). Changes in Antarctic temperature, wind and precipitation in response to the Antarctic Oscillation. *Ann. Glaciol.*, 39, 119–126.

Wang, Z., Turner, J., Wu, Y., and Liu, C. (2019). Rapid Decline of Total Antarctic Sea Ice Extent During 2014-2016 Controlled by Wind-Driven Sea Ice Drift. *J. Climate*, 32(17), 5381–5395.

4 Antarctic Aerosols and Climate
Measurements at a Coastal Antarctic Station

Vimlesh Pant
Indian Institute of Technology Delhi, New Delhi, India

Devendraa Siingh and A.K. Kamra
Indian Institute of Tropical Meteorology, Pune, India

CONTENTS

DOI: 10.1201/9781003203742-4

4.1 ATMOSPHERIC AEROSOLS

Atmospheric aerosols are minute particles suspended in the air. These particles span over a wide size – range of a few nm to around 100 μm (Junge, 1963; Prospero et al., 1983). Aerosols directly interact with solar radiation and play an important role in the Earth's radiation budget. Further, these particles can also act as cloud condensation nuclei (CCN) and influence the radiation budget indirectly. These CCN provide a surface for the water vapor to condense into water droplets and thus involve them in the cloud formation processes (Twomey, 1977). Aerosols' size range and chemical composition are primarily governed by the mechanisms responsible for their generation and dissipation. Natural processes or anthropogenic activities can generate these particles. The design and size of aerosols are modified in the atmosphere by various physical and chemical transformation processes. Direct ejection of particles into the atmosphere leads to 'primary aerosols', mostly of natural origin. Windblown dust and sea-salt particles are examples of primary aerosols generated by the mechanical disintegration processes induced due to wind shear force over the land surface and sea surface, respectively.

On the other hand, the 'secondary aerosols' are formed indirectly by the gas-to-particle conversion processes in the atmosphere. Under favorable meteorological conditions, the gaseous precursor compounds present in the atmosphere, such as nitric oxide and nitrogen dioxide (collectively known as NO_x), sulfur dioxide (SO_2), dimethylsulfide, or DMS (CH_3SCH_3), and hydrocarbons can lead to the formation of secondary aerosols through chemical processes. Aerosols produced by the gas-to-particle conversion mechanisms are generally less than about 0.01 μm diameter in size. Once made, these aerosol particles are transformed in the atmosphere and removed either by 'dry' or 'wet' removal processes (Twomey, 1977). DMS is one of the major precursors for aerosols and CCN in the marine boundary layer over much of the remote ocean, such as the Southern Ocean.

4.2 ANTARCTIC AEROSOLS: COMPOSITION AND SOURCES

The geographically isolated continent of Antarctica serves as a natural laboratory to examine the background aerosols. Antarctica is far from polluted continents and well isolated by the Southern Ocean. The climatological winds over the 50°–70°S latitudes restrict long-range transport from lower margins to the Antarctic continent. In the absence of any long-range atmospheric transport of particles, the aerosols in the Antarctic atmosphere mainly consist of sulphates in the submicrometer range and sodium chloride in the coarse field (Shaw, 1988). While the rough particle concentration is found to be very low (except at coastal sites), the submicrometer aerosols are suspended throughout the continent of Antarctica with significant variations in their engagements during the summer and winter seasons.

The primary sources of aerosol particles in Antarctica are either the nucleation of gaseous precursors to form particles in nucleation mode or the sea-salt particles formed due to the wave-breaking activity at the surrounding ocean surfaces (Hogan et al., 1982; Ito, 1985; Shaw, 1988; Deshpande and Kamra, 2004; Kerminen et al., 2018; Jurányi and Weller, 2019). In addition, the volcanic activity from Mount

Erebus, situated on Ross Island, introduces high fluxes of SO_2 and H_2S, which contribute to the sulfate particles in the Antarctic atmosphere (Radke, 1982; Rose et al., 1985; Chuang et al., 1986; Shaw, 1988). The origin of Aitken particles has been suggested in the middle or upper troposphere, which are brought down to the ice sheet by turbulence (Hogan and Barnard, 1978; Hogan, 1979; Hogan et al., 1982, 1984; Shaw, 1988). The airborne measurements of aerosol constituents over coastal Antarctic stations Syowa (69.00°S, 39.58°E) and Mizuho (70.41°S, 44.19°E) made by Hara et al. (2006) confirm the abundance of sulfur-rich aerosol particles in the free troposphere, whereas sea-salt, mineral, and anthropogenic particles are found as minor constituents. The subsiding air from the mid- or upper-troposphere can suddenly increase the total aerosol concentration over the ice sheet (Jaenicke et al., 1992). These two mechanisms of sub-micrometer particle gen-eration in Antarctica are responsible for several investigators' observed bimodal size distribution (Bigg, 1980; Ito, 1985; Gras and Adriaansen, 1985; Shaw, 1986). Particles in the sub-micrometer size range mainly constitute sulfate in the form of sulphuric acid droplets (Cadle et al., 1968; Bigg, 1980; Ono et al., 1981; Parungo et al., 1981; Ito, 1985). These sulphuric acid droplets are formed due to DMS emissions by surrounding oceans and their subsequent oxidation in the atmosphere by photochemical processes. While in the interior of the Antarctic continent, the NSS-sulfate constitutes about 70% of total mass compared to the 25% contribution from the sea-salt aerosols (Shaw, 1988), the coastal Antarctic aerosols have been observed with higher contribution from the sea-salt particles. Particles in the ac-cumulation mode are aged particles that constitute mainly sulfates and, up to some extent, sea-salt components (Harvey et al., 1991). The sea-salt particles deposited on the fresh snow can also be re-suspended in the atmosphere during periods of high winds (Hogan, 1975; Jaenicke and Matthias-Maser, 1992; Hall and Wolff, 1998). The elemental analysis of aerosol samples collected at a coastal site by Harvey et al. (1991) show sulphur as the only element present in particles of sizes d <0.15 μm, whereas particles in larger size (d > 0.5 μm) classes show a greater proportion of Na and Cl as compared to S, which indicates the sea-salt existence in accumulation and coarse modes. Measurements at the Antarctic Peninsula station King Sejong (62° 13'S, 58° 47'W) show the presence of metals (e.g. Bi, Cd, Co, Cr, Cu, Ni) of crustal and anthropogenic origin in addition to the sea-salt and NSS-sulfate particles (Mishra et al., 2004). The large aerosol concentration during the summer season can be explained by new particle formation events (Kerminen et al., 2018). A year-round measurement at the German coastal Antarctic station Neumayer III highlights both the maritime and continental origin (Jurányi and Weller, 2019). At the same time, the passing cyclones advect marine airmass with the easterly wind, the southern katabatic winds transport particles to the Neumayer III station (Jurányi and Weller, 2019).

Continental air masses freely mix with marine air masses at the Antarctic coastline. The aerosol characteristics of these two types of air masses significantly differ in their generation and growth processes and the chemical transformation they undergo after their generation. Understanding the aerosols in marine air mass is also essential to study the aerosol characteristics at coastal stations. Following is a brief review of the marine aerosols.

Marine aerosols are generated by the action of wind on the sea surface (as primary aerosols) or through the gas-to-particle conversion (as secondary aerosols). The mechanical disruption of the sea surface leads to the formation of sea-salt particles, which are significant contributors to the mass of particulate matter injected into the atmosphere globally (Peterson and Junge, 1971; Andreae et al., 1995; Penner et al., 2001; Lewis and Schwartz, 2004). Marine aerosol contributes significantly to the global aerosol load and has an essential impact on the Earth's albedo and climate (O'Dowd et al., 2004; Reddington et al., 2017). Sea-salt aerosols play a vital role for air-sea interaction and marine atmospheric chemistry (Finlayson-Pitts and Hemminger, 2000; Rossi, 2003; von Glasow and Crutzen, 2004). These sea-salt particles can be generated into different size ranges from a bubble bursting and tearing of the wave crests under high wind conditions. The wind-generated oceanic surface waves entrain air to various depths, forming numerous bubbles that rise to the sea surface. These bubbles give rise to the formation of 'film drops' when they burst. The bubble bursting process starts as soon as the wind speed exceeds 3–4 m s^{-1}. Subsequently, the left-out cavity of each bubble injects a few 'jet drops' into the atmosphere, which is fewer in number than the film drops. Typically, a bow produces tens to hundreds of film drops and only a few jet drops (Blanchard and Woodcock, 1957; Blanchard, 1963). The presence of phytoplankton can alter the organic chemical composition and physical proprieties of marine aerosols (Kuznetsova et al., 2005). Some recent studies reported the generation of sea-salt aerosols by blowing snow over sea ice (Huang et al., 2018; Giordano et al., 2018; Murphy et al., 2019; Yang et al., 2019).

The secondary particles over the marine boundary layer are mostly the non-sea-salt sulfate (NSS-sulfate) (Fiebig et al., 2014; Weller et al., 2015). The NSS-sulfate particles are generated through the gas-to-particle conversion of the organo-sulphur gases emitted by the ocean. The dimethyl sulphide (DMS; CH_3SCH_3) is the primary source of NSS-sulfate over oceans (Andreae et al., 1983, 1986; Cline and Bates, 1983; Fitzgerald, 1991), which constitutes the fine particle mode ($r < 0.25$ μm) over the marine boundary layer. Compared to sea salt, the NSS-sulfate particles are more volatile, i.e. they decompose at lower temperatures than sea salt. Clarke et al. (1987) measured the volatility of particles of 0.1–0.5 μm radius over the tropical South Pacific. They found that about 99% of particles smaller than 0.2 μm radius constitute sulfuric acid or ammonium sulfate/bisulfate. Global emission of DMS from the ocean has been estimated in the range from 17 Tg S yr-1 (Bates et al., 1987) to 40 Tg S yr-1 (Andreae and Raemdonck, 1983). Once emitted into the atmosphere, the DMS is oxidized by OH, NO_3, and IO radicals and leads to the formation of SO_2, methanesulfonic acid (MSA), and H_2SO_4 (Atkinson et al., 1984; Hynes et al., 1986; Yin et al., 1986). Fitzgerald (1991) suggested three principal gas-to-particle conversion processes of the photo-oxidation products of the DMS. These processes are: (i) new particle formation by hetero-molecular homogeneous nucleation involving MSA, H_2SO_4, and other trace gases; (ii) condensation of MSA, H_2SO_4, and other low volatility gas-phase reaction products on newly formed and existing particles; and (iii) aqueous-phase oxidation of SO_2 by O_3 and H_2O_2 in cloud droplets. Once formed, these particles grow by condensation or in-cloud sulfate formation when cycled through non-precipitating clouds (Hoppel et al., 1986).

However, it has been suggested that the number of non-sea-salt sulfate particles over the marine atmosphere is often controlled by the subsidence from the free troposphere (Clarke et al., 1996, 1998b; Bates et al., 1998, 2001; Weber et al., 1998, 2001; Koponen et al., 2002) and less frequently by local nucleation process (Covert et al., 1992; Hegg et al., 1992; Clarke et al., 1998a; O'Dowd et al., 2002; Pirjola et al., 2000; Dall'Osto et al., 2017). Further, the oxidation of biogenically produced organic nitrogen gases from the ocean may be an essential source of nitrate in marine aerosols (Parungo et al., 1986; Savoie et al., 1989; Fitzgerald, 1991). The background concentration of nitrates over remote oceans is in the range of 0.1–0.2 $\mu g\ m^{-3}$. Nitrate is found in larger (> 0.35 μm) particles than the NSS-sulfate (Savoie and Prospero, 1982; Parungo et al., 1986; Fitzgerald, 1991). Therefore, a different gas-to-particle conversion mechanism proposed for the formation of nitrate aerosol is the selective dissolution of NO_2 in the more alkaline sea-salt droplets where it is subsequently oxidized (Gravenhorst, 1978; Parungo and Pueschel, 1980). Various organic acids, amines, carbonyl compounds, and amino acids contribute to the water-soluble organic compounds in marine aerosols (Saxena and Hildemann, 1996). Apart from the sea salt and NSS-sulfate, the mineral dust that originated from continents also constitutes a fraction of marine aerosols. The McMurdo dry valleys in West Antarctica are prominent locations of frequent local dust suspension in the atmosphere (Atkins and Dunbar, 2009; Bullard et al., 2016). The mineral dust aerosols mainly contribute to the coarse mode particles and have large spatiotemporal variability governed by the favorable winds that transport these particles over remote oceans (Raemdonck et al., 1986; Fitzgerald, 1991). Mineral dust particles mainly constitute Si, Fe, Ca, Al, etc. These dust particles can get internally mixed with sea-salt particles in the cloud processes (Andreae et al., 1986). Although typical mineral dust concentrations were found to be less than 0.5 $\mu g\ m^{-3}$ over most oceanic areas (Prospero, 1979; Raemdonck et al., 1986; Savoie et al., 1987), they can sometimes exceed sea-salt concentrations in the vicinity of continents during periods of land-to-ocean winds (Prospero, 1979; Raemdonck et al., 1986; Murugavel et al., 2001; Deshpande and Kamra, 2002; Virkkula et al., 2006). A recent study by Tatlhego et al. (2020) found a positive correlation between dust deposition and chlorophyll concentration in the remote Southern Ocean between 45°S and 65°S. The increase in oceanic primary productivity was associated with the supply of Fe by the dust transport.

Polar sunrise can activate the reactive bromine (BrO_x) related to the ozone depletion in Antarctica (Abbatt et al., 2012; Hara et al., 2018). Additionally, the heterogeneous reactions on sea-salt aerosols is another process through which the BrO_x cycle is found to be activated at the Syowa station (Hara et al., 2018). However, the declining multi-year ice due to climate change has implications on polar regions' atmospheric sea-salt and BrOx cycles (Hara et al., 2018). Antarctic summer particle number size distribution was dominated by aerosols of size <100 nm, which are newly formed particles from the gas phase (Fiebig et al., 2014). Legrand et al. (2016) reported multi-year particulate and gaseous bromine measurements from a central east Antarctic site, Concordia, and a coastal Antarctic location, Dumont d'Urville. Their observations and simulations indicate sea salt as the primary source of gaseous inorganic bromine species during the summer.

Abrahamsson et al. (2018) reported the formation of atmospheric bromine from sea ice during Antarctic winter periods. This new bromine source in the Antarctic atmosphere affects atmospheric chemistry and, thus, has climatic implications. Maffezzoli et al. (2017) analyzed 2 m core samples collected from Talos Dome to GV7 (roughly between 70°S and 72°S, 158 and 159°E) to examine total bromine, iodine, and sodium concentrations. They found seasonality in bromine enrichment with a maximum during Austral spring. However, iodine concentrations do not show a clear seasonal cycle and remain around the average value of 0.04 ppb. The competing effects of air masses originating from the Ross Sea and the Southern Ocean were found to maintain homogeneous air to snow fluxes of bromine, iodine, and sodium (Maffezzoli et al., 2017).

4.3 TEMPORAL (ANNUAL) VARIABILITY OF ANTARCTIC AEROSOLS

Seasonal variability has been observed in the concentration of sub-micrometer particles over the ice sheet of Antarctica by several investigators (Hogan, 1975; Shaw, 1979; Ito, 1985; Gras, 1993; Asmi et al., 2018). Hogan (1975) found sub-micrometer particle concentrations as low as a few cm^{-3} in winter to several thousand cm^{-3} in summer at the South Pole. Similar seasonal variability was observed in the concentration of condensation nuclei (CN) at the coastal stations of Syowa (Iwai, 1979; Ito and Iwai, 1981; Ito, 1983, 1985), George-von-Neumayer (Jaenicke and Stingl, 1984; Jaenicke et al., 1992) and Mawson (Gras and Adriaansen, 1985; Gras, 1993). As sulphur is present in excess all over the continent of Antarctica, higher concentrations of the sub-micrometer particles observed in the summer season are explained by the enhanced gas-to-particle conversion by photochemical reactions during sun-lit months (Ito, 1982; Gras and Adriaansen, 1985; Gras, 1993). Observed annual variations of the CN at Mawson (67.6°S, 62.9°E) by Gras and Adriaansen (1985) show only a few tens of particles cm^{-3} in winter (April to May) and about 300–400 cm^{-3} in summer (October to March). They found a slight increase in total concentration during July–August, which is associated with the penetration of the maritime air from the cyclonic storms passing around the coast during that period.

Ono et al. (1981) observed a significant difference in the annual variation of total concentration from the large particle concentration at the Syowa station. While the total aerosol concentration is low in winter and high in summer, the large (r > 0.15 μm) particle concentration was low in summer with a rapid increase in September month. Based on the variation of the peak values in total and large particle concentrations during different seasons, Ono et al. (1981) concluded that the aerosols in summer months are not of maritime origin but are generated by the photochemical process within the Antarctic atmosphere. At the South Pole, sodium and chloride fractions in aerosol particles were about an order of magnitude higher in winter than summer (Cunningham and Zoller, 1981). A similarly high fraction of sodium and chloride was also observed at the South Pole by Parungo et al. (1981) in the winter month of July. Asmi et al. (2018) measured aerosol particle optical properties at the Marambio station in the Antarctic Peninsula. Their measurements showed the presence of a large fraction of fine mode particles during the summer season. Major constituents of aerosols at the Marambio station were sea salt, sulphate, and crustal soil minerals (Asmi et al., 2018).

4.4 SPATIAL VARIABILITY OF ANTARCTIC AEROSOLS

The physiochemical properties and the size distribution of aerosol particles differ in the coastal or at any interior site on the Antarctic continent. The geographical features, such as surrounding oceans and dome-like topography of the icy continent, result in contrasting climatology of the wind flows at different locations. In addition, the difference in the amount of solar radiation received at the coastal stations and an interior site such as the South Pole impacts the aerosol concentration and their seasonal variations. However, relative contributions of different processes are primarily determined by the prevailing meteorological conditions in the region. Unlike at the South Pole, where NSS-sulfate is the dominant component of particulate mass, particles at the coastal stations constitute a significant fraction of sea salt to the total group. While the total number concentrations range between 100 and 200 cm^{-3} during summer and below about 20 cm^{-3} during winter at the South Pole, their values are found to be 300–2,000 cm^{-3} at the Antarctic coast (Jaenicke et al., 1992; Ito, 1993; Gras, 1993; Koponen et al., 2003). Aerosol measurements made at the South Pole by Park et al. (2004) also show concentrations of the order of 100–300 cm^{-3} during summer when local contamination is neglected.

An occasional increase in total particle concentrations has been observed at coastal Antarctic stations associated with the passage of cyclonic storms circulating the continent (Hogan, 1975; Lal and Kapoor, 1989; Deshpande and Kamra, 2004; Jurányi and Weller, 2019). Coastal stations receive a significant input of sea-salt particles from these cyclonic systems. These sea-salt particles are then modified in the atmosphere by heterogeneous reactions with gaseous sulfur species in summer and reactive nitrogen oxides in winter (Hara et al., 2004). However, size distribution measurements at the South Pole by Parungo et al. (1981) demonstrate some increase in coarse particles associated with the transport of sea salt up to the South Polar Plateau in August. There is a significant difference in absolute concentrations observed at sea level at coastal sites than at the Antarctic plateau (Gras, 1993). Recent measurements of Koponen et al. (2003) at Aboa (73°03'S, 13°25'W) in the Queen Maud Land region of Antarctica show total concentrations of particles in the size range of 0.003–0.8 µm diameter to be 200–2,000 cm^{-3}, with higher average concentrations in the airmass of oceanic/coastal origin than that of continental origin. While similar seasonal variations have been observed at the South Pole and in coastal stations, the total number of concentrations of particles are about an order of magnitude higher at the coastal sites than at the South Pole (Ito, 1993; Koponen et al., 2003). At the King Sejong Station in the Antarctic Peninsula, the total aerosols concentration was in the range of 1,707–83,120 cm^{-3} with the new particle formation rate of 2.79–1.05 cm^{-3} s^{-1} (Kim et al., 2017). Kyrö et al. (2013) reported particle growth events associated with the oxidation of biogenic precursors at the Finnish station Aboa. Giordano et al. (2017) found a consistent mode at 250 nm in the size distribution in coastal Antarctica attributed to sulfate particles. Measurements over the ice-free areas of the Ulu Peninsula (James Ross Island) in the eastern Antarctic

Peninsula showed peak concentrations of particles less than 10 μm (PM_{10}) up to 57 μg m^{-3} when wind speed exceeded 10 ms^{-1} (Kavan et al., 2018).

4.5   ANTARCTIC AEROSOLS AND CLIMATE

The marine aerosol particles influence climate directly by scattering the solar radiation and indirectly by serving as cloud condensation nuclei (CCN), hence modifying cloud properties. Due to their ubiquitous presence over approximately 70% of Earth's surface, sea salt may contribute more than half of the near-surface light extinction, even near the regions influenced by biomass burning (Quinn and Coffman, 1999; Quinn et al., 2001; Shinozuka et al., 2004). Getting aerosol measurements at Antarctica is also essential to check on any substantial change in background aerosol concentration since a small change in it may lead to large changes in global atmospheric processes that contribute to global climatic changes (Charlson et al., 1987). Twomey (1974, 1977) and Twomey et al. (1984) suggested that an increase in aerosol number concentration leads to an increase in cloud droplet number and hence increases cloud albedo (also known as the Twomey effect). This effect is mainly confined to shallow-layer clouds because deep clouds reflect most solar radiation regardless of their drop size (Twomey, 1977). Because of the complex nature of aerosol-CCN-cloud droplet concentration-cloud albedo and the feedback mechanism, the estimation of indirect aerosol effect on climate is one of the most enormous uncertainties in understanding climate change through various global climate models (IPCC, 2007; Boucher et al., 2013). This uncertainty is mainly due to the lack of knowledge of aerosols' formation and physicochemical characteristics (Carslaw et al., 2013; IPCC, 2013).

Calculations show that cloud albedo is sensitive to the concentration of CCN as they have a significant influence on the number and size of cloud drops (Twomey et al., 1984). Higher CCN concentrations result in a higher number of cloud droplets of smaller size, which enhances the albedo and lifetime of the cloud and suppresses the precipitation (Twomey, 1977; Latham and Smith, 1990; Murphy et al., 1998; Lohmann, 2006). Durkee et al. (1988) confirm the higher cloud reflectivity associated with smaller cloud droplets and higher CCN concentrations in marine stratocumulus clouds with the help of simultaneous satellite radiometer and aircraft measurements. The albedo of shallow layer clouds (thickness of the order of several hundred meters), generally found over extratropical marine regions, are more effectively affected by CCN density and therefore significantly contribute to the Earth's radiation budget (Prospero, 1979; Randall et al., 1984; Slingo, 1990; Fitzgerald, 1991). Rosenfeld and coworkers studied the role of aerosol particles in climate change through cloud-aerosol interactions and precipitation formation processes. They explained the role of marine aerosols in deriving the atmospheric circulation via their control on precipitation processes, vertical distribution of latent heat release, and tropospheric lapse rate (Rosenfeld, 1999, 2000, 2006; Rosenfeld et al., 2006), and cleansing of the atmosphere by collecting small pollution nuclei by sea-salt particles (Rosenfeld et al., 2002).

The CCN concentration was believed to be most affected by NSS-sulfate than sea salt over the marine environment. They contribute to most of the sub-micrometer particles and serve as CCN. Recent studies (O'Dowd and Smith, 1993; O'Dowd et al., 1997; Murphy et al., 1998; Clarke et al., 2006) show that sea-salt aerosols can significantly contribute to the sub-micrometer aerosol concentration and therefore have considerable influence on CCN concentration and cloud albedo. Measurements by Meszaros and Vissy (1974) and O'Dowd et al. (1997) show the existence of sub-micrometer sea-salt mode, which dominates the total number concentration of sea salt. Moreover, the sea-salt particles are hygroscopic and get activated at a lower supersaturation compared to NSS-sulfate. Therefore, they act more effectively as CCN. O'Dowd and Smith (1993) suggested that sea salt could not only provide a significant source of CCN but sometimes dominate the actual CCN population. O'Dowd et al. (1999) used a simple cloud model to combine the observed sea-salt and NSS-sulfate size distributions over the North Atlantic and East Pacific Oceans. They found that approximately 80% of the cloud droplets can be formed upon the sea-salt nuclei. Since sea-salt particles larger than 0.05 μm radius are also readily activated into cloud droplets at supersaturations of less than 0.2% (Pruppacher and Keltt, 2010), the concentration of total aerosols larger than 0.05 μm radius can be considered as the particles over a marine atmosphere that can act as CCN (O'Dowd et al., 1999).

A regulatory feedback mechanism (popularly known as CLAW hypothesis) for control of climate by marine phytoplankton affecting the cloud microphysical properties was proposed by Charlson et al. (1987). According to the CLAW hypothesis, the DMS emitted by the ocean is oxidized in the atmosphere to form low vapor pressure species i.e. sulfate and methyl sulfonic acid (MSA), which then form new particles by nucleation and grow the existing aerosol particles by condensation to serve as CCN. The increased population of CCN enhances the number of cloud drops, increasing multiple scattering within clouds and enhancing cloud albedo. The high cloud reflectivity decreases the phytoplankton population and constitutes a regulatory feedback system for biogenic control of the climate. However, a few studies have brought the CLAW hypothesis into question as the MBL nucleation process appears uncommon both as measured (Clarke et al., 1996) and modeled (Katoshevski et al., 1999). The free troposphere sulfate flux may be linked to surface DMS, and only a tiny fraction of the DMS flux participates directly in their formation (Clarke et al., 2006). Quinn and Bates (2011) noted that the cloud response to aerosols formed from dimethyl sulphide is more complex than the simple CLAW hypothesis. Cameron-Smith et al. (2011) utilized the Community Climate System Model (CCSM) to examine the DMS distribution and fluxes using present and future projected atmospheric CO_2 concentrations. They found changes in zonal averaged DMS flux to the atmosphere of over 150% in the Southern Ocean. This change was associated with the concurrent sea ice changes and ocean ecosystem composition shifts caused by changes in temperature and other biogeochemical parameters. The linkage between the sulfur cycle and climate is most extensive in the remote southern ocean, where the oceanic DMS emission dominates over the anthropogenic sulfur emission (Cameron-Smith et al., 2011). The palaeoclimate record from ice-core drilling at the Russian station Vostok in east Antarctica confirms the strong correlation between atmospheric greenhouse gas concentrations and Antarctic temperature (Petit et al., 1999).

The phytoplankton is grown and the micronutrient Fe partially limits biological productivity in remote regions of the Southern Ocean and, thus, making it high-nutrient low-chlorophyll (HNLC) waters (Boyd et al., 2000; Coale et al., 2004). The supply of atmospheric Fe, primarily from desert regions, affects ocean biogeochemical cycles (Jickells et al., 2005; Blain et al., 2007; Boyd et al., 2010; Johnson et al., 2010). Additionally, the Fe is also supplied to the near-surface waters in the Southern Ocean via an upwelling process (Dulaiova et al., 2009; de Jong et al., 2012) and melting of sea ice in coastal Antarctic waters (Sedwick and DiTullio, 1997; Lin et al., 2011).

However, Fitzgerald (1991) pointed out that the dependence of DMS production on sea water temperature and intensity of solar radiation received at the ocean surface has not been well quantified and, therefore, a need for numerical modeling is expressed to quantify the relationship between CCN concentration and DMS flux rate over MBL. Latham and Smith (1990) suggested a negative feedback process associated with the increased wind speeds due to global warming. An increase in wind speed enhances the production of sea-salt particles over the ocean surface, which subsequently act as CCN and increase the number of cloud droplets in marine stratus clouds, enhancing the cloud albedo and producing a cooling effect. It was predicted that a ~10 ms^{-1} increase in average wind speeds might sufficiently improve cloud droplet concentrations in marine stratus to have a cooling comparable with the estimated global warming. Andreae et al. (2005) point out that opposed effects of aerosol cooling and greenhouse gas warming will depend on the present-time aerosol cooling. The Moderate Resolution Imaging Spectrometer (MODIS) has been imaging the globe from space since 1999. It provides essential information to quantify the effective radius of background marine aerosols down to 0.05 μm, covering the full spectrum of CCN. A constellation of satellites can provide an in-depth understanding of the impact of aerosols on clouds, precipitation, and climate through synchronized measurements of aerosol and cloud properties (Rosenfeld, 2006).

4.6 MEASUREMENTS OF AEROSOLS AT A COASTAL ANTARCTIC STATION

Measurements of background aerosols in Antarctica help to estimate the influences of human activities on global atmospheric aerosols. Therefore, measurements of physical and chemical properties of aerosol particles have been made at several scientific stations established by some countries at different geographical positions (e.g. South Pole, Antarctic coast, Antarctic Peninsula, and interior continental Antarctic sites) of the continent to study the characteristics of background aerosol and estimate the effect of aerosols on global climate and climate change. Here, we present measurements of the aerosols made at a coastal Antarctic station, Maitri (70°45'52"S, 11°44'03"E), during the 24th Indian Scientific Expedition to Antarctica (24th ISEA). Measurements of the size distribution of aerosols in the size range of 0.003–20 μm diameter are carried out for 1 January–28 February 2005. The high-resolution size distributions of aerosols are analyzed to investigate the temporal evolution, identification of their sources, and impact of different meteorological events at the coastal Antarctic station.

4.6.1 STUDY REGION AND INSTRUMENTATION

The aerosol measurements were made at the Indian station Maitri, situated in the Schirmacher Oasis in East Antarctica. The Schirmacher Oasis spans ~ 35 km^2 area in the Drowning Maud Land and consists of frozen lakes, cliffs, and sandy soil. It has a vast ice shelf (90 km) in the north with polar ice and ice-rock interface with seasonal fluctuations in the south. The locations of the Schirmacher Oasis, Maitri station, and its detailed layout of laboratories and other facilities are shown in Figure 4.1. During the summer season, the ice-melt regions exist at a distance of about 0.5 km on the northern and southern sides of Maitri. This figure also includes a wind-rose diagram showing the pattern of wind directions during the period of measurements. The persistent southeasterly winds reduced the chances of emissions from Maitri station arriving at sampling location at the Kamet Observatory.

The number size distribution and total concentration of aerosols were measured in the vast particle size range of 0.003–0.7 μm and 0.5–20 μm diameter at Maitri and onboard ship stationed near the ice-shelf region at the Antarctic coast. The micron-size (0.5–20 μm diameter) particles were measured with an Aerodynamic Particle Sizer (APS, TSI Model 3321) and the sub-micrometer size (0.003–1 μm diameter) particles with a Scanning Mobility Particle Sizer (SMPS, TSI Model 3936). Simultaneous measurements with the SMPS and APS provide size distributions in the range of 0.003–20 μm diameter with an overlap of particle sizes between the two instruments. Particles are classified based on their electrical mobility in the SMPS system's Electrostatic Classifier (TSI Model 3080). These particles are then counted in the Ultrafine Condensation Particle Counter (UCPC, TSI Model 3025A) attached to the SMPS. To cover the wide spectrum mobility ranges, SMPS uses two differential mobility analyzers (DMA) i.e. the long DMA (LDMA) for particle sizes of 0.01–1.0 μm diameter and nano DMA (NDMA) for the ultrafine particles in size range of 0.003–0.168 μm diameter. The exact range of particle sizes in a scan depends on the type of DMA, orifice size of the impactor, set values of flow rates, and scan time.

A time-of-flight technique is used to measure the size distribution of accumulation and coarse mode particles (0.5–20 μm) in the Aerodynamic Particle Sizer (APS, TSI Model 3321). Two laser beams focused on an accelerated flow field of air-sample estimate the time-of-flight and detect the aerodynamic size of particles. This method is superior to the light scattering method as the time-of-flight is unaffected by Mie scattering (index of refraction) and accounts for the shape of the particle. After entering the sampling inlet, the air flow splits into two parts, i.e. sheath flow (4.0 L min^{-1}) and sample flow (1.0 L min^{-1}). However, the sheath flow is reunited with the sample flow after passing the orifice. The sheath flow confines the sample air in the central core region for the accurate measurement of particles. Smaller particles move faster than the larger particles within the flow field. The velocity of a particle is measured from the side-scattered light which is converted into electrical pulses by a photodetector. The measured velocity then gives an estimate of the particle size. For the calibration of APS, mono-disperse spherical particles of known density are used.

(a)

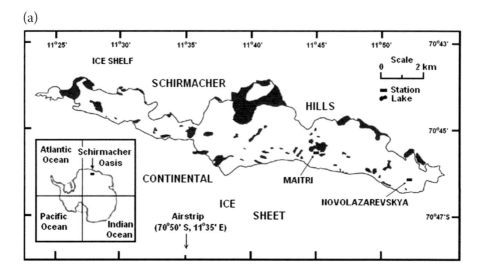

(b)

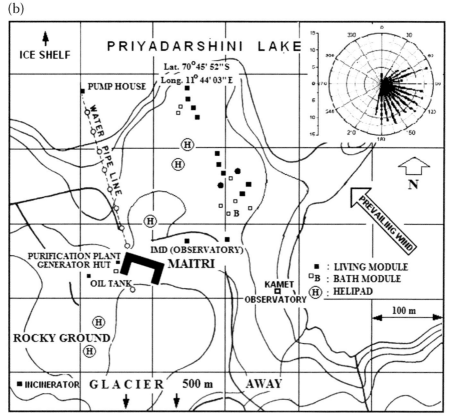

FIGURE 4.1 The map of Antarctica, showing the location of the Maitri station, location of instruments (Kamet Observatory) at Maitri, and the wind rose showing the magnitudes and directions of wind speed during the January to February 2005 period (Adopted from Pant et al., 2010).

Both instruments (i.e. APS and SMPS) are equipped with a dedicated software, 'Aerosol Instrument Manager', that facilitates continuous online data logging, display, and interface to connect with a personal computer for data logging. The aerosol size distribution data were obtained from both APS and SMPS systems simultaneously and a 10-minute average for each sample was stored in a computer during both expeditions. All the 10-minute samples are averaged for 1 hour.

4.6.2 Meteorological Measurements

The meteorological measurements were made at a height of ~5 m above ground level at Maitri. Meteorological parameters such as surface atmospheric pressure, air temperature, wind speed, wind direction, total cloud coverage, etc. were measured at the hourly interval by the observers from the India Meteorological Department (IMD). Figure 4.2 shows the temporal variations in meteorological parameters measured at Maitri during the period of measurements. During this period, the air temperature at Maitri ranged between +8°C and −7°C. The atmospheric pressure varied in the range of 982 hPa to 942 hPa. The eastward propagating low-pressure systems circulating the Antarctic continent between 65°S and 70°S caused significant pressure drops at Maitri on a few occasions. In response to these low-pressure systems, a change in cloud coverage from clear sky to overcast was observed. During the period of measurements, two cyclonic storms were observed on

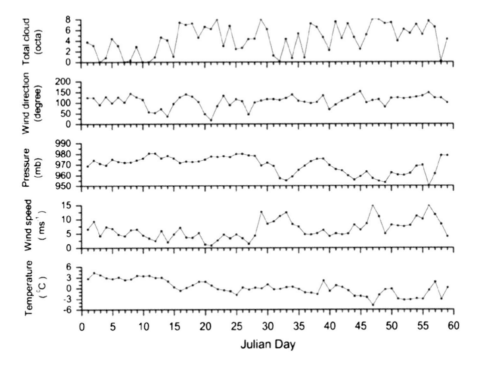

FIGURE 4.2 Daily average values of meteorological parameters at Maitri for Julian days in 2005.

Julian days 33 and 56, which caused a surface pressure drop of 20 and 30 hPa, respectively (Figure 4.2). An analysis of the 5-day back trajectory showed that the air mass was of continental origin before February 16, 2005, and mostly (>50%) changes to oceanic origin after this date. The continental air mass flowed downslope as drainage from the elevated southern latitudes. The passage of cyclonic storms near Maitri and the type of air mass (continental or oceanic origin) influenced the aerosol size distributions at Maitri. Such cases are analyzed in detail and presented in Section 6.3.

4.6.3 Aerosol Measurements

A total of 7,356 number size distributions with the APS system and 6,834 with the SMPS system were measured during this period. Each number size distribution is integrated to separately compute the total number concentrations of micrometer and sub-micrometer particles, respectively. The variations of total number concentrations of micrometer (as measured with the APS system) and sub-micrometer (as measured with the SMPS system) particles are plotted in Figure 4.3. Very low and almost constant values of total number concentrations of 0.1–0.8 particles cm^{-3} characterize our observations of micrometer aerosol particles (0.5–20 μm diameter) during this period. On the other hand, the total number of concentrations of submicrometer particles (0.003–0.7 μm diameter) generally ranges between 100 and 800 particles cm^{-3} in January and between 100 and 2,000 particles cm^{-3} in February. Temporal variations of particles in these two size ranges need not be always similar during this period, indicating different sources for the origin of particles in the two size ranges. Concentrations of condensation nuclei of the same order have also been observed in this season at several other Antarctic coastal stations such as Syowa (69°S, 39.6°E), Mawson (67.6°S, 62.9°E), Ross Island (77.8°S, 166.8°E), and Georg-von-Neumayer (72.6°S, 8.4°W) (Ito, 1983, 1985, 1993; Jaenicke and Stingl, 1984; Gras and Adriaansen, 1985; Shaw, 1986; Lal and Kapoor, 1989; Jaenicke et al., 1992; Gras, 1993; Deshpande and Kamra, 2004). Voskresenskii (1968), however, reports comparatively much lower concentrations at Mirny (66.6°S, 93°E). Measurements of Koponen et al. (2003) at Aboa (73° 03'S, 13° 25'W), a continental station, also show

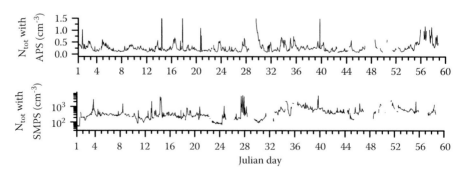

FIGURE 4.3 Time series of total number concentration (numbers cm^{-3}) of aerosols measured using APS (upper panel) and SMPS (lower panel) at Maitri, Antarctica during January–February, 2005.

similar aerosol concentrations. However, concentrations are comparatively smaller by an order of magnitude at the South Pole (Shaw, 1988; Ito, 1993).

Time variations of total number concentrations of particles during this period show some peaks that may sometimes even exceed 1.2 particles cm^{-3} in the case of micrometer particles and 5,000 particles cm^{-3} in the case of sub-micrometer particles. These peaks in micrometer and sub-micrometer particle concentrations, unlike the variations in particle concentrations associated with the passage of cyclonic systems revolving around the continent of Antarctica, do not always occur simultaneously and last from a few tens of minutes to a few hours.

4.6.3.1 Number Size Distributions of Aerosols

The APS and SMPS systems were operated simultaneously throughout the measurement period at the Mairi station. While the APS system measured aerosol size distribution in a uniform range of particle sizes (0.5–20 μm diameter), the size range of sub-micrometer aerosols varied in the SMPS system, depending on the use of either NDMA or LDMA. Further, the LDMA was used with two settings to provide measurements in the size range of 0.016–0.7 μm in January and 0.01–0.4 μm in February. These differences in size ranges were maintained to capture most of the evolving modes of the number size distributions. The NDMA was used with a size range of 0.003–0.16 μm whenever there were incidences of higher ultrafine particles and chances of the presence of <10 nm diameter particles. A comparison of aerosol size distribution measurements at Maitri (January and February 2005) with earlier measurements from different coastal Antarctic stations is shown in Figure 4.4. Although our size distribution curves are similar in shape to those of earlier measurements at Antarctica, there is a large difference in the position of modes and concentration of particles in same size bins, particularly of particles of <0.1 μm diameter. The size distributions measured by LDMA and NDMA (in the SMPS system) are shown separately in the figure. An analysis of modal structure of our number size distributions measured at Maitri showed the presence of nucleation mode (d <0.02 μm), Aitken mode (0.02 <d <0.1 μm), accumulation mode (0.1 <d <1.0 μm), and coarse mode (d > 1 μm) particles during summer. As compared to the Aitken and accumulation modes, the nucleation mode shows a larger variability among measurements from different coastal stations. Large variations in nucleation mode are found since the production and growth rate of ultrafine particles vastly differ under meteorological conditions such as ambient surface temperature, wind speed, relative humidity, and concentration of DMS gases that could greatly vary at different coastal Antarctic stations. Different sources of nucleation mode particles could result in the observed scatter in the distribution of nucleation mode particles. On the other hand, there are only a few measurements in the micron-size particle range. Our data in the coarse mode range (>1 μm) of particles agree well with the measurements of Radke and Lyons (1982) and Harvey et al. (1991). Presented high-resolution size distributions could precisely mark the Aitken mode position over the earlier measurements of aerosols at Maitri by Deshpande and Kamra (2004). Moreover, the open-ended curve in the smallest size end of distribution reported by Deshpande and Kamra (2004) is covered well and closed with the presented measurements.

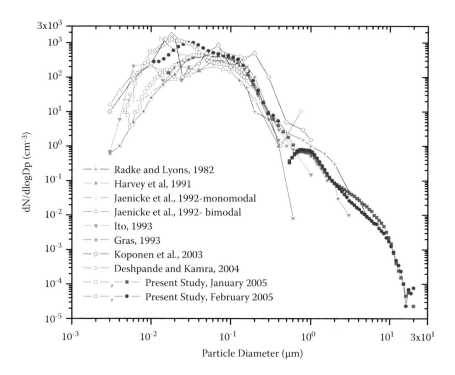

FIGURE 4.4 Number size distribution curves averaged for the whole data simultaneously obtained with the APS and SMPS systems in different size categories during January and February 2005. Hollow and solid labels for the present study represent data obtained with the NDMA and LDMA, respectively. Also shown, for comparison, are the number size distributions observed by various investigators at different coastal Antarctic stations (Modified after Pant et al., 2011).

4.6.3.2 Variability in Aerosol Size Distribution

Our high temporal resolution data reveals variability in different modes of the aerosol size distributions measured during the summer season at an Antarctic coastal site. The fine structure observed in coarse mode (>1 μm), accumulation mode (0.1–1.0 μm), and Aitken mode (<0.1 μm) particles are discussed in this section.

i Accumulation and Coarse Modes

The coarse mode particles generally show a sharp decrease in concentration with their increasing size. However, sometimes the large (>2 μm diameter) particles are observed in greater concentration towards the higher end of the size distribution. Such incidences give rise to higher variability in the concentration of these large particles, compared to the smaller particles (<2 μm) in the coarse and accumulation modes. The coarse mode particles are found to have a maximum of 2.129 ± 0.325 μm diameter under wind speed of 4–6 ms^{-1} on a good sunshine day. The accumulation mode is generally found at 0.777 ± 0.054 μm. However, this accumulation mode shifts to a lower value of 0.723 ± 0.062 μm when a maximum appears in the

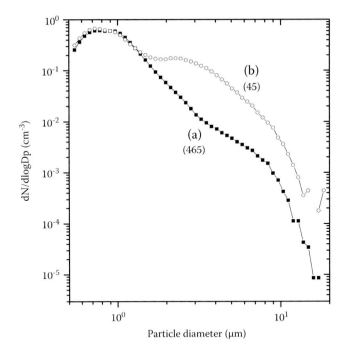

FIGURE 4.5 Mean size distribution curves of mono-modal (a) and bimodal (b) shapes computed from the hourly averaged curves (as measured with the APS system) of respective shapes observed from January to February 2005. Numbers in parentheses show the number of aerosol size distribution curves used to compute the mean curve (Modified after Pant et al., 2011).

coarse mode. On the other hand, in February when the temperatures were low, concentrations of these particles did not attain this peak value, despite stronger winds than in January. In general, the accumulation and coarse modes together constitute either a mono-modal or a bimodal size distribution. Figure 4.5 shows the averaged mono-modal and bimodal size distributions observed throughout measurements. The total number of samples with each type of size distribution is mentioned in parentheses against each curve in the figure. Our observed size distributions agree with the measurements at Butter Point on Ross Island (Harvey et al., 1991) also reported large variations in particles of size >2.0 μm diameter that mainly consisted of sea salt.

ii Aitken Mode

The Aitken mode in the size distribution can be considered as an intermediate stage between the newly generated 'nucleation mode' (ultrafine) particles and the particles that can act as CCN. The monthly averaged size distribution shows a single Aitken mode. However, the higher frequency (10 minutes to hourly) size distributions reveal bimodal characteristics of the Aitken mode that can be classified into Aitken mode I and Aitken mode II.

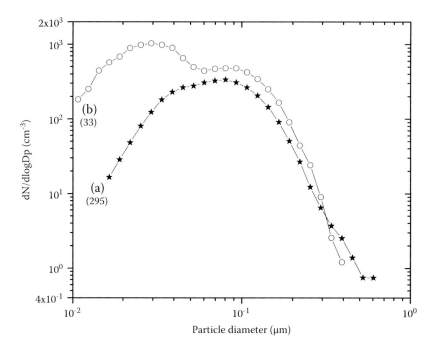

FIGURE 4.6   Mean size distribution curves of mono-modal (a) and bimodal (b) shapes computed from the hourly averaged curves (as measured with the SMPS system using LDMA) of respective shapes observed from January to February 2005. Numbers in parentheses show the number of aerosol size distribution curves used to compute the mean curve (Modified after Pant et al., 2011).

Figure 4.6 shows two curves obtained from all such high-resolution size distributions showing the mono-modal and bimodal character during the period of measurements (i.e. January–February, 2005). The single Aitken mode was found at 0.08 ± 0.011 µm in the monomodal size distribution. The bimodal distribution consists of two modes, the Aitken mode I at 0.029 ± 0.004 µm and Aitken mode II at 0.088 ± 0.006 µm. While the majority of monomodal distributions were observed in January, the bimodal distributions were found to occur mainly in February. Several past observational studies reported a broad mode below 0.1–0.2 µm accompanied by a nucleation mode below 0.02 µm during the summer months (Jaenicke et al., 1992; Gras, 1993; Ito, 1993; Koponen et al. 2003). However, the bimodal characteristics of the Aitken modes were not observed in earlier studies at Antarctica, which could be due to their coarser temporal resolution or larger bin size of aerosol distribution. Jaenicke et al (1992) noted from their measurements at the Georg von Neumayer Antarctic station that the multimodal nature of the distribution is not revealed due to averaging of data.

iii Nucleation Mode
The ultrafine nucleation mode particles in the size range of 0.003–0.16 µm diameter were measured during our study period on a few occasions. Particles in this size

range (<0.01 µm diameter) are frequently observed in our measurements made during the summer months in Antarctica. Observations show the peak concentrations of these particles at about ~0.01 µm or 0.017 µm diameters with large variability in concentration. Throughout measurements, primarily three types of distributions were observed that are shown as curves a, b, and c in Figure 4.7. Curve 'a' shows the maxima at 0.017 µm, while particles less than this size were also measured. In curve 'b', the nucleation mode (at 0.01 µm) is accompanied by an Aitken mode I. The third curve (i.e. curve 'c') bears a nucleation mode at 0.01 µm along with both the Aitken modes (i.e. Aitken modes I and II). It was also noticed that the concentration of Aitken mode particles increases in the coexistence of nucleation mode in the distribution. However, the absence of particles in the lowest

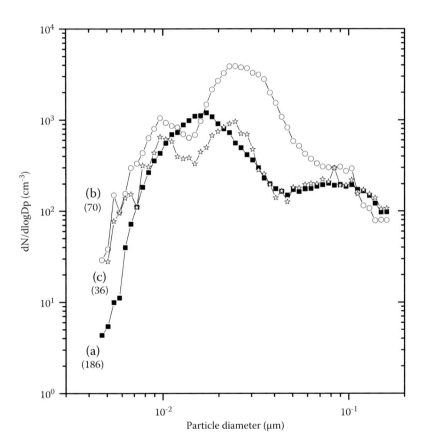

FIGURE 4.7 Mean size distribution curves of mono-modal (a), bimodal (b), and tri-modal (c) shapes computed from the hourly averaged curves (as measured with the SMPS system using NDMA) of respective shapes observed from January to February 2005. Numbers in parentheses show the number of aerosol size distribution curves used to compute the mean curve (Adopted from Pant et al., 2011).

range of size distribution (<0.005 μm) signifies that local new particle formation is not an active process at the Maitri station in east Antarctica.

4.6.3.3   Characteristics of Aerosols in the Oceanic and Continental Air Masses

Being a coastal Antarctic station, Maitri received aerosols of both marine and continental origin. Based on the surface wind direction and history of air mass estimated through the 5-day air back-trajectory analysis (not shown here), the samples are divided into two categories: (i) oceanic air mass and (ii) continental air mass. These size distributions for both the categories are plotted for large size range (APS measured) and small size ranges (SMPS measured) in Figure 4.8. The coarse mode particles show a typical power-law size distribution i.e. the concentration of particles decreases with the increase in their size. However, these particles have a higher concentration in the air mass of oceanic origin than that of continental origin. The peak concentration is found to be around the same size range in accumulation mode, but the oceanic air mass has three times higher concentration than the continental air mass (Figure 4.8a). Back trajectories of air mass suggest that the sea-salt particles generated due to bubble-breaking in the coastal waters off the Antarctic coast get transported to the Maitri station.

In the Aitken range of particle sizes, the distribution was found to be mono-modal (at 0.039 μm) in oceanic air mass, whereas it was bimodal in continental air mass with another peak at 0.085 μm (Figure 4.8b). The two types of size distributions may be associated with air masses of oceanic and continental origin. The

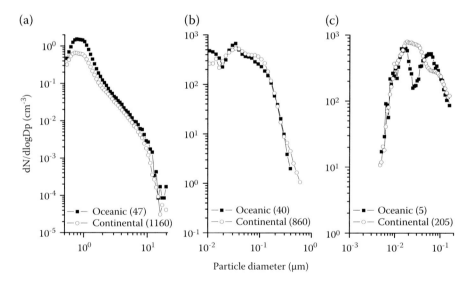

FIGURE 4.8 Mean number size distribution curves for aerosol particles in different size ranges associated with the oceanic and continental air masses. Numbers in parentheses show the number of hourly averaged size distribution curves used to compute the mean curve (Adopted from Pant et al., 2011).

lower cloud amount and SO_2 concentration over the continent than the oceanic region lead to the growth of these particles through cloud processing (Hoppel et al., 1994). Further, the transport of aged particles from the inland locations to the coastal Antarctic site may lead to the second peak in the size distribution. The presence of these aged particles is also noticed in further smaller particle sizes with a maximum of around 0.017 µm (Figure 4.8c). On the other hand, the oceanic air mass shows a sharp nucleation mode at ~ 0.01 µm. Kulmala et al. (2004) pointed out that more than one nucleation process operating in the atmosphere can lead to such large variability in the nucleation mode maxima. Hall and Wolff (1998) found the dominant contribution from the sea-salt particles generated as a result of wave-breaking activity to the aerosol size distributions in coastal Antarctic locations. Moreover, the size distributions measured at Maitri are in good agreement with those observed at the Finnish station, Aboa (Kerminen et al., 2000).

4.6.3.4   Changes in Aerosol Characteristics during Cyclonic Storms

Comparatively higher winds associated with the eastward propagating low-pressure systems circulating the Antarctic continent enhance the bubble-bursting processes over oceans and generate more sea-salt particles. These sea-salt particles of oceanic origin are transported to the coastal locations in Antarctica. Therefore, the passage of cyclonic storms close to the coastline of Maitri is expected to enhance the number concentrations of particles at Maitri. Additionally, the ultrafine particles in Aitken mode are photochemically generated over the continent through sulphuric acid during the summer season (Ito and Iwai, 1981). The physical and chemical properties of Antarctic aerosols are thus modified by the particles of oceanic origin transported by cyclonic storms to the interior of the continent. Many studies have shown that the transport of sea-salt particles is not only limited to coastal stations but also extends to the South Pole (Hogan, 1975; Hogan and Barnard, 1978; Ito and Iwai, 1981). During the summer season at the South Pole, the frontal pressure aloft and dry air subsidence are found to enhance the concentration of small particles (Hogan and Barnard, 1978). Also, an increase in the number of large particles has been observed under the lower level frontal passage and the warm moist air advection from the Weddell Sea (Hogan and Barnard, 1978). At another coastal Antarctic station, Syowa, transport of maritime air mass and descent of acid particles along a cold frontal surface were reported (Ito and Iwai, 1981). In this section, an analysis of the evolution of aerosol number size distribution during the periods of the passage of two cyclonic storms near the Maitri station is presented.

i Cyclonic Storm I

Figure 4.9 shows variations of meteorological parameters recorded at Maitri from the Julian day (JD) 31 to 36, 2005 (January 31–February 5, 2005) during which period an eastward propagating low-pressure area passed close to the coast of Maitri. The low-pressure area caused a pressure drop of ~20 hPa on JD 33 and strong southeasterly winds of 6–16 ms^{-1} during this period at Maitri. Clear skies changed to almost overcast from the morning of JD 33 onwards. An examination of 5-day back trajectories at 0000 UT and 1200 UT during this period showed that the air mass over Maitri was always of continental origin.

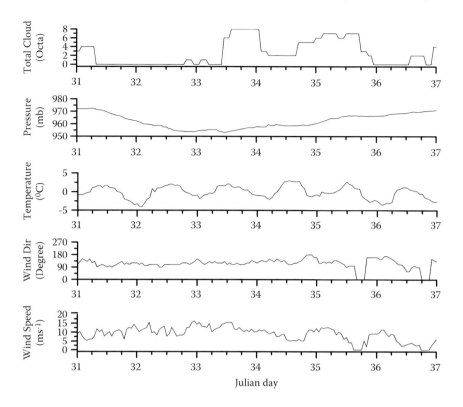

FIGURE 4.9   Variations in meteorological parameters recorded at Maitri during the passage of cyclonic storm from Julian day 31 to 36 (Adopted from Pant et al., 2010).

The total particle concentration of micrometer particles showed several peaks of relatively enhanced concentration exceeding 0.8 particles cm^{-3}. The total particle concentration of sub-micrometer particles, however, steadily increased by about an order of magnitude. Figure 4.10 shows the three-dimensional plot of the variations in number size distributions of micrometer and sub-micrometer particles from JD 31 to 37. Gaps in the data series for sub-micrometer particles are mostly because of cleaning and drying of inlet and making arrangements in the SMPS setup for changing the size range of particles to be measured. Also shown in this figure is a trace of variation in atmospheric surface pressure at Maitri.

The concentration of accumulation mode particles is low when the atmospheric surface pressure at Maitri is minimum but increases and shows some periods of high concentrations before and after this minimum when the surface pressure has either decreasing or increasing tendencies during which period the cyclonic storm is approaching or departing away from the Maitri station. The coarse particles, which normally follow a power-law size distribution, occasionally develop a coarse mode maximum for particles of ~2 μm diameter when wind speeds are ~15 m s^{-1}. On the other hand, although the concentration of the particles in the sub-micrometer range are also low when the surface pressure at Maitri is minimum, their concentration

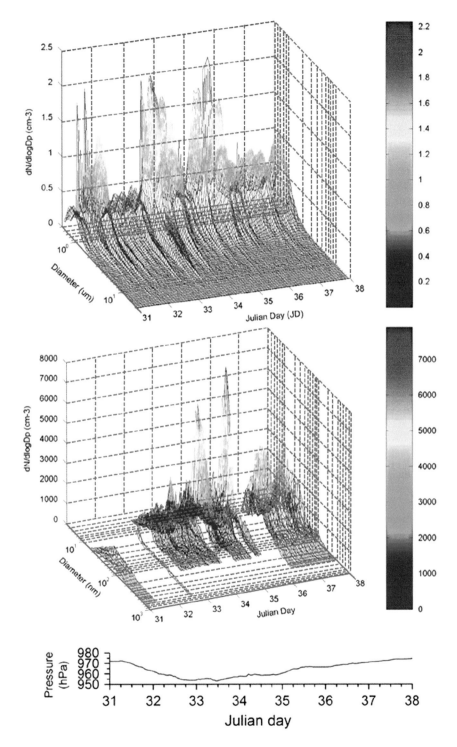

FIGURE 4.10 Three-dimensional plot of number size distributions as measured by APS and SMPS systems for the period of JD 31 to 37, 2005.

increases, and the periods of enhanced concentrations of Aitken and nucleation mode particles appear only when the surface pressure has an increasing tendency. Events of nucleation particles with a mode at 0.01 μm frequently appear during this period. The Aitken mode maximum becomes very prominent in the later period, particularly on JD 35 when the nucleation mode particles appear prominently. During this period, the cyclonic system is departing away from Maitri.

ii Cyclonic Storm II

In the case of the cyclonic storm that caused a pressure drop of 30 hPa on JD 56, Figures 4.11 shows the three-dimensional plot of the variations in size distributions of micrometer and sub-micrometer particles along with that of the atmospheric surface pressure at Maitri for JD 54–58, 2005 (February 23–27, 2005). The data gaps are due to the adverse weather conditions when the measurements could not be continued. A significant difference in this storm is that, unlike in storm I, the air mass at 1200 UT on JD 57, 0000 UT, and 1200 UT on JD 58 originated over the sea and descended from an altitude of 1,000–3,000 m before reaching Maitri. During these periods, Maitri was under the departing end of cyclonic storm II and is thus likely to have to descended atmospheric motions in the region. Unfortunately, there are no data on Julian day 56, the day of minimum surface pressure. However, similar to the case of a storm I, peaks in accumulation mode appear when the surface pressure has either increasing or decreasing tendency but peaks in nucleation mode at <0.015 μm appear and strengthen the Aitken mode maxima at 0.03–0.04 μm only when the surface pressure has an increasing tendency. Nucleation mode is particularly strong when the storm is departing. During this period, 5-day back trajectories show that the air mass originates at the ocean and descends 1,000–3,000 m before reaching Maitri. Accumulation mode maxima are almost twice as high as in the case of storm I and the size of coarse particles often extend to ~10 μm diameter, especially during the period when the surface pressure has an increasing tendency.

4.6.3.5   Aerosol Measurements over Ice-Shelf Zone at the Antarctic Coast

A unique opportunity for aerosol measurements arose while departing from the Maitri station. Just before starting the return cruise into the Southern Ocean, the ship was stationed just beside the large ice-shelf of the Antarctic coast for the period of 1–17 March 2005. The sea surface waters were cold and of low salinity, resulting from the sea-ice melt. The ocean surface did not yet start freezing. Measurements of aerosol size distributions were carried out in the coarse, accumulation, and Aitken modes with the help of APS and SMPS systems during 2–17 March 2005. Unfortunately, however, the routine hourly meteorological measurements were not collected onboard during this period. However, care was taken to keep the air samples free of any exhaust from the ship. The instruments were installed in a front-side cabin on the opposite end of the ship's smokestack. The prevailing wind direction was such that the ship's smoke was not approaching the sampling inlet. Moreover, we filtered short-term (<30 min) spikes in the aerosol concentrations and did not include them in the analysis.

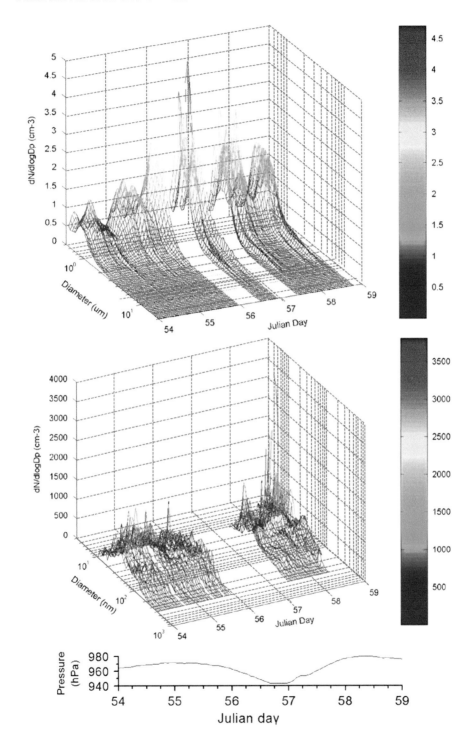

FIGURE 4.11 Three-dimensional plot of number size distributions as measured by APS and SMPS systems for the period of JD 54 to 58, 2005.

i Coarse and Accumulation Mode Particles

The coarse and accumulation mode particles measured by the APS system showed large variability in the number size distribution. The total concentration of particles in the size range of 0.5–20.0 μm varied in the range of 2–40 particles cm^{-3} during the period of measurements at the ice-shelf region. Figure 4.12 shows two types of typical size distributions measured during the period of 2–17 March 2005 near the ice shelf at the Antarctic coast. The accumulation mode was found to be at ~0.8 μm with a peak concentration of about 4 particles cm^{-3} (Figure 4.12a). However, the accumulation mode shifted to 1.0 μm when the peak concentration increased to >10 particles cm^{-3} (Figure 4.12b). In a few measured samples, the accumulation mode was absent and the distribution followed the power law, indicating a high concentration of submicron particles. Larger particles (> 5 μm) towards the end of the measured distribution showed high variability. A broad mode around 7–8 μm with significant variations in its magnitude was observed frequently during this period (Figure 4.12b). The sea-salt particles mainly constitute this broad mode in larger size particles. Further, there are episodes of very high (up to two orders of magnitude higher than normal) concentrations of particles of >5 μm diameter. At the same time, the concentration of accumulation mode particles also increases. A comparison of size distributions measured at a coastal ice-self region to that of the Maitri station (about 80 km inland from the Antarctic coast) shows remarkable differences. The concentration of accumulation mode particles is 4–20 times higher at the coastal ice-shelf region (Figures 4.12a and 4.12b) than at the Maitri station (Figure 4.4).

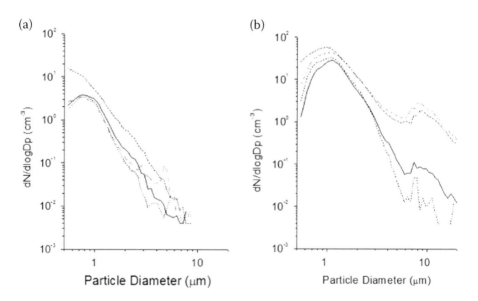

FIGURE 4.12 Typical number size distributions of accumulation and coarse mode particles measured with APS system on 8 March 2005 (a) and 15 March 2005 (b) onboard ship near the ice-shelf region in coastal Antarctica.

ii Aitken Mode Particles

In the size range of 9–430 nm, using LDMA, two distinct shapes of the number size distributions were observed, as shown in Figures 4.13a and 4.13b. The majority of air samples showed a uni-modal distribution with the mode varying in the range of 20–30 nm diameter (Figure 4.13a). On the other hand, some of the air samples did not show any mode in the distribution and remained almost flat, with a marginal decrement in particle numbers with the increasing size (Figure 4.13b). However, a few measurements showed broad maxima appearing at 40–70 nm diameter. The total concentration of particles in this size range was 3×10^2 to 4×10^5 cm^{-3}. Larger concentrations were observed when there appeared a prominent mode centered around 20–30 nm (as in Figure 4.13a) than those in the absence of mode.

The number size distributions of ultrafine particles over the size range of 4–163 nm using NDMA are shown in Figure 4.14. It can be noticed from the figure that size distributions are of mainly three different shapes. The single mode was observed to vary between 25–60 nm diameter. The magnitude of peak concentration (mode) remained higher for the appearance of mode at a smaller size (Figures 4.14a and 4.14b). Apart from these, there are few cases of distributions without a clear mode (Figure 4.14b). The total number concentration of particles in both size ranges measured by SMPS was found to be up to two orders of magnitude higher than those at the Maitri station. These episodes of very high Aitken particles near the ice-shelf region persisted for at least 2 hours. This large number of ultrafine particles could be associated with the secondary particle formation resulting from DMS and other sulfur-containing precursor gases, which get converted into particles. The southern high latitudes and Antarctic coastal waters are known for high DMS emissions (Cameron-Smith et al., 2011; de Jong et al., 2012). Therefore, the high-resolution size distribution measurements made near the ice-shelf region in Antarctic coastal waters provide a signature of gas-to-particle formation in the coastal Antarctic environment.

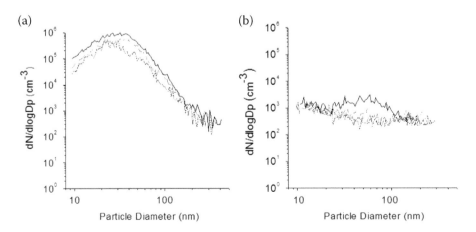

FIGURE 4.13 Typical number size distributions of fine particles in the size range of 9–430 nm using SMPS system with LDMA on 11 March 2005 (a) and 13 March 2005 (b) onboard ship near the ice-shelf region in coastal Antarctica.

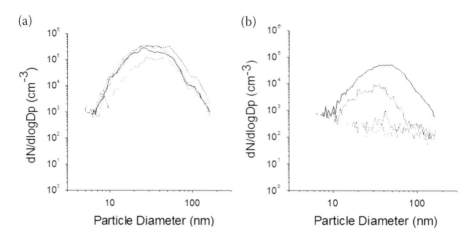

FIGURE 4.14 Typical number size distributions of ultrafine particles in the size range of 4–163 nm measured using SMPS system with NDMA on 8 March 2005 (a) and 12 March 2005 (b) onboard ship near the ice-shelf region in coastal Antarctica.

It calls for further research with a suite of aerosol instruments, along with sophisticated gas analyzers to quantify the contributions from the primary and secondary particles in the background site, such as coastal Antarctica.

4.7   CONCLUSIONS

Past studies on the Antarctic aerosols' characteristics and their effect on climate have been briefly reviewed. Measurements of aerosol number size distributions in the size range of 3 nm–20 μm diameter were made during January–February 2005 at the coastal Antarctic station Maitri. The unique aerosol measurements in this wide size range show a variety of number size distributions with the presence of single, double, and triple modes. The positions of maxima (mode) were found to vary a little (0.72–0.77 μm) in the accumulation range. However, the finer particles in Aitken mode show a larger variability in both the position and magnitude of the modes. Occasionally, large (>2 μm diameter) particles are observed towards the higher end of the size distribution under wind speed of 4–6 ms^{-1}. The total concentrations of sub-micrometer (3–700 nm) and micron-size (0.5–20 μm) particles were found to be in the ranges of 0.1–0.8 particles cm^{-3} and 100–2,000 particles cm^{-3}, respectively at the Maitri station. However, there was more than an order of magnitude increase in the number concentration of coarse mode particles sub-micrometer particles near the ice-shelf regions of coastal Antarctica. The gas-to-particle formation mechanism could be the possible reason for a very high concentration of ultrafine particles in coastal waters around Antarctica in the summer season. A supply of Fe from the interior of the ocean via upwelling or through the long-range atmospheric transport enhances the productivity in nutrient-rich waters off the Antarctic coast. This leads to the high emission of sulfur-containing precursor gases (such as DMS), which leads to the formation of new particles. The sea-salt particles originated from bubble bursting in the surrounding oceanic regions also found to contribute to the total aerosol loading at the Maitri station. The cyclonic storms passing over the higher latitudes of the Southern Ocean were found to significantly enhance the total particle count at Maitri. Given the warming climate and rapid melting of sea ice, the aerosol-climate interactions need to be revisited with enhanced efforts in the modeling and observational fronts.

ACKNOWLEDGMENTS

The authors gratefully acknowledge the National Center for the Antarctic and Polar Research (NCPOR), Goa for participation in the 24th Indian Scientific Expedition to Antarctica, 2004–2005. The support and facilities provided by the crew members of the research vessel 'Emerald Sea' are acknowledged. The meteorological data provided by the India Meteorological Department (IMD) is thankfully acknowledged. The Indian Institute of Tropical Meteorology is supported by the Ministry of Earth Sciences, Govt. of India. One of us (AKK) is thankful for the support under the INSA Emeritus Scientist program.

REFERENCES

Abbatt, J. P. D., Thomas, J. L., Abrahamsson, K., Boxe, C., Granfors, A., Jones, A. E., King, M. D., Saiz-Lopez, A., Shepson, P. B., Sodeau, J., Toohey, D. W., Toubin, C., von Glasow, R., Wren, S. N., and Yang, X. (2012). Halogen activation via interactions with environmental ice and snow in the polar lower troposphere and other regions, *Atmos. Chem. Phys.*, 12, 6237–6271, https://doi.org/10.5194/acp-12-6237-2012.

Abrahamsson, K., Granfors, A., Ahnoff, M., Cuevas, C. A., and Saiz-Lopez, A. (2018). Organic bromine compounds produced in sea ice in Antarctic winter, *Nat. Commun.*, 9, 5291, https://doi.org/10.1038/s41467-018-07062-8.

Andreae, M. O., Charlson, R. J., Bruynseels, F., Storms, H., Vangrieken, R., and Maenhaut, W. (1986). Internal mixture of sea salt, silicates, and excess sulfate in marine aerosols, *Science*, 232(4758), 1620–1623.

Andreae, M. O., Elbert, W., and Mora, S. J. (1995). Biogenic sulfur emissions and aerosols over the tropical South Atlantic 3. Atmospheric dimethylsulfide, aerosols, and cloud condensation nuclei, *J. Geophys. Res.*, 100(D6), 11335–11356.

Andreae, M. O., Jones, C. D., and Cox, P. M. (2005). Strong present-day aerosol cooling implies a hot future, *Nature*, 435, 1187–1190.

Andreae, M. O., and Raemdonck, H. (1983). Dimethyl sulfide in the surface ocean and the marine atmosphere: A global view, *Science*, 221(4612), 744–747.

Asmi, E., Neitola, K., Teinilä, K., Rodriguez, E., Virkkula, A., Backman, J., Bloss, M., Jokela, J., Lihavainen, H., Leeuw, G., Paatero, J., Aaltonen, V., Mei, M., Gambarte, G., Copes, G., Albertini, M., Fogwill, G. P., Ferrara, J., Barlasina, M. E., and Sánchez, R. (2018). Primary sources control the variability of aerosol optical properties in the Antarctic Peninsula, *Tellus B: Chemical and Physical Meteorology*, 70(1), 1–16, https://doi.org/10.1080/16000889.2017.1414571.

Atkins, C. B., and Dunbar, G. B. (2009). Aeolian sediment flux from sea ice into southern McMurdo Sound, Antarctica, *Global Planet. Change*, 69, 133–141. https://doi.org/10.1016/j.gloplacha.2009.04.006.

Atkinson, R. J., Pitts Jr. J. N., and Aschmann, S. M. (1984). Tropospheric reactions of dimethyl sulphide with NO_3 and OH radicals, *J. Phys. Chem.*, 88, 1583–1587.

Bates, T. S., Charlson, R. J., and Gammon, R. H. (1987). Evidence for the climatic role of marine biogenic sulphur, *Nature*, 329, 319–321.

Bates, T. S., Kapustin, V. N., Quinn, P. K., Covert, D. S., Coffman, D. J., Mari, C., Durkee, P. A., DeBruyn, W., and Saltzman, E. (1998b), Processes controlling the distribution of aerosol particles in the lower marine boundary layer during the First Aerosol Characterization Experiment (ACE 1), *J. Geophys. Res.*, 103, 16369–16384.

Bates, T. S., Quinn, P. K., Coffman, D. J., Johnson, J. E., Miller, T. L., Covert, D. S., Wiedensohler, A., Leinert, S., Nowak, A., and Neusub, C. (2001). Regional physical and chemical properties of the marine boundary layer aerosol across the Atlantic during Aerosols99: An overview, *J. Geophys. Res.*, 106, 20767–20782.

Bigg, E. K. (1980). Comparison of aerosol at four base-line atmospheric monitoring stations, *J. Appl. Meteorol.*, 19, 521–533.

Blain, S., Quéguiner, B., Armand, L. et al. (2007). Effect of natural iron fertilization on carbon sequestration in the Southern Ocean, *Nature*, 446, 1070–1074, https://doi.org/ 10.1038/nature05700.

Blanchard, D. C. (1963). The electrification of the atmosphere by particles from bubbles from the sea, *Prog. Oceanogr.*, 1, 71–202.

Blanchard, D. C., and Woodcock, A. H. (1957). Bubble formation and modification in the sea and its meteorological significance, *Tellus*, 9, 145–158.

Boucher, O., Randall, D., Artaxo, P., Bretherton, C., Feingold, G., Forster, P., Kerminen, V. M., Kondo, Y., Liao, H., Lohmann, U., Rasch, P., Satheesh, S., Sherwood, S., Stevens, B., and Zhang, X. (2013). Clouds and aerosols, in Climate change 2013: the physical science basis. Contribution of working group I to the fifth assessment report of the intergovernmental panel on climate change, edited by Stocker, T. F., Qin, D., Plattner, G. K., Tignor, M., Allen, S., Boschung, J., Nauels, A., Xia, Y., Bex, V., and Midgley, P., chap. 7, Cambridge University Press, Cambridge.

Boyd, P. W., Mackie, D. S., and Hunter, K. A. (2010). Aerosol iron deposition to the surface ocean: Modes of iron supply and biological responses, *Mar. Chem.*, 120, 128–143, https://doi.org/10.1016/j.marchem.2009.01.008.

Boyd, P., Watson, A., Law, C. et al. (2000). A mesoscale phytoplankton bloom in the polar Southern Ocean stimulated by iron fertilization, *Nature*, 407, 695–702, https:// doi.org/10.1038/35037500.

Bullard, J. E., Baddock, M., Bradwell, T., Crusius, J., Darlington, E., Gaiero, D., et al. (2016). High-latitude dust in the Earth system, *Rev. Geophys*, 54, 447–485. https:// doi.org/10.1002/2016RG000518.

Cadle, R. D., Fischer, W. H., Frank, E. R., and Lodge, J. P. Jr. (1968). Particles in the Antarctic atmosphere, *J. Atmos. Sci.*, 25, 100–103.

Cameron-Smith, P., Elliott, S., Maltrud, M., Erickson, D., and Wingenter, O. (2011). Changes in dimethyl sulfide oceanic distribution due to climate change, *Geophys. Res. Lett.*, 38, L07704, https://doi.org/10.1029/2011GL047069.

Carslaw, K. S., Lee, L. A., Reddington, C. L., Pringle, K. J., Rap, A., Forster, P. M., Mann, G. W., Spracklen, D. V., Woodhouse, M. T., Regayre, L. A., and Pierce, J. R. (2013). Large contribution of natural aerosols to uncertainty in indirect forcing, *Nature*, 503, 67–71, https://doi.org/10.1038/nature12674.

Charlson, R. J., Lovelock, J. E., Andreae, M. O., and Warren, S. G. (1987). Oceanic phytoplankton, atmospheric sulfer, cloud albedo, and climate, *Nature*, 326, 655–661.

Chuang, R. L., Palais, J., and Rose, W. I. (1986). Fluxes, sizes, morphology and compositions of particles in the Mt. Erebus volcanic plume, December 1983, *J. Atmos. Chem.*, 4, 467–477.

Clarke, A. D., Ahlquist, N. C., and Covert, D. S. (1987). The Pacific marine aerosol: evidence for natural acid sulfates, *J. Geophys. Res.*, 92, 4179–4190.

Clarke, A. D., Davis, D., Kapustin, V. N., Eisele, F., Chen, G., Paluch, I., Lenschow, D., Bandy, A. R., Thornton, D., Moore, K., Mauldin, L., Tanner, D., Litchy, M., Carroll,

M. A., Collins, J., and Albercook, G. (1998a), Particle nucleation in the tropical boundary layer and its coupling to marine sulfur sources, *Science*, 282, 89–92.

Clarke, A. D., Li, Z., and Litchy, M. (1996). Aerosol dynamics in the Pacific marine boundary layer: Microphysics, diurnal cycles and entrainment, *Geophys. Res. Lett.*, 23, 733–736.

Clarke, A.D., Owens, S. R., and Zhou, J. (2006). An ultrafine sea-salt flux from breaking waves: Impactions for cloud condensation nuclei in the remote marine atmosphere, *J. Geophys. Res.*, 111, D06202, https://doi.org/10.1029/2005JD006565.

Clarke, A. D., Varner, J. L., Eisele, F., Mauldin, R. L., and Tanner, D. (1998b), Particle production in the remote marine atmosphere: Cloud outflow and subsidence during ACE 1, *J. Geophys. Res.*, 103, 16397–16409.

Cline, J. D., & Bates, T. S. (1983). Dimethyl sulfide in the Equatorial Pacific Ocean: A natural source of sulfur to the atmosphere. *Geophys. Res. Lett.*, 10. https://doi.org/10.1029/gl010i010p00949.

Coale, K. H., et al. (2004). Southern ocean iron enrichment experiment: Carbon cycling in high- and low-Si waters, *Science*, 304(5669), 408–414, https://doi.org/10.1126/Science.1089778.

Covert, D. S., Kapustin, V. N., Quinn, P. K., and Bates, T. S. (1992). New particle formation in the marine boundary layer, *J. Geophys. Res.*, 97, 20581–20589.

Cunningham, W. C., and Zoller, W. H. (1981). The chemical composition of remote area aerosol, *J. Aerosol Sci.*, 12, 367–384.

Dall'Osto, M., Beddows, D. C. S., Tunved, P., Krejci, R., Ström, J., Hansson, H. C., Yoon, Y. J., Park, K. T., Becagli, S., Udisti, R., Onasch, T., Ódowd, C. D., Simó, R., and Harrison, R. M. (2017). Arctic sea ice melt leads to atmospheric new particle formation, *Sci. Rep.*, 7, 3318, https://doi.org/10.1038/s41598-017-03328-1.

de Jong, J., Schoemann, V., Lannuzel, D., Croot, P., de Baar, H., and Tison, J. L. (2012). Natural iron fertilization of the Atlantic sector of the Southern Ocean by continental shelf sources of the Antarctic Peninsula, *J. Geophys. Res.*, 117, G01029, https://doi.org/10.1029/2011JG001679.

Deshpande, C. G., and Kamra, A. K. (2002), Atmospheric electric conductivity measurements over the Indian Ocean during the Indian Antarctic Expedition in 1996-97, *J. Geophys. Res.*, 107(D21), 4598, https://doi.org/10.1029/2002JD002118.

Deshpande, C. G., and Kamra, A. K. (2004), Physical properties of aerosols at Maitri, Antarctica, Proc. *Indian Acad. Sci. (Earth Planet. Sci.)*, 113(1), 1–25.

Dulaiova, H., Ardelan, M. V., Henderson, P. B., and Charette, M. A. (2009). Shelf-derived iron inputs drive biological productivity in the southern Drake Passage, *Global Biogeochem. Cycles*, 23, GB4014, https://doi.org/10.1029/2008GB003406.

Durkee, P. A., Hudson, J. G., and Mineart, G. M. (1988). Aerosol effects on marine stratocumulus development observed from multispectral satellite measurements, *Lecture Notes in Physics, Atmospheric Aerosols and Nucleation, Springer Berlin / Heidelberg*, 309, 612–613.

Fiebig, M., Hirdman, D., Lunder, C. R., Ogren, J. A., Solberg, S., Stohl, A., and Thompson, R.L. (2014). Annual cycle of Antarctic baseline aerosol: Controlled by photooxidation-limited aerosol formation, *Atmos. Chem. Phys.*, 14, 3083–3093. https://doi.org/10.5194/acp-14-3083-2014.

Finlayson-Pitts, B. J., and Hemminger, J. C. (2000). Physical chemistry of airborne sea salt particles and their components, *J. Phys. Chem.*, 104(49), 11463–11477.

Fitzgerald, J. W. (1991). Marine aerosols: A review, *Atmos Environ*, 25A, 533–545.

Giordano, M. R., Kalnajs, L. E., Avery, A., Goetz, J. D., Davis, S. M., and DeCarlo, P. F. (2017). A missing source of aerosols in Antarctica – beyond long-range transport, phytoplankton, and photochemistry, *Atmos. Chem. Phys.*, 17, 1–20, https://doi.org/10.5194/acp-17-1-2017.

Giordano, M. R., Kalnajs, L. E., Goetz, J. D., Avery, A. M., Katz, E., May, N. W., Leemon, A., Mattson, C., Pratt, K. A., and DeCarlo, P. F. (2018). The importance of blowing snow to halogen-containing aerosol in coastal Antarctica: Influence of source region versus wind speed, *Atmos. Chem. Phys.*, 18, 16689–16711, https://doi.org/10.5194/acp-18-16689-2018.

Gras, J. L. (1993). Condensation nucleus size distribution at Mawson, Antarctica: Microphysics and chemistry, *Atmos. Environ.*, 27, 1417–1425.

Gras, J. L., and Adriaansen, A. (1985). Concentration and size variation of condensation nuclei at Mawson, Antarctica, *J. Atmos. Chem.*, 3, 96–103.

Gravenhorst, G. (1978). Maritime sulphate over the North Atlantic, *Atmos. Environ.*, 12, 707–713.

Hall, J. S., and Wolff, E. W. (1998). Causes of seasonal and daily variations in aerosol sea-salt concentrations at a coastal Antarctic station, *Atmos. Environ.*, 32, 3669–3677.

Hara, K., Iwasaka, Y., Wada, M., Ihara, T., Shiba, H., Osada, K., and Yamanouchi, T. (2006). Aerosol constituents and their spatial distribution in the free troposphere of coastal Antarctic regions, *J. Geophys. Res.*, 111, D15216, https://doi.org/10.1029/2005 JD006591.

Hara, K., Osada, K., Kido, M., Hayashi, M., Matsunaga, K., Iwasaka, Y., Yamanouchi, T., Hashida, G., and Fukatsu, T. (2004). Chemistry of sea-salt particles and inorganic halogen species in the Antarctic regions: Compositional differences between coastal and inland stations, *J. Geophys. Res.*, 109, D20208, https://doi.org/10.1029/2004JD004713.

Hara K., Osada, K., Yabuki, M., Takashima, H., Theys, N., and Yamanouchi, T. (2018). Important contributions of sea-salt aerosols to atmospheric bromine cycle in the Antarctic coasts, *Scientific Reports*, 8, 13852, https://doi.org/10.1038/s41598-018-32287-4.

Harvey, M. J., Fisher, G. W., Lechner, I. S., Issac, P., Flower, N. E., and Dick, A. L. (1991). Summertime aerosol measurement in the Ross Sea region of Antarctica, *Atmos. Environ.*, 25A, 569–580.

Hegg, D. A., Covert, D. S., and Kapustin, V. N. (1992). Modeling a case of particle nucleation in the marine boundary layer, *J. Geophys. Res.*, 97(D9), 9851–9857.

Hogan, A. W. (1975). Antarctic aerosols, *J. App. Met.*, 14, 550–559.

Hogan, A. W. (1979). Meteorological transport of particulate material to the south polar plateau, *J. Appl. Meteorol.*, 18, 741–749.

Hogan, A. W., and Barnard, S. (1978). Seasonal and frontal variation in Antarctic aerosol concentrations, *J. Appl. Meteorol.*, 17, 1458–1465.

Hogan, A. W., Barnard, S., Samson, J., and Winters, W. (1982). The transport of heat, water vapor, and particulate material to the south polar plateau, *J. Geophys. Res.*, 87, 4287–4292.

Hoppel, W. A., Frick, G. M., and Larson, R. E. (1986). Effect of nonprecipitating clouds on the aerosol size distribution in the marine boundary layer, *Geophys. Res. Lett.*, 13, 125–128.

Hoppel, W. A., Frick, G. M., Fitzgerald, J. W., and Larson, R. E. (1994). Marine boundary layer measurements of new particle formation and the effect which non-precipitating clouds have on the aerosol size distribution, *J. Geophys. Res.*, 99, 14443–14459.

Hogan, A. W., Kebschull, K., Townsend, R., Murphy, B., Samson, J., and Barnard, S. (1984). Particle concentrations at the south pole, on meteorological and climatological time scales; is the difference important? *Geophys. Res. Lett.*, 11, 850–853.

Huang, J., Jaeglé, L., and Shah, V. (2018). Using CALIOP to constrain blowing snow emissions of sea salt aerosols over Arctic and Antarctic sea ice, *Atmos. Chem. Phys.*, 18, 16253–16269, https://doi.org/10.5194/acp-18-16253-2018.

Hynes, A. J., Wine, P. H., and Semmes, D. H. (1986). Kinetics and mechanism of OH reactions with organic sulfides, *J. Phys. Chem.*, 90, 4148–4156.

IPCC (2007). *Climate Change 2007: The Physical Science Basis. Contribution of the Working Group I to the Fourth Assessment Report of the Intergovernmental Panel on*

Climate Change, edited by Solomon, S., D. Qin, M. Manning, Z. Chen, M. Marquis, K. B. Averyt, M. Tignor, and H. L. Miller, Cambridge University Press, Cambridge, United Kingdom, and New York, NY, USA, p. 996.

IPCC (2013). *Climate change 2013: The physical science basis, Intergovernmental Panel on Climate Change*, Cambridge University Press, New York, USA, pp. 571–740.

Ito, T. (1982). On the size distribution of submicron aerosols in the Antarctic atmosphere, *Antarctic Record*, 76, 1–19.

Ito, T. (1983). Study on properties and origin of aerosol particles in the Antarctic atmosphere (in Japanese, English abstract), *Pap. Meteorol. Geophys.*, 34 (3), 151–219.

Ito, T. (1993). The size distribution of Antarctic submicron aerosols, *Tellus*, 45B, 145–159.

Ito, T., (1985). Study of background aerosols in the Antarctic troposphere, *J. Atmos. Chem.*, 3, 69–91.

Ito, T., and Iwai, K. (1981). On the sudden increase in the concentration of Aitken particles in the Antarctic atmosphere, *J. Meteorol. Soc. Japan*, 59, 262–271.

Iwai, K. (1979). Concentration of Aitken particles observed at Syawo station, Antarctica (in Japanese), *Nankyoku Shiryo*, 67, 172–179.

Jaenicke, R., Dreiling, V., Lehmann, E., Koutsenoguii, P. K., and Stingl, J. (1992). Condensation nuclei at the German Antarctic station 'Georg von Neumayer', *Tellus*, 44B, 311–317.

Jaenicke, R., and Matthias-Maser, S. (1992). Natural sources of atmospheric aerosol particles, In Fifth International Conference on Precipitation, Scavenging and Atmos., Surface Exchange Processes.

Jaenicke, R., and Stingl, J. (1984). Aitken particle size distribution in Antarctica, In Proceedings of International Conference on Aerosols and Ice Nuclei, Budapest, Hungary, University of Budapest Press.

Jickells, T. D., et al. (2005). Global iron connections between desert dust, ocean biogeochemistry, and climate, *Science*, 308(5718), 67–71, https://doi.org/10.1126/Science.1105959.

Johnson, M. S., Meskhidze, N., Solmon, F., Gasso, S., Chuang, P. Y., Gaiero, D. M., Yantosca, R. M., Wu, S. L., Wang, Y. X., and Carouge, C. (2010). Modeling dust and soluble iron deposition to the South Atlantic Ocean, *J. Geophys. Res.*, 115, D15202, https://doi.org/10.1029/2009JD013311.

Junge, C. E. (1963). *Air chemistry and radioactivity*, Academic Press, New York.

Jurányi, Z., and Weller, R. (2019). One year of aerosol refractive index measurement from a coastal Antarctic site, *Atmos. Chem. Phys.*, 19, 14417–14430, https://doi.org/10.5194/acp-19-14417-2019.

Katoshevski, D., Nenes, A., and Seinfeld, J. H. (1999). A study of processes that govern the maintenance of aerosols in the marine boundary layer, *J. Aerosol Sci.*, 30, 503–532.

Kavan, J., Dagsson-Waldhauserova, P., Renard, J. B., Láska, K., and Ambrožová, K. (2018). Aerosol concentrations in relationship to local atmospheric conditions on James Ross Island, Antarctica, Front. *Earth Sci.*, 6, 207, https://doi.org/10.3389/feart.2018.00207.

Kerminen, V. -M., Teinila, K., and Hillamo, R. (2000). Chemistry of sea-salt particles in the summer Antarctic atmosphere, *Atmos. Environ.*, 34, 2817–2825.

Kerminen, V.-M., Chen, X., Vakkari, V., Petäjä, T., Kulmala, M., and Bianchi, F. (2018). Atmospheric new particle formation and growth: A review of field observations, *Environ. Res. Lett.*, 13, 103003, https://doi.org/10.1088/1748-9326/aadf3c.

Kim, Jaeseok, Yoon, Y. J., Gim, Y., Choi, J. H., Kang, H. J., Park, K., Park, J., and Lee, B. Y. (2017). New particle formation events observed at King Sejong Station, Antarctic Peninsula – Part 1: Physical characteristics and contribution to cloud condensation nuclei, *Atmos. Chem. Phys.*, 19, 7583–7594, https://doi.org/10.5194/acp-19-7583-2019.

Koponen, I. K., Virkkula, A., Hillamo, R., Kerminen, V.-M. and Kulmala, M. (2002). Number size distributions and concentrations of marine aerosols: observations during a cruise between the English Channel and the coast of Antarctica, *J. Geophys. Res.*, 107 (D24), 4753, https://doi.org/10.1029/2002JD002533

Koponen, I. K., Virkkula, A., Hillamo, R., Kerminen, V. -M., and Kulmala, M. (2003). Number size distributions and concentrations of the continental summer aerosols in Queen Maud Land, Antarctica, *J. Geophys. Res.*, 108 (D18), 4587, https://doi.org/10.1029/2002JD002939.

Kulmala, M., Vehkamaki, H., Petaja, T., Dal Maso, M., Lauri, A., Kerminen, V. -M., Birmili, W., and McMurry, P. H. (2004). Formation and growth rates of ultrafine atmospheric particles: A review of observations, *J. Aerosol Sci.*, 35(2), 143–176.

Kuznetsova, M., Lee, C., and Aller, J. (2005). Characterization of the proteinaceous matter in marine aerosols, *Mar. Chem.*, 96, 359–377, https://doi.org/10.1016/j.marchem.2005.03.007.

Kyrö, E.-M., Kerminen, V.-M., Virkkula, A., Dal Maso, M., Parshintsev, J., Ruíz-Jimenez, J., Forsström, L., Manninen, H. E., Riekkola, M.-L., Heinonen, P., and Kulmala, M. (2013). Antarctic new particle formation from continental biogenic precursors, *Atmos. Chem. Phys.*, 13, 3527–3546, https://doi.org/10.5194/acp-13-3527-2013.

Lal, M., and Kapoor, R.K. (1989). Certain meteorological features of submicron aerosols at schirmacher oasis, East Antarctica. *Atmos. Environ.*, 23(4), 803–808. https://doi.org/10.1016/0004-6981(89)90484-8.

Latham, J., and Smith, M. H. (1990). Effect on global warming of wind-dependent aerosol generation at the ocean surface, *Nature*, 347, 372–373.

Legrand, M., Yang, X., Preunkert, S., and Theys, N. (2016). Year-round records of sea salt, gaseous, and particulate inorganic bromine in the atmospheric boundary layer at coastal (Dumont d'Urville) and central (Concordia) East Antarctic sites, *J. Geophys. Res. Atmos.*, 121, 997–1023, https://doi.org/10.1002/2015JD024066.

Lewis, E. R., and Schwartz, S. E. (2004). *Sea salt aerosol production: Mechanisms, methods, measurements, and models: A critical review*, American Geophysical Union, Washington, DC, pp. 413.

Lin, H., Rauschenberg, S., Hexel, C. R., Shaw, T. J., and Twining, B. S. (2011). Free-drifting icebergs as sources of iron to the Weddell Sea, *Deep Sea Res.*, Part II, 58(11–12), 1392–1406, https://doi.org/10.1016/J.Dsr2.2010.11.020.

Lohmann, U. (2006). Aerosol effects on clouds and climate, *Space Sci. Rev.*, 125, 129–137.

Maffezzoli, N., Spolaor, A., Barbante, C., Bertò, M., Frezzotti, M., and Vallelonga, P. (2017). Bromine, iodine and sodium in surface snow along with the 2013 Talos Dome–GV7 traverse (northern Victoria Land, East Antarctica), *The Cryosphere*, 11, 693–705, https://doi.org/10.5194/tc-11-693-2017.

Meszaros, A., and Vissy, K. (1974). Concentration, size distribution and chemical nature of atmospheric aerosol particles in remote ocean areas, *J. Aerosol Sci.*, 5, 101–109.

Mishra, V. K., Kim, K. H., Hong, S., and Lee, K. (2004). Aerosol composition and its sources at the King Sejong Station, Antarctic Peninsula, *Atmos. Environ.*, 38(24), 4069–4084.

Murphy, D. M., Froyd, K. D., Bian, H., Brock, C. A., Dibb, J. E., DiGangi, J. P., Diskin, G., Dollner, M., Kupc, A., Scheuer, E. M., Schill, G. P., Weinzierl, B., illiamson, C. J., and Yu, P. (2019). The distribution of sea-salt aerosol in the global troposphere, *Atmos. Chem. Phys.*, 19, 4093–4104, https://doi.org/10.5194/acp-19-4093-2019.

Murphy, D. M., Anderson, J. R., Quinn, P. K., McInnes, L. M., Brechtelk, F. J., Kreidenweisk, S. M., Middlebrook, A. M., Posfai, M., Thomson, D. S., and Buseck, P. R. (1998). Influence of sea-salt on aerosol radiative properties in the southern ocean marine boundary layer, *Nature*, 392, 62–65.

Murugavel, P., Pawar, S. D., and Kamra, A. K. (2001). The size distribution of submicron aerosol particles over the Indian Ocean during IFP-99 of INDOEX, *Curr. Sci.*, 80, 123–127.

O'Dowd, C. D., and Smith, M. H. (1993). Physico-chemical properties of aerosol over the northeast Atlantic: Evidence for wind speed related submicron sea salt aerosol production, *J. Geophys. Res.*, 98, 1137–1149.

O'Dowd, C. D., et al. (2002). A dedicated study of new particle formation and fate in the coastal environment (PARFORCE): An overview of objectives and achievements, *J. Geophys. Res.*, 107 (D19), 8108, https://doi.org/10.1029/2001JD000555.

O'Dowd, C. D., Facchini, M. C., Cavalli, F., Ceburnis, D., Mircea, M., Decesari, S., Fuzzi, S., Yoon, Y. J., and Putaud, J. P. (2004). Biogenically driven organic contribution to marine aerosol, *Nature*, 431, 676–680, https://doi.org/10.1038/nature02959.

O'Dowd, C. D., Lowe, J. A., Smith, M. H., and Kaye, A. D. (1999). The relative importance of NSS-sulphate and sea-salt aerosol to the marine CCN population: An improved multi-component aerosol-cloud droplet parameterization, *Q. J. R. Meteorol. Soc.*, 125(556), 1295–1313.

O'Dowd, C. D., Smith, M. H., Consterdine, I. E., and Lowe, J. A. (1997). Marine aerosol, sea salt, and the marine sulphur cycle: A short review, *Atmos. Environ.*, 31, 73–80.

Ono, A., Ito, T., and Iwai, K. (1981). A note on the origin and nature of the Antarctic aerosol, Nat. Inst. *Polar Res.*, 19, 141–151.

Park, J., Sakurai, H., Vollmers, K., and McMurry, P. H. (2004). Aerosol size distribution measured at the south pole during ISCAT, *Atmos. Environ.*, 38, 5493–5500.

Pant V., Siingh, D., and Kamra, A. K. (2010). Concentrations and size distribution of aerosol particles at Maitri, during the passage of cyclonic storms revolving around the continent of Antarctica, *J. Geophys. Res.*, 115, D17202, https://doi.org/10.1029/2009JD013481.

Pant V., Siingh, D., and Kamra, A. K. (2011). The Size distribution of atmospheric aerosols at Maitri, Antarctica, *Atmos. Environ.*, 45, 5138–5149.

Parungo, F. P., Bodhaine, B., and Bortnaik, J. (1981). Seasonal variation in Antarctic aerosol, *J. Aerosol Sci.*, 12, 491–504.

Parungo, F. P., Nagamoto, C. T., Rosinski, J., and Haagenson, P. L. (1986). A study of marine aerosols over the Pacific Ocean, *J. Atmos. Chem.*, 4, 199–226.

Parungo, F. P., and Pueschel, R. F. (1980). Conversion of nitrogen oxide gases to nitrate particles in oil refinery plumes, *J. Geophys. Res.*, 85(C8), 4507–4511.

Penner, J. E., et al. (2001). *Aerosols, their direct and indirect effects, in Climate Change 2001: The Scientific Basis. Contribution of working group I to the Third Assessment Report of the Intergovernmental Panel on Climate Change*, Cambridge University Press, pp. 289–348.

Peterson, J. T., and Junge, C. E. (1971). Sources of particulate matter in the atmosphere, in *Man's Impact on the Climate*, W. J. Matthews, W. W. Kellogg, and G. D. Robinson (eds.), MIT Press, Cambridge, MA, pp. 310–320.

Petit, J. R., et al. (1999). Climate and atmospheric history of the past 420,000 years from the Vostok ice core, Antarctica, *Nature*, 399, 429–436.

Pirjola, L., O'Dowd, C. D., Brooks, I. M., and Kulmala, M. (2000). Can new particle formation occur in the clean marine boundary layer? *J. Geophys. Res.*, 105(D21), 26531–26546.

Prospero, J. M., (1979). Mineral and sea salt aerosol concentrations in various ocean regions, *J. Geophys. Res.*, 84 (C2), 725–731.

Prospero, J. M. et al. (1983). The atmospheric aerosol system—an overview, *Reviews of Geophysics and Space Physics*, 21, 1607–1629.

Pruppacher, H., and Klett, J. (2010). Growth of cloud drops by collision, coalescence and breakup, in *Microphysics of Clouds and Precipitation,Atmospheric and Oceanographic Sciences Library*, vol 18, Springer, Dordrecht. https://doi.org/10.1007/978-0-306-48100-0_15.

Quinn, P. K., and Bates, T. S. (2011). The case against climate regulation via oceanic phytoplankton sulphur emissions, *Nature*, 480, 52–56, https://doi.org/10.1038/nature10580.

Quinn, P. K., and Coffman, D. J. (1999). Comment on "Contribution of different aerosol species to the global aerosol extinction optical thickness: Estimates from model results" by Tegen et al., *J. Geophys. Res.*, 104(D4), 4241–4248.

Quinn, P. K., Coffman, D. J., Bates, T. S., Miller, T. L., Johnson, J. E., Voss, K., Welton, E. J., and Neusub, C. (2001). Dominant aerosol chemical components and their contribution to extinction during the Aerosols99 cruise across the Atlantic, *J. Geophys. Res.*, 106, 20783–20809.

Radke, L. F. (1982). Sulphur and sulphate from Mt. Erebus, *Nature*, 299, 710–712.

Radke, L. F., and Lyons, J. H. (1982). Airborne measurements of particles in Antarctica, In 2nd symposium on the composition of the non-urban troposphere, American Met. Soc. Massachusetts, USA, 159–163.

Raemdonck, H., Maenhaut, W., and Andreae, M. O. (1986). Chemistry of marine aerosols over the tropical and equatorial Pacific, *J. Geophys. Res.*, 91, 8623–8636.

Randall, D. A., Coakley, Jr. J. A., Fairall, C. W., Kropfli, R. A., and Lenschow, D. H. (1984). Outlook for research on subtropical marine stratiform clouds, *Bull. Amer. Meteor. Soc.*, 65, 1290–1301.

Reddington, C., Carslaw, K., Stier, P., Schutgens, N., Coe, H., Liu, D., Allan, J., Pringle, K., Lee, L., and Yoshioka, M. (2017). The Global Aerosol Synthesis and Science Project (GASSP): Measurements and modeling to reduce uncertainty, *B. Am. Meteorol. Soc.*, 98(9), 1857–1877.

Rose, W. I., Chuan, R. L., and Kyle, P. R. (1985). Rate of sulphur dioxide emission from Erebus volcano, Antarctica, December 1983, *Nature*, 316, 710–712.

Rosenfeld, D. (1999). TRMM observed first direct evidence of smoke from forest fires inhibiting rainfall, *Geophys. Res. Lett.*, 26(20), 3105–3108.

Rosenfeld, D. (2000). Suppression of rain and snow by urban and industrial air pollution, *Science*, 287 (5459), 1793–1796.

Rosenfeld, D. (2006). Aerosols, clouds, and climate, *Science*, 312, 1323–1324.

Rosenfeld, D., Kaufman, Y., and Koren, I. (2006). Switching cloud cover and dynamical regimes from open to closed Benard cells in response to aerosols suppressing precipitation, *Atmos. Chem. Phys.*, 6, 2503–2511

Rosenfeld, D., Lahav, R., Khain, A. P., and Pinsky, M. (2002). The role of sea-spray in cleansing air pollution over ocean via cloud processes, *Science*, 297, 1667–1670.

Rossi, M. J. (2003). Heterogeneous reactions on salts, *Chem. Rev.*, 103, 4823–4882.

Savoie, D. L., and Prospero, J. M. (1982). the Particle size distribution of nitrate and sulphate in the marine atmosphere, *Geophys. Res. Lett.*, 9, 1207–1210.

Savoie, D. L., Prospero, J. M., and Nees, R. T. (1987). Nitrate, non-sea-sulfate, and mineral aerosol over the northwestern Indian Ocean, *J. Geophys. Res.*, 92, 933–942.

Savoie, D. L., Prospero, J. M., Merill, J. T., and Uematsu, M. (1989). Nitrate in the atmospheric boundary layer of the tropical South Pacific: Implications regarding sources and transport, *J. Atmos. Chem.*, 8, 391–415.

Saxena, P., and Hildemann, L. M. (1996). Water-soluble organics in atmospheric particles: A critical review of the literature and application of thermodynamics to identify candidate compounds, *J. Atmos. Chem.*, 24, 57–109, https://doi.org/10.1007/bf00053823.

Sedwick, P. N., and DiTullio, G. R. (1997). Regulation of algal blooms in Antarctic shelf waters by the release of iron from melting sea ice, *Geophys. Res. Lett.*, 24(20), 2515–2518, https://doi.org/10.1029/97GL02596.

Shaw, G. E. (1979). Consideration on the origin and properties of the Antarctic aerosol, *Rev. Geophys. Space Phys.*, 17, 1983–1998.

Shaw, G. E. (1986). On the physical properties of aerosol at Ross Island, Antarctica, *J. Aerosol. Sci.*, 17, 937–945.

Shaw, G. E. (1988). Antarctic aerosols: A review, *Rev. Geophys.*, 26, 89–112.

Shinozuka, Y., Clarke, A. D., Howell, S. G., Kapustin, V. N., and Huebert, B. J. (2004). Sea-salt vertical profiles over the Southern and tropical Pacific oceans: Microphysics, optical properties, spatial variability, and variations with wind speed, *J. Geophys. Res.*, 109, D24201, https://doi.org/10.1029/2004JD004975.

Slingo, A. (1990). Sensitivity of earth's radiation budget to changes in low clouds, *Nature*, 343, 49–51.

Tatlhego, M., Bhattachan, A., Okin, G.S., and D'Odorico, P. (2020). Mapping areas of the Southern Ocean where productivity likely depends on dust-delivered iron, *J. Geophys. Res. Atmospheres*, 125, https://doi.org/10.1029/2019JD030926.

Twomey, S. (1974). Pollution and the planetary albedo, *Atmos. Environ.*, 8, 1251–1256.

Twomey, S. (1977). *Atmospheric Aerosols*, Elsevier, New York, p. 302.

Twomey, S., Piepgrass, M., and Wolfe, T. L. (1984). An assessment of the impact of pollution on global cloud albedo, *Tellus*, 36B, 1356–1366.

Virkkula, A., Teinila, K., Hillamo, R., Kerminen, V. -M., Saarikoski, S., Aurela, M., Viidanoja, J., Paatero, J., Koponen, I. K., and Kulmala, M. (2006). Chemical composition of boundary layer aerosol over the Atlantic Ocean and at an Antarctic site, *Atmos. Chem. Phys.*, 6, 3407–3421.

von Glasow, R., and Crutzen, P. J. (2004). Tropospheric halogen chemistry, edited by R. F. Keeling, *The Atmosphere*, Elsevier, Amsterdam, pp. 21–64.

Voskresenskii, A. I. (1968). Condensation nuclei in the Mirny region, *Sov. Antark. Eksped. Trudy*, 38, 149–198.

Weber, R. J., K. Moore, V. Kapustin, A. Clarke, R. L. Mauldin, E. Kosciuch, C. Cantrell, F. Eisele, B. Anderson, and L. Thornhill (2001). Nucleation in the equatorial Pacific during PEM-Tropics B: Enhanced boundary layer H2SO4 with no particle production, *J. Geophys. Res.*, 106(D23), 32,767–32,776.

Weber, R. J., McMurry, P. H., Mauldin, L., Tanner, D. J., Eisele, F. L., Brechtel, F. J., Kreidenweis, S. M., Kok, G. L., Schillawski, R. D., and Baumgardner, D. (1998). A study of new particle formation and growth involving biogenic and trace gas species measured during ACE 1, *J. Geophys. Res.*, 103, 16385–16396.

Weller, R., Schmidt, K., Teinilä, K., and Hillamo, R. (2015). Natural new particle formation at the coastal Antarctic site Neumayer. *Atmos. Chem. Phys.*, 15, 11399–11410. https://doi.org/10.5194/acp-15-11399-2015.

Yang, X., Frey, M. M., Rhodes, R. H., Norris, S. J., Brooks, I. M., Anderson, P. S., Nishimura, K., Jones, A. E., and Wolff, E. W. (2019). Sea salt aerosol production via sublimating wind-blown saline snow particles over sea ice: Parameterizations and relevant microphysical mechanisms, *Atmos. Chem. Phys.*, 19, 8407–8424, https://doi.org/10.5194/acp-19-8407-2019.

Yin, F., Grosjean, D., and Seinfeld, J. H. (1986). Analysis of atmospheric photooxidation mechanisms for organosulfur compounds, *J. Geophys. Res.*, 91, 14417–14438.

5 Transient Variations in En Route Southern Indian Ocean Aerosols, Antarctic Ozone Climate, and Its Relationship with HO$_x$ and NO$_x$

S.M. Sonbawne, P.R.C. Rahul and K.K. Dani
High Altitude Cloud Physics Laboratory (HACPL) and Cloud-Aerosol Interaction and Precipitation Enhancement Experiment (CAIPEEX)
Indian Institute of Tropical Meteorology, Pune, India

P.C.S. Devara
Centre of Excellence in Ocean-Atmospheric Science and Technology (ACOAST) & Environmental Science and Health (ACESH)
Amity University Haryana (AUH), Gurugram, India

CONTENTS

DOI: 10.1201/9781003203742-5

5.1   INTRODUCTION

Ozone is one of the key constituents of the atmosphere and it plays an important role in the energy budget and dynamics of the atmosphere (e.g. Brasseur and Solomon, 1968; World Meteorological Organization [WMO], 2003, 2007). Atmospheric ozone protects life on Earth by absorbing UV radiation. Since the discovery of the Antarctic 'ozone hole' by Farman et al. (1985), it is observed and studied that the Antarctica continent that has suffered from the largest variations in total ozone column, various studies were made to understand the chemical, dynamic, and radiative processes, which are responsible for these ozone losses, and efforts were also made on predicting the future of polar ozone. It is now well established that concentrations of certain gases, such as the highly reactive chlorine compounds released into the atmosphere as chlorofluorocarbons (CFCs), have a critical effect on ozone levels over the polar latitudes (World Meteorological Organization [WMO], 2003, 2007; UNEP, 1989, 1998). In recent years, even though the concentrations of ozone-depleting gases have stabilized due to international measures, variations and interplay of chemistry, dynamics, and meteorological conditions have become important in influencing the behavior of ozone over the polar regions.

Energetic particles, namely, "protons" emitted by the solar flare during Solar Proton Events (SPEs), correspond to solar coronal mass ejections (CME) and enter the Earth's atmosphere, which is guided by the magnetic field when they reach the magnetosphere and they made an impact over polar cap regions at >60° geomagnetic latitude (e.g. Jackman and McPeters, 2004). Studies were made over SPEs between 1976 and 2006, which show the flux and energies were very different from one to another.[1] Even though their frequency is low over a long period, a single SPE can have the most extreme effects on low-Earth orbit space systems as well as on the Earth's atmosphere (Jackman et al., 1999, 2001). The high-energy protons of 10 MeV can deposit their energy at different altitudes and latitudes of the Earth while traveling into the Earth's atmosphere, such as the mesosphere and stratosphere and also in the upper troposphere. These SPEs provide extreme force on the middle atmosphere (e.g. Seppala et al., 2004), causing ionization in the polar atmosphere, particularly during the solar maxima. Earlier studies were reported by Reid et al. (1991), Jackman et al. (2001), Seppala et al. (2004), and Verronen et al. (2005) show the short-long-term effect on the ozone has been observed after each SPE event. Large SPEs can change the chemical composition of the atmospheric important constitute of HO_x and NO_x which can produce directly or via photochemical reaction and by ion-pair production through energized particles, particularly in the polar region that influences the ozone variation in the Earth atmosphere. These SPE lead to produce odd hydrogen HO_x by associated chemistry of ion-pair and NO_x constituent by dissociation of molecular nitrogen via charged particle impact and ion chemistry (Solomon et al., 1981; Rusch et al., 1981) are important to understand their impact on mesospheric and stratospheric ozone and even on the upper tropospheric atmosphere. NO_x plays a key role

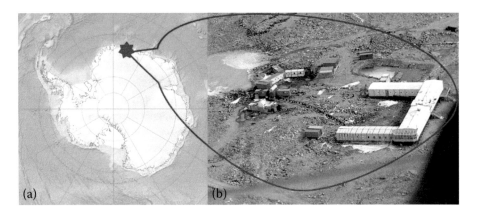

FIGURE 5.1 (a) Closed red circle indicates the location of the Indian station in the Antarctica region, and (b) aerial view of the Indian station, Maitri, situated at Queen Maud Land Region, Antarctica.

in the ozone balance of the middle atmosphere because it destroys odd oxygen through catalytic reactions (Grenfell et al., 2006), which leads to short- and long-term ozone destruction with associated chemistry and through the well-known NO_x catalytic ozone loss cycle in the atmosphere.

In this chapter, ground-based total column ozone measurements (Ozonometer and Brewer spectrophotometer) made over the Indian Antarctic station (Figure 5.1) Maitri (70.76°S, 11.74°E) and also satellite (TOMS) data of total ozone during the period 2004–2006 have been examined in this study for a possible influence of Solar Proton Events (data from GOES 10 and GOES 11) on short-term ozone depletion and the results are discussed.

5.2 MATERIALS AND METHOD OF ANALYSIS

As part of the 24th Indian Antarctica Expedition, from both sun-photometer and ozone-monitor versions of MICROTOPS-II, high-resolution solar radiometric observations of aerosol optical depth (AOD) at six wavelengths, covering UV to NIR (380, 440, 500, 675, 870, and 1,020 nm), total column ozone (TCO) and precipitable water content (PWC) have been estimated. The differential optical absorption (DOAS) method has been followed for the measurement of TCO (300, 305.5, and 312.5 nm) and PWC (940 and 1020 nm). An important advantage with this radiometer, as compared to many complementary instruments, lies in its portability and online data acquisition and analysis so that it yields instantaneous and simultaneous estimates of aerosol and pre-cursor gaseous optical depths. More details about the observational scheme, calibration, and data retrieval procedures for this radiometer have been published in literature (Devara et al., 2001; Ichoku et al., 2002). These extensive data sets, collected during 9, 11, and 12 March 2005, have been utilized to investigate the aerosol optical, physical, and their interface with simultaneously measured gases such as total column ozone, water vapor, and surface-level meteorological parameters over the southern Indian Ocean.

Total column ozone measurements were carried out using an Ozonometer (Micro tops-II, Solar Light Inc., USA) and the Brewer spectrophotometer (Brewer) at the Indian Antarctic station, "Maitri" (70.76°S, 11.74°E). This research station is situated in the Droning Maud Land of East Antarctica. The station area is small in size, 2 km width and 1.8 km length in an east-west direction with a height of 117 m from average mean sea level and is ice-free, mainly covered by sandy and loamy sand types of soil. During the observational period, December 2004 to February 2005, the meteorological condition of the station is the average temperature was −0.75°C, with a variation of +1.41 to −8.20°C. In the December 2006 to February 2007 summer observational period, the average temperature was +0.07°C with a variation of +4.50 to −4.80°C. Observed surface temperature and relative humidity show a decreasing trend in both observational campaigns.

High temporal resolution Ozonometer observations have been undertaken for the summer periods as part of the 24th and 26th Indian Antarctic Expeditions (IAEs) in the years 2004–2005 and 2006–2007. The compact, hand-held, microprocessor-based calibrated ozone meter was installed on a sturdy tripod stand for shake-free direct solar (sunlight) observations. For the observation of the ozone meter, the solar zenith angle at Maitri for optical remote sensing measurements varied from 48° to 88° during the current observational period (Sonbawne et al., 2009). It gives total column ozone (TCO) in DU at three wavelengths of light (300, 305.5, and 312.5 nm). The global positioning system (GPS) receiver is attached to this instrument, which provides the geographical coordinates for the observation location, which are required for estimating the local air mass. The accuracy of the measurement is ~1%. More details of the instrument and measurement accuracy can be found in Morys et al. (2001), Ichoku et al. (2002), and Ernest Raj et al. (2004), etc. Ozonometer observations of TCO were made from morning to evening at 10-minute intervals in the early morning hours and late evening hours due to the large variation in solar zenith angle, and in the afternoon, 30-minute interval observations were made for a total of more than 30–35 sets of observations of direct solar radiation collected on each clear-sky experimental day. To study the short-term variation in total column ozone, data sets were collected at a close interval of 10–15 minutes on clear-sky days.

A Brewer spectrophotometer was installed at Maitri in July 1999, and observations were made simultaneously and used at the same interval for this study. The Brewer measures column ozone at five wavelengths using five slits allowing the simultaneous measurement of SO_2 (Staehelin et al., 2003) and in NO_2 mode with 0.85 nm resolution (Kulandaivelu and Venkateshwara, 2004) with FOV of 3°. This instrument was calibrated regularly with the side-by-side operation of the two types of instruments calibrated from two independent sources, which allows assessment of the reliability of the long-term stability of the involved instrument and network.

Daily data of TCO from the TOMS satellite have also been collected and Ozonemeter observations were performed for the same geographical location during solar proton events to see the effects on atmospheric ozone influence by solar proton events on satellite-derived ozone and their comparisons with ground-based observations. It is to be noted that the differences in the ground observation–derived ozone levels and TOMS ozone levels is 2–4% (Thompson et al., 2003). The difference in Ozonometer-derived ozone values is found to exceed by 2% that of TOMS, as expected (Aculinin, 2006) and the difference between Brewer spectrometer values and TOMS values is 4%, as expected (Evtushevsky et al., 2008). Also, we used the SCIMACHY NO_2 profiles during 12th December 2006 to corroborate the implications of the SPEs. Solar proton fluxes are being continuously measured by several satellites in orbit around the Earth. The National Oceanic and Atmospheric Administration (NOAA) Geostationary Operational Environmental Satellite (GOES) measured data of proton fluxes[2] have been used in the current study. Data from GOES satellites are used for the proton fluxes in the years 2004–2006; GOES-10 for the period 1 January 2005 and 15 January 2005; and GOES-11 for the period October–December 2003, October–November 2004, September 2005, and December 2006. The differential energy spectra of solar protons, derived using the GOES-10 and GOES-11 data in the integral energy channels > 1 MeV, > 5 MeV, > 10 MeV, > 30 MeV, > 50 MeV, and > 100 MeV, has been used to derive the variations in proton fluxes during the SPEs.

5.3   RESULTS AND DISCUSSION

5.3.1   SPECTRAL VARIATION OF MARINE AOD

The daily mean aerosol optical depth (AOD) at six sensing wavelengths of 380, 440, 500, 675, 870, and 1,020 nm is plotted in Figure 5.2. Table 5.1 presents the

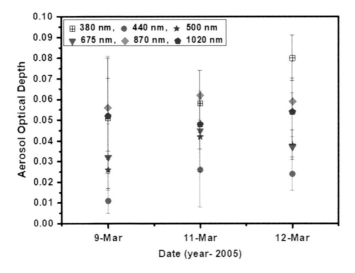

FIGURE 5.2   Day-to-day mean variation in AOD over southern Indian Ocean. The vertical bars denote standard error in the estimation.

TABLE 5.1

Range of Aerosol and Pre-Cursor Gas Parameters Measured during the Study Period

Parameter	Minimum	Maximum	Average
AOD (at 500 nm)	0.026	0.042	0.035
Angstrom Exponent (Alpha)	0.002	0.035	0.017
TCO (DU)	268.6	293	279.9
CWV (cm)	0.20	0.37	0.32

minimum, maximum, and average values of aerosols and pre-cursor gas para-meters observed on 9, 11, and 12 March 2005. The daily mean AOD was found to be 0.026 with an average Angstrom exponent (an indicator of aerosol size dis-tribution) of 0.17, revealing the abundance of coarse-mode particles. In other words, AOD increases with a decrease in wavelength. The day-to-day variation in total column ozone and water vapor exhibited an inverse relationship on particular days associated with a low temperature (Figure 5.3). The results show lower ozone for low water vapor during the initial phase of the experiment (at the coast). This trend continued until the ship reached the mid-latitude region around 11 March 2019, and, thereafter, they both showed an inverse relationship on ap-proaching Antarctica. This feature may partly be due to heterogeneous effects wherein aerosols grow (increase in surface area, acts as a catalyst) at the expense of available water vapor.

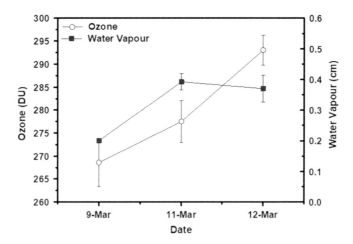

FIGURE 5.3 Daily mean variations in TCO and PWC (H_2O) during the voyage. The vertical bars denote standard error in the estimation.

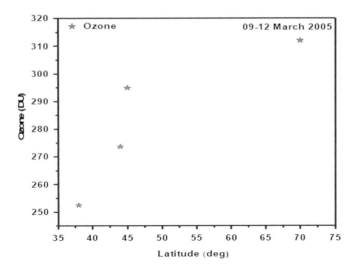

FIGURE 5.4   Latitudinal variation of ozone over southern Indian Ocean.

5.3.2   LATITUDINAL VARIATION OF OZONE

The latitudinal variation of columnar ozone over the oceanic region during 9–12 March 2005 is plotted in Figure 5.4. It is clear from the figure that total column ozone over the oceanic region in the Southern Hemisphere increased from about 252 DU around 38°S to about 312 DU at 70°S, showing a gradual increase in ozone with increasing latitude. Variability in ozone on the daily scale during the period of observation was less by 4% over the Antarctic region.

5.3.3   TIME VARIATION OF OZONE OVER THE SEA REGION

The diurnal variation of columnar ozone over the sea region on three typical experimental days is plotted in Figure 5.5. The results exhibit low coastal ozone on 9 March 2005, whereas ozone is observed to be high over mid-latitude (11 March 2005) and polar latitudes (12 March 2005). Moreover, the ozone variability appears to be high in the middle and polar latitude regions compared to coastal regions.

5.3.4   SPECTRAL DEPENDENCE OF MARINE AOD

The AOD variations at different wavelengths observed over the ocean on 9, 11, and 12 March 2005 are depicted in Figure 5.6. A clear-cut wavelength dependency of AOD can be seen with some typical undulations instead of a monotonic variation. One can note from the figure that AOD was more at the 340 nm wavelength and it decreased at 440 nm and thereafter it increased up to 940 nm and then decreased up

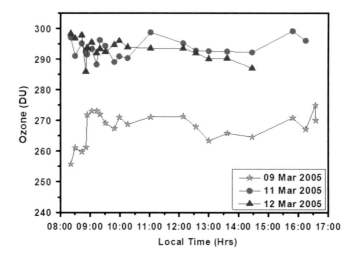

FIGURE 5.5   Temporal variations in TCO over the southern Indian Ocean.

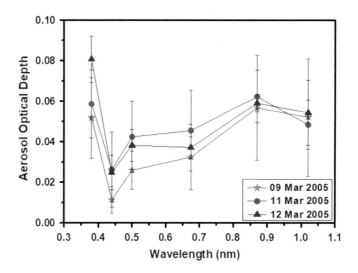

FIGURE 5.6   Wavelength dependence of AOD over the southern Indian Ocean.

to 1,020 nm. The initial decrease of AOD is according to power-law size distribution. AOD increases from 500 nm up to 870 nm could be attributed to the enhancement of secondary aerosol particles due to bubble-bursting and gas-to-particle conversion processes. The decrease in AOD from 940 to 1,020 nm could be due to water vapor absorption (Figure 5.7).

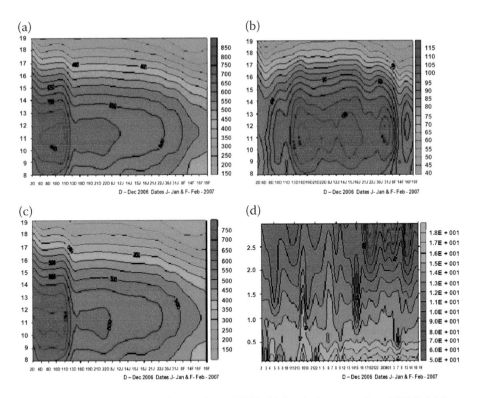

FIGURE 5.7   (a) Flux at top-of-atmosphere (TOA), (b) flux in the atmosphere (ATM), (c) flux at bottom of atmosphere (BOA), and (d) aerosol size distribution during 26th IAE 2006-07.

It is clear from the figure that photon flux cools the TOA. The photon flux varies from 500 to 850 W/m^2 and appears to be similar to that of BOA. It could be due to abundant contribution from gases. Photon flux is found to decrease from summer to winter, which is consistent and indicates from warming to cooling. The flux at BOA is found to vary between 500 and 750 W/m^2. The frame (d) denotes the aerosol size distribution matching from summer to winter. The results further indicate smaller particle-mode dominates in summer and larger particles dominate in the winter months. This feature can be ascribed to be due to variations in the boundary layer height. Normally, lower mixing depths are observed to be higher during summer and lower in the winter months.

5.4   INTERACTIONS BETWEEN SOLAR PROTON EVENTS AND ATMOSPHERIC OZONE

Solar proton events of X-class start bombarding on 16 January 2005, with an intensity of X2.6 solar flare, with a peak of 0.1–0.8 nm. This event solar flares continued to increase for another few days with associated Coronal Mass Ejection (CME) was observed. The solar flare intensity was on 17 January was X3.8, whereas on 20 January, it was even more intense with a X7 solar flare. Solar proton flux (Particle Flux Unit, particle cm^2 sec^{-1} sr^{-1}) measurements made by GOES-11 for the period 1 January–15 February 2005 for the energy channels of > 1 MeV, > 5 MeV, > 10 MeV, > 30 MeV, and > 50 MeV have been taken and daily mean flux values have been computed, and their day-to-day variations are plotted in Figure 5.8. It is seen that there is a substantial and abrupt enhancement in proton flux after 15 January that continued for nearly 10 days. The highest proton fluxes were reported to be observed around midnight on 17/18 January, and around noon on 20 January (Verronen et al., 2007). This flare marked the start of an extraordinary solar proton storm with the flux of extremely high-energy solar protons (>100 MeV), which was of the same order as in the well-documented October 1989 SPE (Reid et al., 1991; Jackman et al., 1999). The lower energy fluxes remained at moderate levels, making the January 2005 event the hardest and most energetic proton event of Solar Cycle 23 so far (Seppala et al., 2006).

Figure 5.8 also shows the day-to-day variation in the daily mean total column ozone from the ground-based Ozonemeter and Brewer pectrometer TCO over the Indian Antarctic station and also the daily TCO from the TOMS satellite for the same region during 1 January–15 February 2005. It can be readily seen that a substantial reduction in TCO (~35 DU) over the Antarctic region takes place after 16 January, when the SPE began and it recovers in 10–12 days to its value when the proton force subsides. Seppala et al. (2006) from their model results have shown that the SPEs that began on 16 January 2005, led to large HO$_x$ and NO$_x$ enhancements in the mesosphere that in turn led to the destruction of ozone through

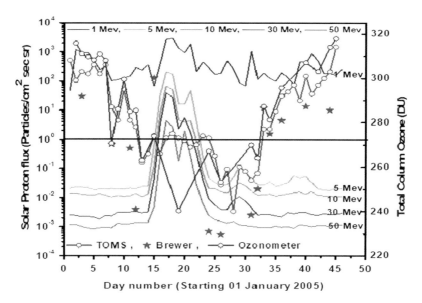

FIGURE 5.8   Day-to-day variations in solar proton flux at 5 energy levels and total column ozone derived from ground-based ozonometer and Brewer spectrometer at the Indian Antarctic station Maitri and from TOMS satellite over the region during the period from 1 January to 15 February 2005.

chemical reactions, and also that the greatest ionization takes place in between 17 and 18 January and 20 January when the ionization rate peak is below 50 km altitude are also shown by Seppala et al. (2006). SPE particle precipitate in the atmosphere produced a HO_x constituent as a result of an increase in ionization, which has a relatively short lifetime of a few days, and destroyed the ozone. An observed decrease in ozone content at the Indian Antarctic station was recorded for about 12 days when it recovered to normal value. To ascertain short-term fluctuations, temporal variation of solar proton flux at a 1-minute interval on 15 January 2005, during the period 0700–2400 hours is plotted in Figure 5.9, along with a temporal variation of ozone meter-derived TCO measured on that day at the Indian Antarctic station. At shorter time intervals within a day, one can also see a decrease in ozone whenever a sudden increase in bursts in proton flux (especially at > 1 MeV) occur.

In another case study, during early December 2006, two SPEs of Halo-type X-class with an intensity of X9 and X3 solar flare occurred on 6 December and 13 December, associated with CME. Daily mean solar proton flux measurements made by GOES-11 for the period 1–31 December 2006 for the energy channels of > 1 MeV, > 5 MeV, > 10 MeV, > 30 MeV, and > 50 MeV as in the previous case, are plotted in Figure 5.10. One can see that after 5 December, there are two distinguishable peaks in proton flux correspond to the two SPEs that occurred in this month. Day-to-day variation in the daily mean total column ozone from the ground-based Ozonometer, Brewer spectrophotometer, and also from the TOMS satellite over the Indian Antarctic station for the period 1–31 December 2006 are shown in Figure 5.10. Here, because of the seasonal increasing trend, the effect

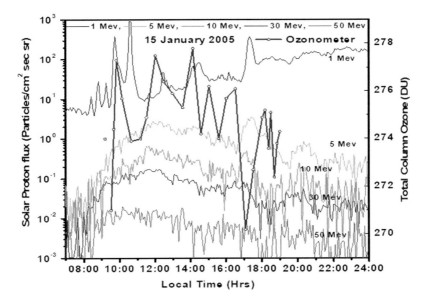

FIGURE 5.9 Temporal variations in solar proton flux at 5 energy levels and total column ozone from Ozonometer at Maitri on 15 January 2005.

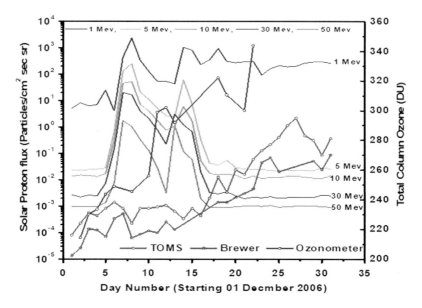

FIGURE 5.10   Same as Figure 5.8 but for the period 1–31 December 2006.

due to a SPE is evident but not dramatic. Also, Ozonometer-derived TCO seems to be overestimating the other two measurements. The linear trend in all three measurements is removed and the day-to-day variation in the deviations only is plotted in Figure 5.11. Now one can see the immediate response of decreasing

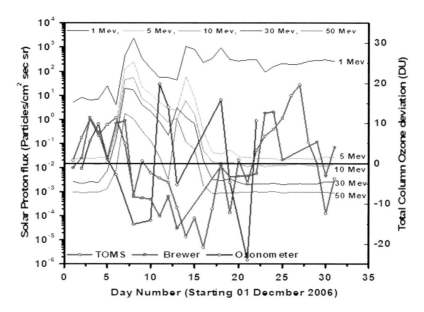

FIGURE 5.11   Daily variation of total column ozone and deviation during December–February 2007.

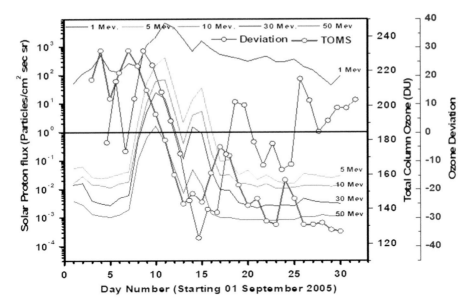

FIGURE 5.12   Day-to-day variations in solar proton flux at 5 energy levels and total column ozone derived from TOMS and trend removed ozone deviations (DU) during the period 1–30 September 2005.

ozone content in all three measurements. The Ozonometer measurement of TCO shows a temporary recovery in content during the intermittent period of the above-mentioned two SPEs. All three measurements of TCO, especially those by TOMS and ground-based Brewer spectrometer, showed that ozone content recovered to normal values after 20 December 2006.

A strong SPE of X-class solar flare (X17) associated with CME occurred on 8 September 2005. Daily mean solar proton flux measurements made by GOES-11 for the period 1–30 September 2005, for the energy channels of > 1 MeV, > 5 MeV, > 10 MeV, > 30 MeV, and > 50 MeV are plotted in Figure 5.12. During this event only, TOM's total column ozone data for the Antarctic region is available and thus the daily TCO values are also plotted in Figure 5.12. Since there was a steady decreasing trend in TCO, the linear trend is removed and only the deviations from the linear trend are also plotted in the same figure to bring out the variations better. There is an abrupt and substantial increase in solar proton fluxes after 7 September at all the energy levels and continued to be high until 16 September. Almost simultaneously, the ozone content decreased by 35–40 DU in a few days and recovered to pre-SPE day's value after 16 September. Thus, this case also shows the short-term depletion in ozone over the Antarctic region due to the influx of highly charged solar particles. The SPE-produced HO_x could have led to the short-term catalytic ozone destruction in the lower mesosphere and stratosphere.

Ozone variation during yet another SPE that occurred in November 2004 is also studied. A large SPE that occurred on 7 November 2004, was an X2-class solar flare of halo-type CME, with maximum flux energy of 2.07E + 09 protons cm^2 sec sr (> 10 MeV). The daily mean solar proton flux with energy > 1 Mev, > 5 Mev, > 10 Mev,

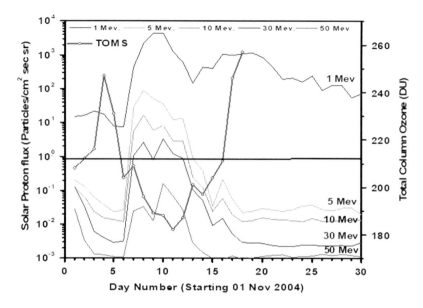

FIGURE 5.13 Day-to-day variations in solar proton flux at 5 energy levels and total column ozone derived from TOMS during the period 1–30 November 2004.

> 30 Mev, and > 50 Mev, as measured by GOES-10 and GOES-11, is plotted in Figure 5.13 for the period 1–30 November 2004. Simultaneously, the TOMS-measured daily TCO for the Antarctic location is also plotted and shown in the same figure. Coincident with the increase in solar proton flux after 7 November, the ozone content also decreased and after 15 November as the intensity of proton flux subsides, ozone content also started to be restored to its pre-SPE value.

In Figures 5.14 and 5.15, we show the monthly ozone depletion (in Dodson Units) time series over the Antarctica region encompassing the considered 2004–2006 period of this study. The main purpose of looking at this ozone variability time series is to identify the period in which the ozone hole season was active. The observational results reported here are during 1–15th January 2005, and 1–30th December 2006; it is important to note that January 2005 is not in the usual annual ozone depletion/ozone hole formation season whereas during December 2006, it is clear from Figure 5.14 that the stratospheric ozone had recovered completely. Now, if we consider the two other cases in which we used only TOM's ozone data during the solar proton events of November 2004 and September 2005, it is understood and observed in Figure 5.15 that these periods coincided with the usual annual ozone depletion cycle over Antarctica. In other words, we have two contrasting conditions in which the solar proton events have occurred: (1) During the annual ozone- hole occurring periods (cases of November 2004 and September 2005) and (2) during the normal periods (cases of January 2005 and December 2006). Hence, it becomes clear that the variation in the TCO during January 2005 and December 2006 was not definitely because of the annually occurring ozone hole formation methods; instead, the change in the TCO stands to be accounted for the observed solar proton events

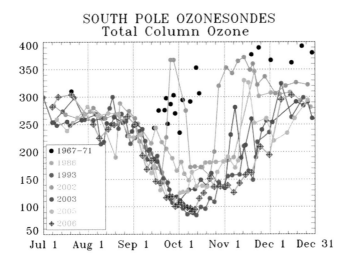

FIGURE 5.14 Monthly total ozone variations during 2005 and 2006 at the Southern Hemisphere.

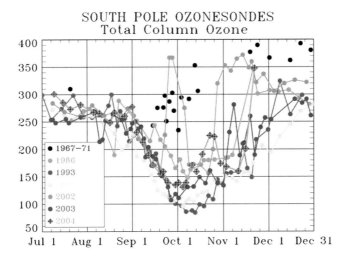

FIGURE 5.15 Same as Figure 5.14 but also incorporating 2004.

that had occurred during these periods. To further consolidate this reason, we have also plotted the SCIMACHY NO_2 (NO_x family) profiles during 12th December 2006 (which happened to be the peak time during which the SPE occurred) over the Southern Hemisphere, encompassing the Antarctica region (Figure 5.16). From these plots, it can be seen that the satellite passes over the entire Antarctica domain, and also reveals enhanced loading of the NO_2. Since NO_2 is primarily present below 50 km, these plots reveal that the NO_2 was enhanced as a result of the SPE and they were present in the atmospheric layer between 12 and 47 km. Some of the especially large SPEs have been documented to have a substantial influence on chemical

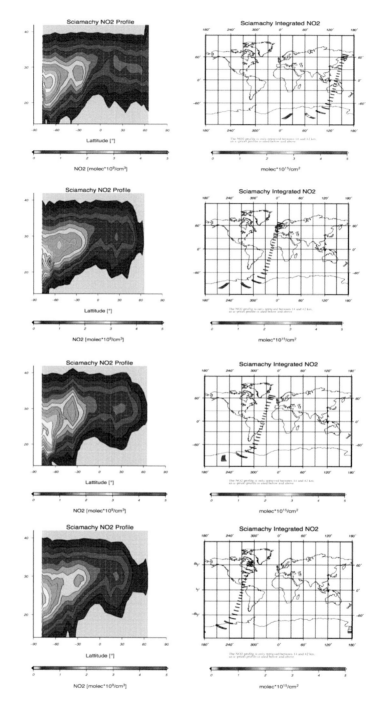

FIGURE 5.16   SCIMATCHY NO_2 profiles (between 14 and 42 km) during 12 December 2006 over the Southern Hemisphere.

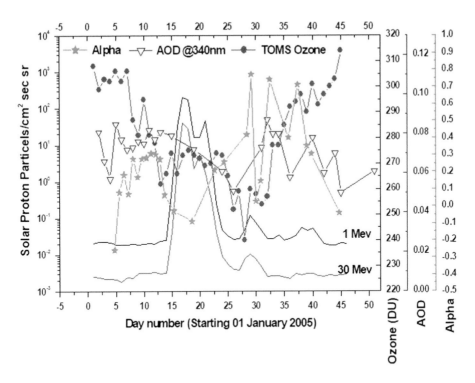

FIGURE 5.17   Day-to-day variations in solar proton flux at 2 energy levels and total column ozone derived from TOMS satellite and AOD at 380 nm with alpha value over the region, Indian Antarctic station Maitri during 1 January–15 February 2005.

constituents in the polar middle atmosphere, especially NO_x and ozone (Weeks et al., 1972; Heath et al., 1977; Reagan et al., 1981; McPeters et al., 1981; Thomas et al.,1983; McPeters and Jackman, 1985; McPeters, 1986; Jackman and McPeters, 1987; Zadorozhny et al., 1992; Jackman et al., 1999; 2001; 2005; Randall et al., 2001; Seppala et al., 2004; 2006; Lopez-Puertas et al., 2005a; 2005b; von Clarmann et al., 2005; Orsolini et al., 2005; Degenstein et al., 2005; Rohen et al., 2005; Verronen et al., 2006). Here we have a very rare case where the SPE was short term but did affect decreas of TCO, through the enhanced loading of NO_2 that destructs the ozone (Figure 5.17).

   In the polar cap region, where the energetic partiles mainly precipitate, enhance the constitutes of both NOy and HO_x, resulting in ozone depletion. During the polar night, dark photolysis of water vapor decreases, thus reducing the HO_x in the atmosphere. In SPEs on polar nights, a large amount of HO_x is produced in terms of enhancing the HO_x catalytic cycles on Ox loss, whereas in sunlight, the mesosphere ozone loss is less severe, which can be explained by differences in the ambient HO_x production (see Rohan et al., 2005). The SPEs enhanced ion production rates, which can lead to the formation of NOy when the charged particles collide and dissociate N_2. These produce the other NOy constituent in the lower stratosphere and upper troposphere leads to the Ox loss through NO_x catalytic cycle. Thus, all the case studies during the years 2004–2006 shown previously indicate a strong influence of precipitating charged particles during SPEs on the ozone content in the Antarctic region.

5.5   SUMMARY AND CONCLUSIONS

- Day-to-day variations in ozone and water vapor exhibited an inverse relationship on certain days, revealing heterogeneous chemical processes wherein aerosol acts as a catalyst for the ozone destruction mechanism.
- Aerosol size distribution showed an abundance of fine-mode particles in summer and course-mode particles in winter.
- Spectral dependence of AOD suggests enhancement of secondary aerosol particles due to bubble-bursting/gas-to-particle conversion/new particle formation phenomenon.
- Atmospheric ozone showed an increase with increase in latitude.
- Studies during the years 2004–2006 indicate a strong influence of precipitating charged particles during SPEs on the ozone content in the Antarctic region.
- Results indicate that even a moderate SPE can, under certain conditions, be important to ozone loss in the middle atmosphere.
- Short-term depletions in ozone content take place due to enhanced loading of NO_2 below 50 km of the atmospheric profile. These effects are transient, as ozone seems to recover to quiescent levels within days.
- The principal reason for the short-lived ozone decreases in the Antarctic region is due to the short lifetimes of the HO_x constituents and NOy catalytic ozone loss cycle.
- Short-term ozone depletion of the type observed here may also have a significant impact on the Earth's biosphere, especially at high latitudes.

NOTES

1 Details of these SPEs can be seen at (http://umbra.nascom.nasa.gov/SEP/).
2 (http://www.ngdc.noaa.gov/stp/GOES/goes.html).

ACKNOWLEDGMENTS

The authors wish to thank the Director, IITM, Pune, for continuous encouragement and support. Thanks, are also due to the Authorities of the Amity University Haryana (AUH), Gurugram, for cooperation and valuable suggestions. We thank Sanjoy K. Saha for his untiring help throughout the study. Support from Director and scientists of NCPOR, Goa is gratefully acknowledged. The authors express their sincere gratitude to the Ministry of Earth Sciences (MoES), Govt of India, for facilitating the participation of one of the authors (SMS) in two Indian summer expeditions (24th and 26th) Antarctica. The valuable comments and suggestions from anonymous Reviewers, which improved the scientific content of the Chapter, are acknowledge with thanks.

REFERENCES

Aculinin A. (2006). Variability of TCO content measured at Chisinau site, Republic of Moldova. *Moldavian Jr of the Physical Sci*, 5, N2.

Brasseur P., and Solomon S. (1968). *Aeronomy of the Middle Atmosphere*, Springer.

Degenstein D. A., Lloyd D., Bourassa A. E., Gattinger R. L., and Llewellyn E. J. (2005). Observations of mesospheric ozone depletion during the October 28, 2003, solar proton event by OSIRIS. *Geophys Res Lett*, 32, L03S11. https://doi.org/10.1029/2004 GL021521.

Devara P. C. S., Maheskumar R. S., Ernest Raj P., Dani K. K., and Sonbawne S. M. (2001). Some features of aerosol optical depth, ozone, and precipitable water content observed over land during the INDOEX-IFP99. *MeteorologischeZeirschrift*, 10, 901–908.

Ernest Raj P., Devara P. C. S., Pandithurai G., Maheskumar R. S., Dani K. K., Saha S. K., and Sonbawne S. M. (2004). Variability in sun photometer derived ozone over a tropical urban station. *J Geophys Res*, 109(1-8), D08309. https://doi.org/10.1029/2003 JD004195.

Evtushevsky O., Milinevsky G., Grytsai A., Kravchenko V., Grytsai Z., and Leonov M. (2008). Comparison of ground-based Dobson and satellite EP-TOMS total ozone measurements over Vernadsky station, Antarctica, 1996–2005. *Inter J of Remote Sens*, 29(9), 2675–2683.

Farman J. C., Gardiner B. G., and Shanklin J. D. (1985). Large losses of total ozone in Antarctica reveal seasonal CLO_x/NO_x interaction. *Nature*, 315, 207–210.

Grenfell J. L., Lehmann R., Mieth P., Langematz U., and Steil B. (2006). Chemical reaction pathways affecting stratospheric and mesospheric ozone. *J Geophys Res*, 111(311), D17. https://doi.org/101029/2004JD005713.

Heath D. F., Krueger A. J., and Crutzen P. J. (1977). Solar proton event: influence on stratospheric ozone. *Science*, 197, 886–889.

Ichoku C., Robert Levy, Kaufman Y. J., Remer L. A., Rong-Rong, Li, Martins V. J., Holben B. N., Abuhassan N., Slutsker I., Eck T. F., and Pietras C. (2002). Analysis of the performance characteristics of the five-channel Micro tops II sun photometer for measuring aerosol optical thickness and precipitable water. *J. Geophys. Res.*, 107, D13. https://doi.org/101029/2001/JD001302.

Jackman C. H., DeLand M. T., Labow G. J., Fleming E. L., Weisenstein D. K., Ko M. K. W., Sinnhuber M., and Russell J. M. (2005). Neutral atmospheric influences of the solar proton events in October–November 2003. *J Geophys Res*, 110, A09S27. https://doi.org/101029/2004JA010888.

Jackman C. H., Fleming E. L., Francis M., Vitt F. M., and Considine D. B. (1999). The influence of solar proton events on the ozone layer. *Advances in Space Res*, 24, 625–630.

Jackman C. H., and McPeters R. D. (1987). Solar proton events as tests for the fidelity of middle atmosphere models. *Physica Scripta*, T18, 309–316.

Jackman C. H., and McPeters R. D. (2004). The effect of solar proton events on ozone and other constituents. *Geophys Monograph*, 141, 305–319.

Jackman C. H., McPeters R. D., Labow G. J., Fleming E. L., Praderas C. J., and Russell J. M. (2001). Northern Hemisphere atmospheric effects due to the July 2000 solar proton event. *Geophys Res Lett*, 28, 2883–2886.

Kulandaivelu E., and Venkateshwara R. (2004). Measurement of Total Ozone, D-UV Radiation, Sulphur Dioxide, and Nitrogen Dioxide through Brewer Spectrophotometer at Maitri Antarctica during the year 2000, 19th Indian Expedition of Antarctica. *Scientific Report DOD, Technical Publication*, No. 17, 147–163.

Lopez-Puertas M., Funke B., Gil-Lopez S., Tsidu G. M., Fische, H., and Jackman C. H. (2005b). HNO3, N2O5, and ClONO2 enhancements after the October–November 2003 solar proton events. *J Geophys Res*, 110, A09S44. https://doi.org/101029/2005JA 011051.

Lopez-Puertas M., Funke B., Gil-Lopez S., von Clarmann T., Stiller G. P., Hopfner M., Kellmann S., Fischer H., and Jackman C. H. (2005a). Observation of NO_x enhancement and ozone depletion in the Northern and Southern Hemispheres after the October–November 2003 solar proton events. *J Geophys Res*, 110, A09S43. https://doi.org/101029/2005JA011050.

McPeters R. D. (1986). A nitric oxide increase observed following the July 1982 solar proton event. *Geophys Res Lett.*, 13, 667–670.

McPeters R. D., and Jackman C. H. (1985). The response of ozone to solar proton events during solar cycle 21: the observations. *J Geophys Res*, 90, 7945–7954.

McPeters R. D., Jackman C. H., and Stassinopoulos E. G. (1981). Observations of ozone depletion associated with solar proton events. *J Geophys Res*, 86, 12071–12081.

Morys M., Mims F. M., Hagerup S., Anderson S. E., Baker A., Kia J., and Walkup T. (2001). Design, calibration and performance of Micro tops II handheld ozone monitor and sunphotometer. *J Geophys Res*, 106, 14573–14582.

Orsolini Y. J., Manney G. L., Santee M. L., and Randall C. E. (2005). An upper stratospheric layer of enhanced HNO3 following exceptional solar storms. *Geophys Res Lett*, 32, L12S01. https://doi.org/101029/2004GL021588.

Randall C. E., Siskind D. E., and Bevilacqua R. M. (2001). Stratospheric NO_x enhancements in the southern hemisphere polar vortex in winter and spring of 2000. *Geophys Res Lett*, 28, 2385–2388.

Reagan J. B., Meyerott R. E., Nightingale R. W., Gunton R. C., Johnson R. G., Evans J. E., Imhof W. L., Heath D. F., and Krueger A. J. (1981). Effects of the August 1972 solar particle events on stratospheric ozone. *J Geophys Res*, 86, 1473–1494.

Reid G. C., Solomon S., and Garcia R. R. (1991). Response of the middle atmosphere to the solar proton events of August-December 1989. *Geophys Res Lett*, 18, 1019–1022.

Rohen G., von Savigny C., Sinnhuber M., Llewellyn E. J., Kaiser J. W., Jackman C. H., Kallenrode M. B., Schroter J., Eichmann K. U., Bovensmann H., and Burrows J. P. (2005). Ozone depletion during the solar proton events of Oct/Nov 2003 as seen by SCIAMACHY. *JGeophys Res*, 110, A09S39, https://doi.org/101029/2004JA010984.

Rusch D. W., Gerard J. C., Solomon S., Crutzen P. J., and Reid G. C. (1981). The effect of particle precipitation events on the neutral and ion chemistry of the middle atmosphere: I Odd nitrogen. *Planet Space Sci*, 29, 767–774.

Seppala A., Verronen P. T., Kyrola E., Hassinen S., Backman L., Hauchecorne A., Bertaux J. L., and Fussen D. (2004). Solar proton events of October–November 2003: Ozone depletion in the Northern Hemisphere polar winter as seen by GOMOS/Envisat. *Geophys Res Lett*, 31, L19107, https://doi.org/101029/2004GL021042.

Seppala A., Verronen P. T., Sofieva V. F., Tamminen J., Kyrola E., Rodger C. J., and Clilverd M. A. (2006). Destruction of tertiary ozone maximum during a solar proton event. *Geophys Res Lett*, 33, L07804, https://doi.org/101029/2005GL025571.

Solomon S., Rusch D. W., Gerard J. C., Reid G. C., and Crutzen P. J. (1981). The effect of particle precipitation events on the neutral and ion chemistry of the middle atmosphere: II Odd hydrogen. *Planet Space Sci*, 29, 885–893.

Sonbawne S. M., Ernest Raj P., Devara P. C. S., and Dani K. K. (2009). Variability in sun photometer derived summertime total column ozone over the Indian station Maitri in the Antarctic region. *Inter Jr of Remote Sens*, 30(4331–4341), 15–16.

Staehelin J., Kerr J., Evans R.,and Vanice kK (2003). Comparison of Total Ozone measurements of Dobson and Brewer Spectrophotometers and Recommended transfer functions. *WMO TD No 1147*, No 149, March 2003.

Thomas R. J., Barth C. A., Rottman G. J., Rusch D. W., Mount G. H., Lawrence G. M., Sanders R. W., and Thomas G. E., Clemens L. E. (1983). Mesospheric ozone depletion during the solar proton event of 13 July 1982, 1, Measurements. *Geophys Res Lett*, 10, 257–260.

Thompson A. M. et al. (2003). Southern Hemisphere Additional Ozonesondes (SHADOZ) 1998–2000 tropical ozone climatology, Comparison with Total Ozone Mapping Spectrometer (TOMS) and ground-based measurements. *J Geophys Res*, 108(D2), 8238, https://doi.org/101029/2001JD000967.

UNEP (1989). United Nations Environment Programme: annual report of the Executive Director, 1989. x, 285 p.: ill, charts, graphs, tables. ISBN/ISSN 9280712195 1010-1268.

UNEP(1998). United Nations Environment Programme UNEP annual report 1998. 24 p.: ill., charts, graphs. ISBN 9280717499.

Verronen P. T., Rodger C. J., Clilverd M. A., Pickett H. M., and Turunen E. (2007). Latitudinal extent of the January 2005 solar proton event in the Northern Hemisphere from satellite observations of Hydroxyl. *Ann Geophys*, 25, 2203–2215.

Verronen P. T., Seppala A., Clilverd M. A., Rodger C. J., Kyrola E., Enell C. F., Ulich T., and Turunen, E. (2005). Diurnal variation of ozone depletion during the October-November 2003 solar proton events. *J Geophys Res*, 110, A09S32, https://doi.org/101029/2004JA010932.

Verronen P. T., Seppala A., Kyrola E., Tamminen J., Pickett H. M., and Turunen E. (2006). Production of odd hydrogen in the mesosphere during the January 2005 solar proton event. *Geophys Res Lett*, 33, L24811, https://doi.org/101029/2006GL028115.

von Clarmann T., Glatthor N., Hopfner M., Kellmann S., Ruhnke R., Stiller G. P., and Fischer H. (2005). Experimental evidence of perturbed odd hydrogen and chlorine chemistry after the October 2003 solar proton events. *J Geophys Res*, 110, A09S45, https://doi.org/101029/2005JA011053.

Weeks L. H., CuiKay R. S., and Corbin J. R. (1972). Ozone measurements in the mesosphere during the solar proton event of 2 November 1969. *J AtmosSci*, 29, 1138–1142.

World Meteorological Organisation (WMO) (2003). Scientific Assessment of Ozone Depletion 2002, WMO Global Ozone Research and Monitoring Project, Report No 47, Geneva, Switzerland.

World Meteorological Organisation (WMO) (2007). Scientific Assessment of Ozone Depletion 2006, WMO Global Ozone Research and Monitoring Project. Report No 50, Geneva, Switzerland.

Zadorozhny A. M., Tuchkov G. A., Kikhtenko V. N., Lastovicka J., Boska J., and Novak A. (1992). Nitric oxide and lower ionosphere quantities during solar particle events of October 1989 after rocket and ground-based measurements. *J AtmosTerrPhys*, 54, 183–192.

6 Physicochemical Properties of Antarctic Aerosol Particles

Vikas Goel

Environmental Sciences and Biomedical Metrology Division, CSIR-National Physical Laboratory, New Delhi, India

Sumit Kumar Mishra

Academy of Scientific and Innovative Research (AcSIR), Kamla Nehru Nagar, Ghaziabad, Uttar Pradesh, India

CONTENTS

6.1 INTRODUCTION

Aerosol particles are tiny solid or liquid particles suspended in a gaseous medium. The size of the aerosol particles generally varies from 1 nm to 100 µm. Aerosol particles less than the size of 2.5 µm, 5 µm, and 10 µm in aerodynamic diameter are classified as $PM_{2.5}$, PM_5, and PM_{10}, respectively. These particles can be directly emitted from various sources or can be formed in the atmosphere through a gas-to-particle conversion. Particles emitted directly from the sources are called primary aerosol particles and the particles formed in the air are called secondary aerosol particles (Seinfeld and Pandis, 2006). Primary aerosol particles are generally bigger in size and secondary aerosol particles are smaller. Particles can emit from various

DOI: 10.1201/9781003203742-6

sources ranging from natural to anthropogenic. Although aerosol particles have a very short lifetime and comprise only a very small fraction of the atmosphere, they have attained great attention because of their role in determining Earth's radiation budget. Not only can they affect the radiation budget, but they also determine the atmosphere's chemical composition, properties of clouds (by acting as cloud condensation nuclei), visibility, and are also found to have acute health effects (Bond et al., 2013; Ching et al., 2019; Formenti et al., 2011; Fuzzi et al., 2015; Goel et al., 2018a; Kim et al., 2015; Lelieveld et al., 2015; Mishra and Tripathi, 2008; Seinfeld and Pandis, 2016). The particles can be removed from the atmosphere by two mechanisms: first is the gravitational settling on the Earth's surface called dry deposition and the second is the incorporation of aerosols into cloud droplets and the washout effect of rain called wet deposition. Because of wet and dry deposition, aerosol particles have a very short lifetime and their concentration varies widely throughout the globe depending upon the sources and meteorological conditions of the region.

Once airborne, aerosol particles interact with the incoming shortwave solar radiations and affect the radiation budget directly and indirectly. An an indirect effect, the particles interact directly with the radiation and cause warming or cooling of the atmosphere (depending on their physicochemical properties); whereas, in indirect interaction, the particles act as Cloud Condensation Nuclei (CCN) and form clouds. The indirect effect is an overall process by which aerosols perturb the Earth's energy budget by altering the cloud amount and albedo. Particles rich in sulfate, nitrate, sea salt, and calcium carbonate can readily act as CCN because of their hydrophilic property. In and indirect effect, particles rich in black carbon (BC), iron, copper, and chromium absorb the solar radiation and cause positive radiative forcing (warming); whereas the particles rich in silicon, nitrate, sulfate, aluminum, and calcium scatter the solar radiation and cause negative radiative forcing (cooling) (Ahlawat et al., 2019; Chaubey et al., 2011; Goel et al., 2018a, 2018b; Mishra and Tripathi, 2008; Zhang et al., 2015). The role of aerosol particles in determining Earth's radiation can be better studied by simulating optical properties like Single Scattering Albedo (SSA) and asymmetric parameter (g).

Antarctica is the coldest, driest, and widest place on Earth. It is the Earth's southernmost continent that covers the South Pole. It is a unique site for the study of natural aerosol particles. Antarctica allows the whole scientific community to study the pristine environment, untouched from direct anthropogenic activities. It is Earth's fifth-largest continent, always covered in snow and barely has any vegetation. But due to human interventions and global warming, the Antarctic ice is melting. It raises the sea level, which increases flooding events in low-altitude regions. According to a report, published in *Nature*, Antarctica alone could result in a 3-foot rise in sea level by the end of this century. This will lead to colossal devastation to mankind. Many low-altitude regions like Maldives, Las Vegas, and Mauritius will be submerged under the sea. Therefore, it becomes an area of prime importance to study the physicochemical properties of Antarctic aerosols and their optical properties.

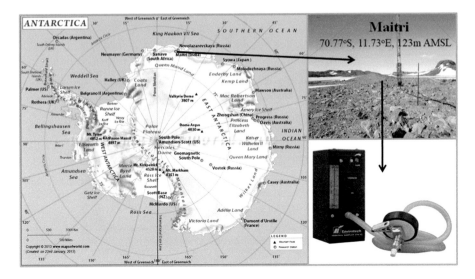

FIGURE 6.1 Geographical location of the sampling site, Indian Antarctic station Maitri and image of the Envirotech APM 801 sampler used for the collection of PM$_5$ (figure adopted from Goel et al., 2018a).

6.2 PARTICLE COLLECTION

PM$_5$ particles were collected on 99.99% pure tin substrates with the help of the Envirotech APM801 hand-held sampler from the Indian Antarctic station Maitri (70.77°S, 11.73°E) (Figure 6.1). During sample collection, the flow rate of the sampler was maintained at 2.5 lpm. The sampler was kept at the height of nearly 2 meters above the ground to prevent contamination from the loose ground soil. The sampling was done from December 2013 to February 2014. After collection, the tin substrates were kept in the small biological tubes and stored in a desiccator to prevent contamination. For physicochemical analysis, the samples were brought to the National Physical Laboratory, New Delhi.

6.3 PHYSICOCHEMICAL CHARACTERIZATION

Physicochemical characterization of aerosol particles give vital information of physical [morphology (size and shape) and mixing state] and chemical properties (chemical composition), which are essential inputs to the optical models. Precise information of physicochemical properties will help in more accurate optical property estimations. A scanning electron microscope (SEM, ZEISS EVO MA-10, Germany), coupled with energy dispersive X-ray spectroscopy (EDS, Oxford Link ISIS 300, England), gives detailed information on the physicochemical properties of aerosol particles at the individual particle level. The instrument uses an electron gun to generate a focused beam of electrons which, on interaction with the specimen, generates secondary electrons, backscattered electrons, auger electrons, and characteristic X-rays. The SEM detector installed in the sample chamber detects the

FIGURE 6.2   SEM-EDS instrument used for the physicochemical analysis of aerosol particles collected from Antarctica.

secondary electrons and gives a clear and focused image of aerosol particles well spaced from each other. The EDS detector is also installed in the sample chamber. It detects the characteristic X-rays and gives important information of the elemental composition of the specimen. The continuous bombardment of electrons on the specimen results in charged surfaces that affect the results; therefore, the sample collection was done on the conducting tin substrate. It prevents the charge accumulation on the sample surface by transferring the charge to the ground. The SEM monographs give vital information of the particle shape and size, and the EDS gives the elemental composition information of the aerosol particles. The SEM-EDS instrument used in this study is shown in Figure 6.2.

6.4   PHYSICOCHEMICAL PROPERTIES

The optical properties of the aerosol particles are governed by their physicochemical properties (Goel et al., 2020). The SEM images give information on the physical properties of the aerosols particles and the EDS gives elemental composition information. The morphological database provided by the National Institute of Standards and Technology (NIST) is used to identify the shape of the particles. Antarctic aerosol particles were observed to be rich in calcium, aluminum, silicon, iron, titanium, nitrogen, sodium, chloride, carbon, and oxygen. The Al- and Si-rich particles are generally classified as alumino-silicate or quartz-rich particles (Figures 6.3a and 6.3b). These particles are triangular and pentagonal. Earth's crust is composed of alumino-silicate particles (more than 70% by weight). The calcium-rich particles are generally identified as calcite ($CaCO_3$) and dolomite ($CaMg$ $(CO_3)_2$) particles and these particles are generally observed in single particle analysis (Pachauri, 2013; Shao et al., 2006). These particles are also found in the

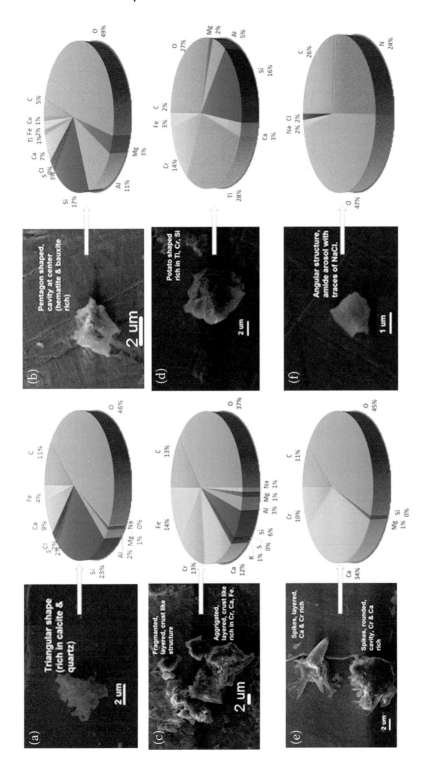

FIGURE 6.3 Elemental compositions of the Antarctic particles and their morphology. The morphology was determined using the morphology glossary provided by NIST (figure adopted from Goel et al., 2018a).

Antarctic environment (Figures 6.3c and 6.3e) with rounded shapes decorated with spikes and cavities. Antarctica is a place where one can find meteorites easily due to the contrast between white snow and the dark-colored meteorite. In the present work, particles rich in Ti and Fe are observed with a characteristic composition of the meteorite. Every year a lot of meteorites fall over Antarctica due to its thinner atmosphere (Klekociuk et al., 2005). The Ti-rich particle was observed in a potato shape with a good amount of Si, Cr, and trace amounts of Fe, Ca, Al, and Mg (Figure 6.3d). Antarctica is surrounded by the sea, which is a source of NaCl particles and in the present work, a NaCl particle was observed (Figure 6.3f). The particle was angular in shape with a good amount of N, O, and C. Some black carbon particles were also observed in the analysis. These particles can be easily identified with their grape-like structure made of carbon monomers. These particles can be emitted from the diesel generators used for the generation of electricity.

6.4.1 MORPHOLOGICAL PARAMETERS

Morphology parameters (aspect ratio, AR, and circulatory factor, CIR) of the aerosol particles can be calculated using the SEM monographs. The frequency distribution of AR and CIR of Antarctic particles is shown in Figure 6.4. AR is a ratio of the maximum projection to the width of the particle and CIR is calculated with the help of the area and perimeter of the particle. Both AR and CIR show the extent of particles' non-sphericity. AR = CIR = 1 indicate spherical particle and the

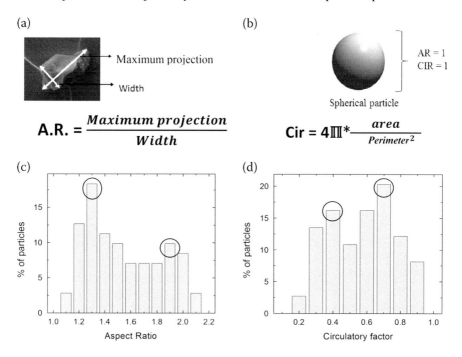

FIGURE 6.4 Frequency distribution of aspect ratio and circulatory factor of Antarctic aerosols (figure adopted from Goel et al., 2018a).

non-sphericity increases with AR > 1 and CIR < 1. The frequency distribution of AR and CIR was observed to be bimodal with their mode peaks at 1.3, 1.9, and 0.4, 0.7, respectively. The median AR and CIR of Antarctic particles was calculated to be 1.4 and 0.52, which is similar to that observed for the Chinese desert and Asian dust (Okada et al., 2001). For cylindrical, spiked, triangular, and smooth-shaped particles, the AR and CIR values were observed to be 2.4 and 0.52, 1.5 and 0.18, 2.22 and 0.65, and 1.07 and 0.82, respectively. Generally, in the satellite- and ground-based observations, the aerosol particles are assumed to be homogeneous spheres (i.e. AR = CIR = 1) (Mishra et al., 2017, 2008). In a realistic environment, the particles are highly non-spherical, as observed in the present study. The observed AR and CIR data will be useful to minimize the uncertainty associated with optical properties simulations.

6.5   SPECTRAL VARIATION OF REFRACTIVE INDEX

Refractive index is a function of wavelength and chemical composition, which are obtained from the EDS analysis (Sokolik and Toon, 1996). Individual particle analysis allows investigating the elemental composition of the single aerosol particles well spaced from each other. In the optical and radiative properties calculations, the aerosol particles are considered homogeneous spheres. But realistically, these particles can have variable composition, e.g. particles from the crust origin are rich in Al, Si, Mg, Ca, and Fe, and particles from vehicular emissions are rich in carbon, nitrogen, and sulfur. A particle of different chemical compositions interacts differently with solar radiation. Particles rich in carbon absorb the radiation and cause warming, whereas the particles rich in Si scatter the solar radiation and cause cooling. The absorption and cooling property of the aerosol particle can be better understood with the help of spectral variation of its refractive index (RI). It is a complex mixture of the real and imaginary parts (eq. 6.1). For the refractive index calculations, the elemental composition of the individual aerosol particles is converted into their respective oxides based on the available literature, e.g. Fe into Fe_2O_3, Al in Al_2O_3, Si into SiO_2, etc. The conversion was done using the approach used by Agnihotri et al. (2015) the effective refractive index of the particle was calculated using the effective volume mixing rule (eqs. 6.2 and 6.3) (Mishchenko et al., 2000).

$$RI = n \mp ik \qquad\qquad 6.1$$

Here, RI is the refractive index, n is the real part of RI, and k is the imaginary part of RI.

$$n_{eff} = \sum_{i=1}^{n} n_i f_i \qquad\qquad 6.2$$

$$k_{eff} = \sum_{i=1}^{n} k_i f_i \qquad\qquad 6.3$$

Here, n_{eff} and k_{eff} are the effective real and imaginary parts of refractive index, n_i and k_i are real and imaginary values of the ith species and f_i is a volume fraction of ith species.

In the spectral refractive index analysis, Fe_2O_3- and Cr_2O_3-rich particles were observed with a very high value of imaginary part of refractive index (k) and Al_2O_3-rich particles were observed to have nearly a zero value of k (Figures 6.5a, 6.5c, and 6.5d). In the case Fe_2O_3 and Cr_2O_3, the k value was observed to be very high at the shorter wavelength (0.38 μm), which decreases on increasing wavelength. In the case of $CaCO_3$-rich particles with a trace amount of Fe_2O_3 the value of k was observed to be highest at 0.38 μm, which decreases to zero at 0.67 μm onwards (Figure 6.5b). The Al_2O_3-rich particle shows no spectral variation; the value of k remained zero at all wavelengths (0.38−1.2 μm). In this analysis, particles rich in Fe and Cr were observed to have a high k value, whereas particles rich in Al were observed to have zero k value.

6.6   OPTICAL PROPERTIES

When a beam of light falls on a particle, the electric charges in the particle are excited and start performing an oscillating motion. These excited charges irradiate the absorbed energy in all directions, which is called scattering. Some part of the absorbed energy is converted into thermal energy, which is called absorption by the particle (Seinfeld and Pandis, 2016). The light absorption property of the aerosol particles is determined by its Single Scattering Albedo (SSA), which varies between 0 and 1. SSA that equals 1 shows that the particle is scattering in nature and the SSA that equals 0 shows that the particle is absorbing in nature. Mathematically, it is represented as a ratio of the fraction of the scattering efficiency (Q_{sca}) to extinction efficiency (Q_{ext}) (eq. 6.4).

$$SSA = \frac{Q_{sca}}{Q_{ext}} \qquad 6.4$$

Another important optical property is the asymmetry parameter (g). It is defined as an intensity-weighted average of the cosine of the scattering angle (eq. 6.5). It gives information on the direction of the scattered light. g = 1 shows that the total scattering takes place in the forward direction, g = −1 shows that scattering takes place in the backward direction, and g = 0 shows isotropic scattering. It is mathematically represented as:

$$g = \frac{1}{2} \int_0^\pi cos\theta \, P(\theta) sin\theta \, d\theta \qquad 6.5$$

Spectral variation of SSA and g of Antarctic aerosol particles are shown in Figures 6.6 and 6.7. The SSA and g were calculated using the T-matrix code (Mishchenko and Travis, 1998) initially developed by Dr. Waterman (Waterman, 1971). The code is publically available on the Internet at http://www.giss.nasa.gov/~crmim. The simulations were done at individual particles using the chemical composition data obtained from the EDS analysis. Based on the EDS analysis, four different particle compositions were considered while simulating the optical properties: (a) Cr_2O_3- and Fe_2O_3-rich

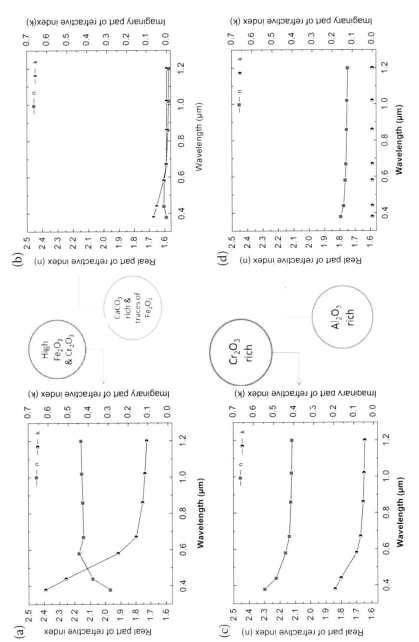

FIGURE 6.5 Spectral variation of the real and imaginary part of the refractive index of the Antarctic aerosol particles (figure adopted from Goel et al., 2018a).

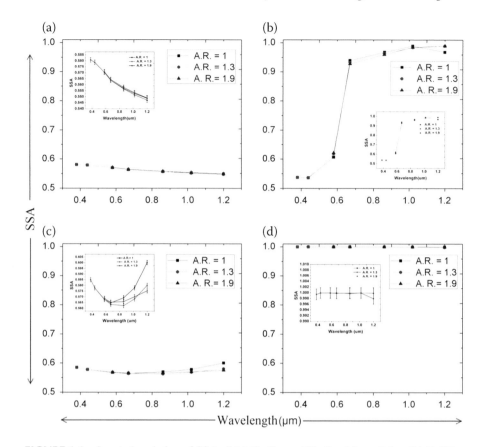

FIGURE 6.6 Spectral variation of SSA of (a) Cr_2O_3- and Fe_2O_3-rich particles, (b) $CaCO_3$-rich particle with a trace amount of Fe_2O_3, c) Cr_2O_3-rich particle, and d) Al_2O_3-rich particle (figure adopted from Goel et al., 2018a).

particles, (b) $CaCO_3$-rich particle with a trace amount of Fe_2O_3, (c) Cr_2O_3-rich particle, and d) Al_2O_3-rich particle. The morphology sensitivity studies were also performed by simulating the SSA and g for the observed particles for the three different model shapes with AR = 1, 1.4, 1.9. Here, we considered only smooth-shaped particles [sphere (AR = 1) and spheroid (AR = 1.3 and 1.9)] due to the model limitations.

6.6.1 Spectral Variation of SSA of the Considered Model Shapes

The spectral variation of SSA of the observed particles discussed in the previous section is shown in Figure 6.6. The difference in SSA between particles of different AR was found to be highest for the particles rich in Cr_2O_3 and $CaCO_3$ and the same was observed to be highest at a longer wavelength (1.2 μm). The effect of shape on SSA was observed to be negligible in Fe_2O_3- and Cr_2O_3-rich particles and Al_2O_3-rich particle. The chemical composition of the aerosol particles is an important

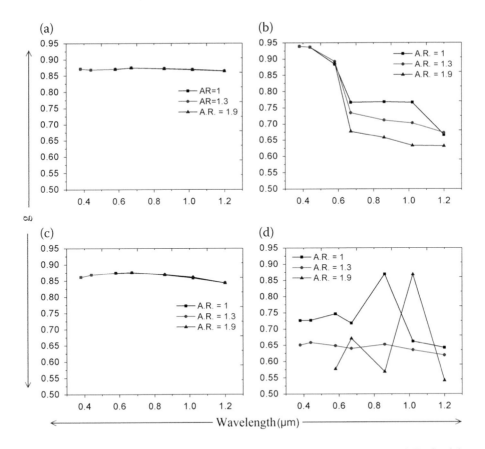

FIGURE 6.7 Spectral variation of asymmetry parameter of (a) Cr_2O_3- and Fe_2O_3-rich particles, (b) $CaCO_3$-rich particle with a trace amount of Fe_2O_3, (c) Cr_2O_3-rich particle, and (d) Al_2O_3-rich particle (figure adopted from Goel et al., 2018a).

input to the optical model that affects most of the results. The difference between SSA of particles having different chemical compositions is visible in Figure 6.6. The SSA of (a) Fe_2O_3- and Cr_2O_3-rich particles was observed to vary between 0.57 and 0.54, (b) the $CaCO_3$-rich particle with a trace amount of Fe_2O_3 varies between 0.53 and 0.98, (c) the Cr_2O_3-rich particle varies between 0.58 and 0.60, and (d) the Al_2O_3-rich particle was observed to be nearly one (~0.99). The spectral variation of SSA was observed to be highest in case c where $CaCO_3$ is having nearly zero value of k but Fe_2O_3 has a high value of k at a shorter wavelength, which gradually decreases with increasing wavelength. The spectral variation in k value of Fe_2O_3 resulted in the strong spectral variation of SSA in case c. SSA gives information on the light scattering and absorption nature of the particle. The particles rich in Fe_2O_3 and Cr_2O_3 were observed to have the lowest SSA, which shows that the particles can absorb the radiation efficiently, whereas the particle rich in Al_2O_3 was observed to have very high SSA (~0.99), which indicated that the particle scatters the solar radiation. SSA of the particle rich in $CaCO_3$ with a trace amount of Fe_2O_3 was observed to vary from 0.53 to 0.93; this indicates that the particle absorbs the solar radiation at a shorter wavelength (0.38 μm) and scatter at the longer wavelength (1.2 μm). Globally, black carbon and Fe_2O_3 are considered important species responsible for global warming, but in the present study, $Cr+O_3$ is also found to be an important element that can absorb solar radiation efficiently and may help in global warming.

6.6.2 Spectral Variation of Asymmetry Parameter of the Considered Model Shapes

The spectral variation of g of the Antarctic aerosol particles is shown in Figure 6.7. The shape effect on g was observed to be highest on the particles rich in $CaCO_3$ and Al_2O_3, and lowest on the particles rich in Cr_2O_3 and Fe_2O_3. From this observation, it can be concluded that the effect of shape on g is highest on the light-scattering particles and lowest on the light-absorbing particles. The g was observed to be different for the particles with a different composition. g was observed to be more sensitive to the particle's shape compared to that of SSA. The highest spectral variation in g was observed for the $CaCO_3$-rich particle.

ACKNOWLEDGMENTS

The authors would like to thank director NPL for the constant support. The authors would also like to thank NCAOR, Goa for the logistical support.

REFERENCES

Agnihotri, R., Mishra, S.K., Yadav, P., Singh, S., Rashmi, R., Prasad, M.V.S.K., Sharma, C., and Arya, B.C., (2015). Bulk level to individual particle level chemical composition of atmospheric dust aerosols (PM5) over a semi-arid urban zone of Western India (Rajasthan). *Aerosol. Air Qual. Res*, 15, 58–71. https://doi.org/10.4209/aaqr.2013.08.0270

Ahlawat, A., Mishra, S.K., Goel, V., Sharma, C., Singh, B.P., and Wiedensohler, A., (2019). Modeling aerosol optical properties over urban environment (New Delhi) constrained with balloon observation. *Atmos. Environ.*, 205, 115–124. https://doi.org/10.1016/j.atmosenv.2019.02.006

Bond, T.C., Doherty, S.J., Fahey, D.W., Forster, P.M., Berntsen, T., DeAngelo, B.J., Flanner, M.G., Ghan, S., Kärcher, B., Koch, D., Kinne, S., Kondo, Y., Quinn, P.K., Sarofim, M.C., Schultz, M.G., Schulz, M., Venkataraman, C., Zhang, H., Zhang, S., Bellouin, N., Guttikunda, S.K., Hopke, P.K., Jacobson, M.Z., Kaiser, J.W., Klimont, Z., Lohmann, U., Schwarz, J.P., Shindell, D., Storelvmo, T., Warren, S.G., and Zender, C.S. (2013). Bounding the role of black carbon in the climate system: A scientific assessment. *J. Geophys. Res. Atmospheres*, 118, 5380–5552. https://doi.org/10.1002/jgrd.50171

Chaubey, J.P., Krishna Moorthy, K., Suresh Babu, S., and Nair, V.S. (2011). The optical and physical properties of atmospheric aerosols over the Indian Antarctic stations during southern hemispheric summer of the International Polar Year 2007–2008. *Ann. Geophys*, 29, 109–121. https://doi.org/10.5194/angeo-29-109-2011

Ching, J., Adachi, K., Zaizen, Y., Igarashi, Y., and Kajino, M. (2019). Aerosol mixing state revealed by transmission electron microscopy pertaining to cloud formation and human airway deposition. *Npj Clim. Atmospheric Sci.*, 2, 22. https://doi.org/10.1038/s41612-019-0081-9

Formenti, P., Schütz, L., Balkanski, Y., Desboeufs, K., Ebert, M., Kandler, K., Petzold, A., Scheuvens, D., Weinbruch, S., and Zhang, D. (2011). Recent progress in understanding physical and chemical properties of African and Asian mineral dust. *Atmospheric chem. Phys.*, 11, 8231–8256. https://doi.org/10.5194/acp-11-8231-2011

Fuzzi, S., Baltensperger, U., Carslaw, K., Decesari, S., Denier van der Gon, H., Facchini, M.C., Fowler, D., Koren, I., Langford, B., Lohmann, U., Nemitz, E., Pandis, S., Riipinen, I., Rudich, Y., Schaap, M., Slowik, J.G., Spracklen, D.V., Vignati, E., Wild, M., Williams, M., and Gilardoni, S. (2015). Particulate matter, air quality, and climate: lessons learned and future needs. *Atmospheric Chem. Phys.*, 15, 8217–8299. https://doi.org/10.5194/acp-15-8217-2015

Goel, V., Mishra, S.K., Ahlawat, A., Kumar, P., Senguttuvan, T.D., Sharma, C., and Reid, J.S. (2020). Three-dimensional physicochemical characterization of coarse atmospheric particles from the urban and arid environment of India: An insight into particle optics. *Atmos. Environ*, 117338. https://doi.org/10.1016/j.atmosenv.2020.117338

Goel, V., Mishra, S.K., Lodhi, N., Singh, S., Ahlawat, A., Gupta, B., Das, R.M., and Kotnala, R.K. (2018a). Physico-chemical characterization of individual Antarctic particles: Implications to aerosol optics. *Atmos. Environ*, 192, 173–181. https://doi.org/10.1016/j.atmosenv.2018.07.046

Goel, V., Mishra, S.K., Sharma, C., Sarangi, B., Aggarwal, S.G., Agnihotri, R., and Kotnala, R.K. (2018b). A non-destructive FTIR Method for the determination of ammonium and sulfate in urban PM2.5 samples. *MAPAN*. https://doi.org/10.1007/s12647-018-0253-9

Kim, K.-H., Kabir, E., and Kabir, S. (2015). A review on the human health impact of airborne particulate matter. *Environ. Int.*, 74, 136–143. https://doi.org/10.1016/j.envint.2014.10.005

Klekociuk, A.R., Brown, P.G., Pack, D.W., ReVelle, D.O., Edwards, W.N., Spalding, R.E., Tagliaferri, E., Yoo, B.B., and Zagari, J. (2005). Meteoritic dust from the atmospheric disintegration of a large meteoroid. *Nature*, 436, 1132–1135. https://doi.org/10.1038/nature03881

Lelieveld, J., Evans, J.S., Fnais, M., Giannadaki, D., and Pozzer, A. (2015). The contribution of outdoor air pollution sources to premature mortality on a global scale. *Nature*, 525, 367–371. https://doi.org/10.1038/nature15371

Mishchenko, M.I., Hovenier, J.W., and Travis, L.D. (2000). *Light scattering by nonspherical particles: theory, measurements, and applications*. Academic Press, San Diego.

Mishchenko, M.I., and Travis, L.D., (1998). Capabilities and limitations of a current FORTRAN implementation of the T-matrix method for randomly oriented, rotationally symmetric scatterers. *J. Quant. Spectrosc. Radiat. Transf.*, 60, 309–324. https://doi.org/10.1016/S0022-4073(98)00008-9

Mishra, S.K., Dey, S., and Tripathi, S.N. (2008). Implications of particle composition and shape to dust radiative effect: A case study from the Great Indian Desert. *Geophys. Res. Lett.*, 35. https://doi.org/10.1029/2008GL036058

Mishra, S.K., Saha, N., Singh, S., Sharma, C., Prasad, M.V.S.N., Gautam, S., Misra, A., Gaur, A., Bhattu, D., Ghosh, S., Dwivedi, A., Dalai, R., Paul, D., Gupta, T., Tripathi, S.N., and Kotnala, R.K., (2017). Morphology, mineralogy, and mixing of individual atmospheric particles over Kanpur (IGP): Relevance of homogeneous equivalent sphere approximation in radiative models. *MAPAN*, 32, 229–241. https://doi.org/10.1007/s12647-017-0215-7

Mishra, S.K., and Tripathi, S.N. (2008). Modeling optical properties of mineral dust over the Indian Desert. *J. Geophys. Res.*, 113. https://doi.org/10.1029/2008JD010048

Okada, K., Heintzenberg, J., Kai, K., and Qin, Y. (2001). The shape of atmospheric mineral particles collected in three Chinese arid regions. *Geophys. Res. Lett.*, 28, 3123–3126. https://doi.org/10.1029/2000GL012798

Pachauri, T. (2013). SEM-EDX characterization of individual coarse particles in Agra, India. *Aerosol Air Qual. Res.*, https://doi.org/10.4209/aaqr.2012.04.0095

Seinfeld, J.H., and Pandis, S.N. (2016). *Atmospheric chemistry and physics: From air pollution to climate change*, 3rd ed. John Wiley & Sons, Hoboken, New Jersey.

Seinfeld, J.H., and Pandis, S.N. (2006). *Atmospheric chemistry and physics: From air pollution to climate change*, 2nd ed. Wiley, Hoboken, NJ.

Shao, L., Shi, Z., Jones, T.P., Li, J., Whittaker, A.G., and BéruBé, K.A. (2006). Bioreactivity of particulate matter in Beijing air: Results from plasmid DNA assay. *Sci. Total Environ.*, 367, 261–272. https://doi.org/10.1016/j.scitotenv.2005.10.009

Sokolik, I.N., and Toon, O.B. (1996). Direct radiative forcing by anthropogenic airborne mineral aerosols. *Nature*, 381, 681–683. https://doi.org/10.1038/381681a0

Waterman, P.C. (1971). Symmetry, unitarity, and geometry in electromagnetic scattering. *Phys. Rev. D*, 3, 825–839. https://doi.org/10.1103/PhysRevD.3.825

Zhang, X. L., Wu, G. J., Zhang, C. L., Xu, T. L., and Zhou, Q. Q. (2015). What's the real role of iron-oxides in the optical properties of dust aerosols? *Atmospheric Chem. Phys. Discuss.*, 15, 5619–5662. https://doi.org/10.5194/acpd-15-5619-2015

7 Impact of Near-Earth Space Environmental Condition to the Antarctic Sub-Auroral Upper Atmospheric Region

Rupesh M. Das
Environmental Sciences and Biomedical Metrology Division,
CSIR-National Physical Laboratory, New Delhi, India

CONTENTS

7.1 Introduction...151
7.2 Facilities for Monitoring the Sub-Auroral Ionosphere 153
    7.2.1 Canadian Advanced Digital Ionosonde (CADI) 153
    7.2.2 Global Ionospheric Scintillation and TEC Monitoring
        (GISTM) System 154
7.3 Earth's Geomagnetic Perturbations due to the Solar
    Wind–Magnetospheric Coupling Process...154
7.4 Response of Sub-Auroral High-Latitude Ionosphere
    to the Geomagnetic Disturbances 157
7.5 Longitudinal Ionospheric Behavior over the
    Antarctic Region during Adverse Space Weather Condition....................162
7.6 Latitudinal Ionospheric Response to the Changing Near-Earth Space
    Environment System 167
Acknowledgments..171
References...172

7.1 INTRODUCTION

The Earth's magnetosphere is very sensitive to the changing solar wind velocity as well as the associated proton density, especially during the directional or magnitude change in the interplanetary magnetic field. In general, when the solar wind strikes

DOI: 10.1201/9781003203742-7

the Earth's magnetosphere, properties of both vary in terms of slowing down solar wind velocity and changes in shape and size of the magnetosphere. However, dramatic changes in the Earth's ionosphere have been experienced when magnetospheric outbursts enter the ionosphere. The changes can be observed in terms of ring current intensification, brightening of auroras, and enhancements in ionization at different latitudes. The polar latitudes are most sensitive and significantly affected due to slight changes in space weather conditions. The nearly open or vertical geomagnetic field lines over the polar region allow the solar wind along with the associated high-energy particles to penetrate up to lower altitudes. The charged particle precipitation produces aurora with a simultaneous increase in ionospheric conductivities. Such effects are not homogenous throughout the polar region, hence the polar region is further divided into three parts, i.e. sub-auroral, auroral, and polar cap region. The auroral zone is a region of continuous and intense precipitation of energetic particles emanating from the Earth's plasma sheet (Eather et al., 1976; Lui et al., 1977). However, in the sub-auroral region, the energetic particles can be stored and accelerated to very high energies (10 KeV to several MeV) before eventual precipitation into the middle atmosphere (in the altitude range between 50 and 100 km). On the other hand, in the polar caps, the geomagnetic field lines are generally thought to be open and connected to the interplanetary medium. This permits direct access for energetic particles of solar or galactic origin. Though the three regions have different behaviors and responses to the changing space weather condition, a combination of the three decides the electrodynamic properties of the polar region. The electric field mapping from the magnetosphere into the ionosphere, along with magnetic field lines, establishes the dawn-to-dusk field in the polar cap; the field-aligned currents connect the currents flowing in the magnetosphere and ionosphere, forming a three-dimensional current system. Hence, the upper atmosphere exhibits various stormy features, affecting mainly the auroral regions. However, the excessive amount of energy deposition over the high-latitude region causes the phenomenon known to be Joule's heating phenomenon. Such a phenomenon is further responsible to modulate the conventional wind direction and strength. The modulated wind pattern is responsible for the transportation of Travelling Ionospheric Disturbances (TIDs) from a high-latitude region to a low-latitude region. The TIDs are often associated with molecular enriched chemical composition and perturbs lower-latitude ionospheric regions within a few hours of the actual commencement of geomagnetic storms.

To study the impact of changing the near-Earth space environment system to the Antarctic sub-auroral upper atmospheric region, CSIR-National Physical Laboratory, New Delhi established a state-of-the-art Indian Polar Space Physics Laboratory (IPSPL) at Indian Permanent Research Base Maitri, Antarctica, during International Polar Year. The facility provided an opportunity to monitor the various ionospheric parameters continuously and on a real-time basis to address the scientific interest of high-latitudinal ionospheric consequences caused by the modulation of near-Earth space environmental conditions.

7.2 FACILITIES FOR MONITORING THE SUB-AURORAL IONOSPHERE

The IPSPL is equipped with state-of-the-art ground- as well as space-based ionospheric probing instruments like the Canadian Advanced Digital Ionosonde (CADI) and Global Ionospheric Scintillation and TEC Monitoring (GISTM) Systems.

7.2.1 CANADIAN ADVANCED DIGITAL IONOSONDE (CADI)

The CADI system (see Figure 7.1) is a simple high-frequency (HF) radar that utilizes radio pulses to detect and range the plasma density in the bottom-side ionosphere. The CADI measurement technique is based on the ionosonde Doppler drift or imaging Doppler interferometry (IDI) technique. CADI provides a sounding capability using high-power radio frequency pulses at vertical incidence. Different plasma parameters of the ionosphere are collected and recorded properly and continuously. Observable quantities include: echo delay (height) versus frequency; phase and amplitude of echo; angle of arrival; polarization of the echo; real-time monitoring of magnetic disturbances; and increased ionization in the ionosphere. The frequency of the vertically transmitted wave determines the plasma density from which the pulse is reflected in the ionosphere. The time delay between transmission and reception of the reflected pulse on the ground is a measure of the height of the ionospheric layer from which the pulse was reflected. An ionosonde operates by stepping the frequency of the pulses from, say, 1–20 MHz, thereby producing a map of the bottom-side ionosphere. The result is an ionogram. Ionograms are recorded tracings of reflected high-frequency radio pulses generated by an ionosonde. Unique relationships exist between the sounding frequency and

FIGURE 7.1 Canadian Advanced Digital Ionosonde (CADI) Laboratory, Maitri, Antarctica.

FIGURE 7.2   Global Ionospheric Scintillation and TEC Monitoring (GISTM) system, Maitri, Antarctica.

the ionization densities which can reflect it. The ionogram measurements were synchronized on 10-minute boundaries, beginning at 00 minute of the hour, and the fixed frequency measurements were nominally synchronized to the nearest minute.

7.2.2   GLOBAL IONOSPHERIC SCINTILLATION AND TEC MONITORING (GISTM) SYSTEM

A Novatel makes a dual-frequency 12-channel GISTM system (Figure 7.2) that has been installed for real-time and around-the-clock monitoring of Ionospheric Total Electron Content (ITEC), along with the L-band phase and amplitude scintillation that occurs due to various types of ionospheric irregularities. The system is also able to find out the accurate position of the location of the monitoring station with a maximum error of 1 meter. For minimizing the multipath reflection, a chock ring antenna system is used.

7.3   EARTH'S GEOMAGNETIC PERTURBATIONS DUE TO THE SOLAR WIND–MAGNETOSPHERIC COUPLING PROCESS

It is often observed that the changing solar-related parameters cause modulation of the near-Earth space environment condition, which is further responsible to perturb the Earth's magnetic field/ionospheric conditions. The geo-potential changing near-Earth space environment depends upon the orientation of the Interplanetary Magnetic Field (IMF), which can be realized in terms of IMF-Bz. The long duration of unstable nature (i.e. continuous north-south orientation) causes episodic energy loading–dissipation

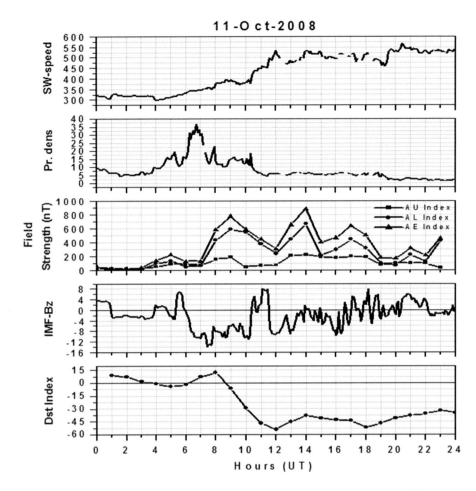

FIGURE 7.3   Geophysical condition along with AE-indices on 11th October 2008.

cycles termed *magnetospheric sub-storms* (as suggested by Baker et al., 1996; Sergeev et al., 1996b). Figure 7.3 shows a sample case observed on 11 October 2008, when IMF-Bz turned southward at around 0530 UT along with increasing proton density and solar wind speed, causing solar wind energy to be transferred to the Earth's magnetosphere through dayside magnetic reconnection. This leads to the storage of a large amount of energy in the magneto-tail, which corresponds to the initial phase (also known as the growth phase) of a sub-storm. The consequences of the growth phase have been analyzed by using the outputs derived with help of the GUMIC model. A similar fact suggested by Tanskanen et al. (2002) also stated that the growth phase typically begins in the quiescent period at the time of the southward turning of the IMF. The model-derived output shows that the progress of the growth phase is responsible for thinning as well as the intensification of the near-Earth current sheet, which very much agrees with earlier works reported by Baker and Pulkkinen (1991). The model output (Figure 7.4) also shows that the plasma sheet becomes very thin at

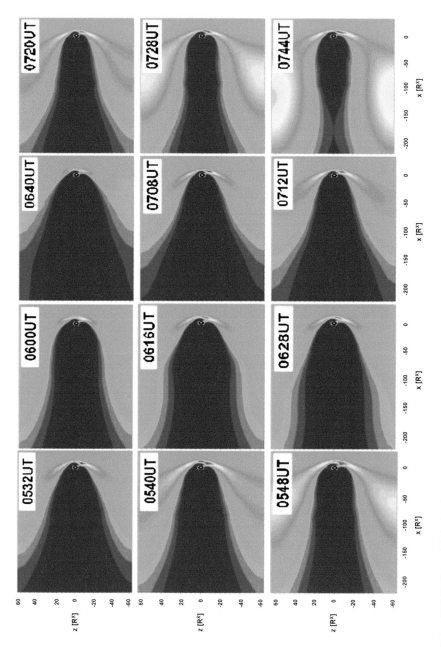

FIGURE 7.4 GUMICS model-derived magnetospheric condition on 11th October 2008.

around 0540 UT, which triggered the next step of the sub-storm cycle (expansion phase) by a rapid release of stored energy. Such effects were also examined by Baker and McPherron (1990) and stated that the cross-tail current sheet close to the Earth became extremely thin during the late growth phase. The extended work reported by Baker et. al. (1996) revealed that the sub-storm expansion began with the onset of the closed field lines reconnection in the center of the plasma sheet, due to which the near-Earth tail plasma-sheet became extremely thin and a so-called X-type neutral point formed within the plasma sheet. Similarly, several sub-storm researchers (Heppner et al., 1967; Siscoe and Cummings, 1969; Lui and Burrows, 1978; Tanskanen et al., 1987; Rothwell et al. 1988; Galperin and Feldstein, 1989; Elphinstone et al., 1990; Lopez and Lui, 1990; Lui, 1991) explained the sub-storm phases and suggested in their work that a sub-storm expansion disturbance is initiated in the near-Earth tail region and propagates to further downstream distances as the sub-storm progresses.

According to this theory, the cross-tail current is disrupted at the neutral point and diverted to the closed field line, magnetic flux being annihilated and plasma energized at the neutral point. A burst of highly energetic particles are injected into the inner magnetosphere, with some being precipitated in the auroral zones (electrojets) and some trapped as an initially asymmetric addition to the radiation belts (ring current).

Therefore, the simultaneous changes should be observed in both AE and Dst indices, as can be seen in Figure 7.3. However, the GUMIC model output (Figure 7.4) clearly shows that the said plasma sheet thinning process has been occurring three times (i.e. at around 0540UT, 0616UT, and 0712UT) during the period of southward turned IMF-Bz (from 0530 UT to 0100 UT) followed by a burst of energy release towards Earth. This is evident in the sequential occurrence of three sub-storm events within a very short frame of time. On the other hand, the cumulative effect of the three sub-storms has been observed in terms of occurrence of a moderate-type geomagnetic storm with the smooth negative excursion of Dst index (−55 nT) just after 0800UT, followed by a long recovery phase. The previously explained phenomenon agrees with the work reported by Gonzalez et al. (1994). Gonzalez et al. (1994) explained that if the energy input continues significantly longer (≥3 hours), a magnetic storm develops. Figure 7.3 reveals that IMF-Bz remained southward for about 4 hours, which is direct evidence of the presence of a magnetic storm. Such storms have a wide impact on the Earth's ionosphere and are responsible for redistribution of ionization based upon the nature of modulation of background electrodynamic conditions.

7.4 RESPONSE OF SUB-AURORAL HIGH-LATITUDE IONOSPHERE TO THE GEOMAGNETIC DISTURBANCES

The high-latitude ionosphere is broadly divided into three parts, i.e. sub-auroral, auroral, and polar cap regions. The boundaries of these three regions are variable and depend upon the background electrodynamic conditions. The magnetic reconnection between IMF and a geomagnetic field has been established during the southward turning of IMF-Bz, which produces open field lines. The open field lines allow mass, energy, and momentum to be transferred from the solar wind to the

Earth's magnetosphere, (as first suggested by Dungey, 1961). This results in a decrease in the horizontal component (H) of Earth's geomagnetic field. Also, the ionospheric conductivities are increased with the increase in precipitation of high-energy particles at a high-latitude region (Behera et al., 2015). The combined effect of a decrease in the H-field and increase in ionospheric conductivity are responsible to modulate the background electrodynamic conditions and consequences have been observed as an equator-ward expansion of auroral oval. The boundary of the auroral oval can be estimated with the quantitative change in Auroral Electrojet (AE-index) values. The AE is mainly driven by particle precipitation and the field-aligned current system, therefore, any changes in the near-Earth space environment system will cause expansion of auroral oval in both equator or poleward directions. Such an expansion of auroral oval changes the regional characteristics of sub-auroral and polar cap dynamics and directly perturbs the ionization over the high-latitude region. Also, the open magnetic field lines allow the solar wind and accompanied high energetic particles to interact with deep atmospheric layers to enhance the joule heating. The enhanced joule heating uplifts the molecular-rich air to the higher altitudes and changes the thermospheric compositions, which are responsible for changing the direction of the neutral wind from poleward to equator-ward (Mayr and Harris, 1978; Burns et al. 1991; Fuller-Rowell et al. 1994). The combined effects attribute a negative ionospheric response over the sub-auroral region, as shown in Figure 7.5.

Arun Kumar (Singh et al., 2019) suggested that the nature of ionospheric responses to the geomagnetic disturbances not only depended upon the status of high-latitudinal electrodynamic processes but was also influenced by the seasonal variations. The study suggested that the combination of equator-ward plasma transportation along with ionospheric compositional changes cause a negative ionospheric impact, especially during summer and equinox seasons. However, the combination of poleward contraction of the oval region, along with particle precipitation, may lead to an exhibit of positive ionospheric response during the winter season (see Figure 7.6).

The results reveal that the summer and equinoctial negative ionospheric response to the geomagnetic disturbances are mainly due to the combined effect of compositional changes and equator-ward displacement of the trough region. On the other hand, winter positive ionospheric response is caused by the combined effect of composition changes and poleward contraction of the auroral oval region. As already explained, the expansion and contraction of the auroral oval region are manifestations of changes in the background electrodynamic conditions due to enhancement in particle precipitation at higher latitudes and deep at ionospheric lower altitudes i.e. up to E-layer. The evidence of such deep particle precipitation impacts are visible in terms of abrupt enhancement in E-layer electron density observed in the ionograms recorded by CADI installed at Maitri, Antarctica (see Figures 7.7a, 7.7b, and 7.7c).

The abrupt enhancement in E-layer is further responsible for the enhancement in auroral electrojet strength and field-aligned current system along with electron/ion temperature. The ionograms in Figures 7.7(a) and 7.7(c) are also evidence of increased ionization at around the E-layer with a simultaneous decrease at higher altitudes (F-region) with upward lifting. This might be due to a change in O^+ and

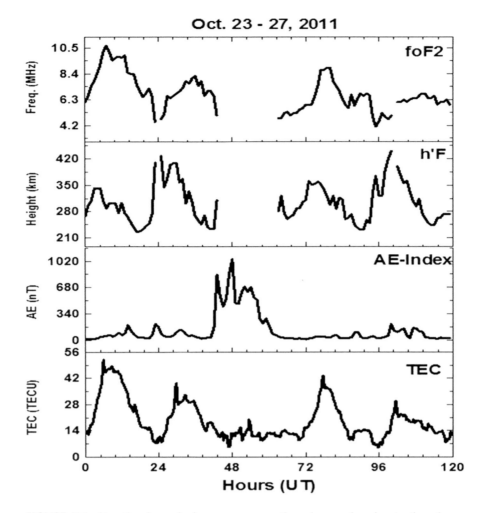

FIGURE 7.5 Negative ionospheric response over the sub-auroral region to the adverse near-Earth space environment condition.

NO$^+$ composition by conversion of O$^+$ into NO$^+$ due to the elevated ion temperature and the lowering of O$^+$ peak height. The lowering of O$^+$ peak height is responsible for a faster [O] recombination than the production rate, hence the ratio of atomic oxygen concentration [O] and molecular nitrogen concentration [N2] decreases. This is further responsible for a decrease in the electron density at peak height and shows a negative ionospheric storm effect at the sub-auroral region. The negative storms are primarily explained by the decrease of the neutral density ratio O/N$_2$ leading to an ion loss rate enhancement (Prölss and Jung, 1978). On the other hand, enhancement in auroral electrojet is responsible for the equator-ward expansion of the oval, hence the lifted F-layer ionization drifted towards the low-mid latitude region along with magnetic field lines. Therefore, the combination of equator-ward plasma transportation, along with ionospheric compositional changes, causes a

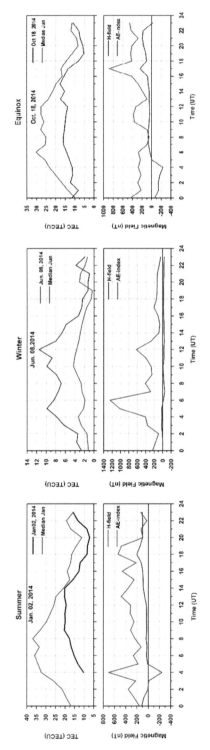

FIGURE 7.6 Seasonal response of the sub-auroral ionosphere to adverse near-Earth space environment system.

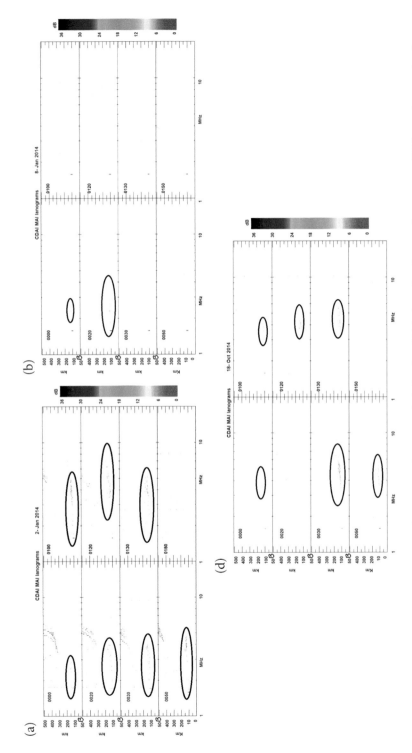

FIGURE 7.7 (a), (b), and (c): Ionograms of 2 January, 8 June, and 18 October 2014 show the abrupt enhancement of E-layer electron density.

negative ionospheric impact during summer and equinox over the sub-auroral latitudes, as can be seen in Figure 7.6.

However, a positive ionospheric effect has been observed during the winter season due to sudden orientation change in IMF-Bz from southward to northward leads to the modulation of the H-field from positive to negative hence the convection pattern also changes from anti-sunward to sunward (Ruohoniemi and Greenwald, 1998). This is responsible for pole-ward contraction of the auroral oval region as described by Kamide et al. (1974). During the wintertime, there is very little EUV available to ionize O so maximum production of O^+ is because of highly energetic particle precipitation. This increases the O^+ concentration with a decrease in N_2 concentration which results in enhancement of the O/N_2 ratio. The depletion of the N_2 densities causes reduced electron loss rates and can contribute to the enhanced electron densities that have been observed in the part of the winter hemisphere (Burns et al., 1995). Therefore, the combination of poleward contraction of the oval region along with particle precipitation may lead to the exhibition of the positive ionospheric response to geomagnetic disturbance observed during the winter month of 2014.

7.5 LONGITUDINAL IONOSPHERIC BEHAVIOR OVER THE ANTARCTIC REGION DURING ADVERSE SPACE WEATHER CONDITION

As explained previously that the boundaries of high-latitudinal, sub-auroral, auroral, and polar cap are not fixed and change according to the nature of events and can be estimated through a proxy like AE-index (Auroral Electrojet). The ionospheric impact on the changing near-Earth space environment system has also been different and modulated accordingly. To evaluate the nature of ionospheric perturbations at these three regions, the data from four Antarctic-based GPS stations (i.e. Maitri, Syowa, Mawson, Casey) has been analyzed. The geographic locations of these mentioned GPS stations are shown in Figure 7.8. The required ITEC parameter over Maitri has been acquired with help of the Novatel Make GPS receiver, installed by the National Physical Laboratory, Council of Scientific and Industrial Research, India, and in operation since November 2007. The other three stations' ITEC data has been derived by utilizing the GPS raw data file downloaded from the website www.sonel.org/ and https://github.com/stevieyu/stevieyu.github.io/blob/master/gps.html. Table 7.1 presents the geographic and geomagnetic coordinates of all the considered GPS stations. Among these stations, Syowa and Mawson permanently come under the auroral oval region, while Maitri and Casey, respectively, come under the sub-auroral and polar cap region. The GPS stations Maitri and Casey lie just outside the auroral oval boundary, which enables the scientific communities to study the ionospheric responses over both the station during equator-ward or poleward expansion of the oval.

Figure 7.9 shows the equator-ward expansion of the auroral oval region with the highest observed AE-index value is 887 nT (Figure 7.3) at around 1400 UT on 11 October 2008. This might be since during the growth phase of a sub-storm, an increasing amount of closed magnetic flux is removed from the dayside, which then

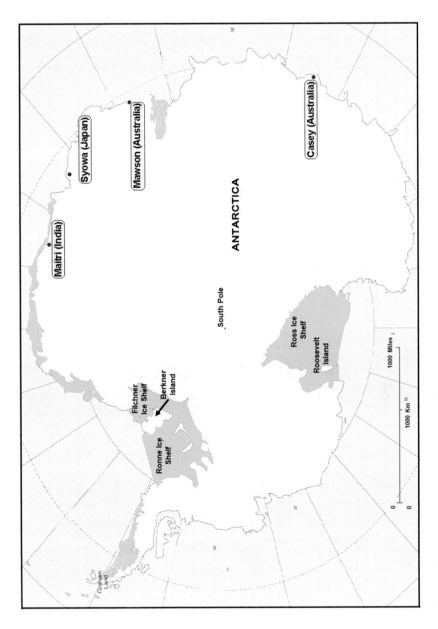

FIGURE 7.8 Geographic locations of considered Antarctic-based stations.

TABLE 7.1
Location Details of Considered Antarctic-Based Stations

GPS Stations		Geographic		Geomagnetic	
Station Name	Station Code	Latitude	Longitude	Latitude	Longitude
Maitri	MAI	70.65S	11.45E	67.16S	58.57E
Syowa	SYOG	69.006S	39.58E	70.33S	84.44E
Mawson	MAWI	67.604S	62.87E	73.09S	111.07E
Casey	CASI	66.28S	110.53E	76.22S	175.61W

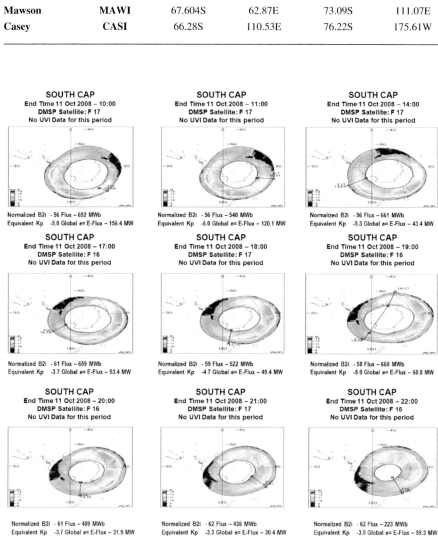

FIGURE 7.9   Movement of the auroral oval on the event day.

appears in the lobes as open flux and particles are energized at the sub-storm onset time (DeForest and McIlwain, 1971; McIlwain, 1974; Sauvaud and Winckler, 1980). These changes appear in the ionosphere as an increase in the size of the polar cap (Akasofu, 1968; Baker et al., 1994a). The ionospheric impact has been seen on the recorded ITEC values observed over four Antarctic-based GPS stations (Figure 7.10). The observation clearly shows that the first ionospheric impact has been observed over conventional auroral stations (i.e. Syowa and Mawson) in terms of enhancement in ITEC with a time lag of 1 hour. The ITEC enhancement at SYOWA (0900 UT) has led by 1 hour from the enhancement observed at MAWSON (1000 UT).

The enhancements in ITEC have been observed just after 0600 UT, which coincides with the enhancement in AE-index. Hence, it is evident that the southward turning of Bz-IMF at around 6 UT leads to the precipitation of energetic particles into the auroral region. However, no abnormal activity has been observed over Maitri, which is mainly because it falls outside the auroral boundary and comes under the sub-auroral region. This is evidence that the effect of particle precipitation has been confined to the auroral oval region only. The particle precipitation event

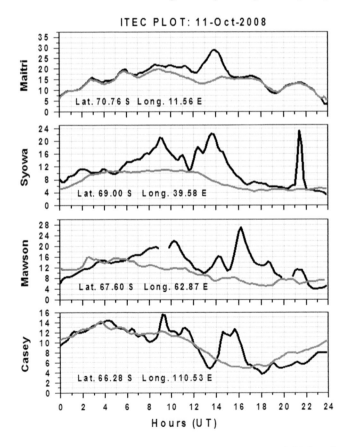

FIGURE 7.10 ITEC variations observed over four Antarctic-based GPS stations.

further causes perturbed polar region electrodynamics, which is reflected in terms of enhancement in the AE index parameter. The dominance of Auroral Electrojet (AE), especially the westward component of auroral electrojet, i.e. AL index is noticed during the event, which implies the expansion of the high-latitude plasma convection equator-ward. Due to this equator-ward expansion, MAITRI comes under the auroral oval region at around 10 UT, due to which an abrupt ITEC enhancement has also been observed over MAITRI along with Syowa and Mawson. The next or second-highest peak was observed at SYOWA at around 1330 UT, while the same is observed at around 1600 UT over MAWSON. Figure 7.11 shows the solar zenith angle variation at four mentioned GPS service stations, from which it can be seen that at around 1330 UT, it was 'afternoon' at SYOWA as well as at MAITRI. Therefore, the time of occurrence of elevated ITEC over Syowa and Maitri coincides with each other. However, the second ITEC enhancement over Mawson has been delayed by 02:30 hours, which is evident of formation and anti-sunward transportation of TOI (Tongue of Ionization). The phenomenon is explained well by Knudsen (1974). In general, the southward turning of IMF-Bz leads to drifting or transportation of solar-produced high-density plasma into the polar cap, which then convects anti-sunward, results in the formation of TOI (Tongue of Ionization) (Sato and Rourke, 1964). Figure 7.3 clearly shows that IMF-Bz was not stable for a while, rather it shows a baylike structure during the complete event, which allowed the TOI to be broken into polar cap patches. However, at CASEY, the ionospheric structures of enhanced TEC (2–3 times relative to quiet time ITEC) were observed during the event (shown in Figure 7.10). These structures can probably be attributed to polar cap patches. Hence, despite the low solar activity, it can be seen that the enhancement or fluctuations in the ITEC are a characteristic feature of the polar ionosphere.

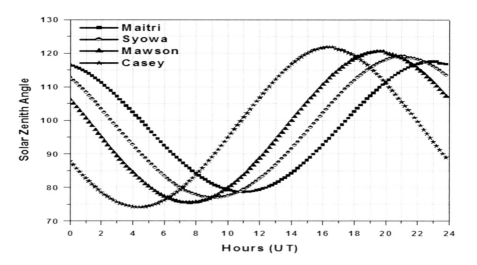

FIGURE 7.11   Solar zenith angle variation at four considered GPS service stations.

7.6 LATITUDINAL IONOSPHERIC RESPONSE TO THE CHANGING NEAR-EARTH SPACE ENVIRONMENT SYSTEM

In this section, an attempt has been made to explain the latitudinal ionospheric response to the changing near-Earth space environment system over the Southern Hemisphere. It is well known that the ionospheric impact of the changing near-Earth space-weather environment majorly depends upon latitudinal locations due to various factors, e.g. EIA, neutral winds, and so on. Until now, very few works reported the latitudinal (low to high) behavior of Earth's ionosphere for a single event. However, studies related to ionospheric response at specific latitudes have been well explained by various researchers (Kikuchi et al., 1978; Prölss and Jung, 1978; Kelley et al., 2004; Mannucci et al., 2005; Balan et al., 2009; Balan et al., 2010, 2011; Vijaya Lekshmi et al., 2011). In line with this, data from five GPS stations (Table 7.2) operated in the South American continent and Antarctic peninsula region have been analyzed to study the spatial and temporal behavior of the ionosphere over the Southern Hemisphere during the two geomagnetic storms (see Figures 7.12a, 7.12b, 7.12c, 7.12d, and 7.12e). The impact analysis has been done to explore the spatial distribution of ionization under the changing equatorial and high-latitudinal electrodynamic conditions.

The investigation is based on the spatial and temporal variation of Earth's ionosphere during two different geomagnetic storms observed on 7th and 9th March 2012. Given this, GPS-derived ITEC values of event days observed at different geographic locations over the South American continent and Antarctic Peninsula region have been utilized. The Dst index (Figure 7.13) shows that event that occurred on 9th March 2012 is more severe than that of the event observed on 7th March 2012.

TABLE 7.2
Position of the Selected GPS Service Stations

GPS Stations		Geographic		Geomagnetic		Remarks
Station Name	Station Code	Latitude	Longitude	Latitude	Longitude	
Fortaleza	BRFT	3.87S	38.42W	3.80N	33.65E	Equatorial region
Salvador	SSAI	12.97S	38.51W	5.25S	32.63E	Low-latitude region
Imbituba	IMBT	28.23S	48.65W	19.10S	21.80E	Low- to mid-latitude region
Bahia Blanca	VBCA	38.70S	62.62W	29.24S	8.85E	Mid-latitude region
Vernadsky	VNAD	65.24S	64.24W	55.45S	5.88E	High-latitude region

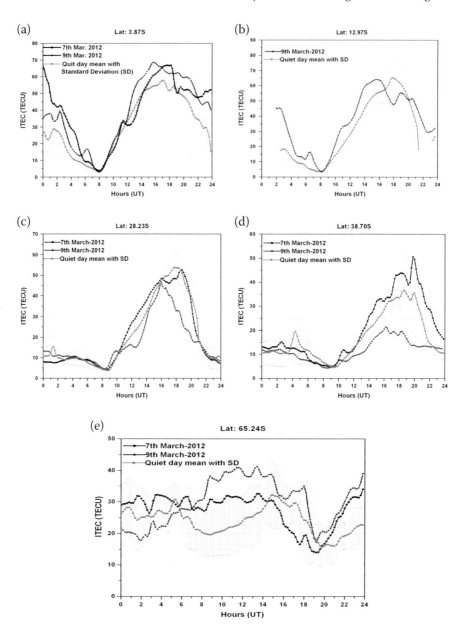

FIGURE 7.12 ITEC variation over different latitudes along with the five quiet days' mean and standard deviation.

The consequences are well observed on the Southern Hemispheric ionosphere according to the intensity of the geomagnetic storms. The results reveal that the combination of equatorial and high-latitudinal electrodynamic processes play important roles in the latitudinal distribution of ionized particles. Here, the considered cases provided an opportunity to have two different types of combinations, i.e. (1) Under

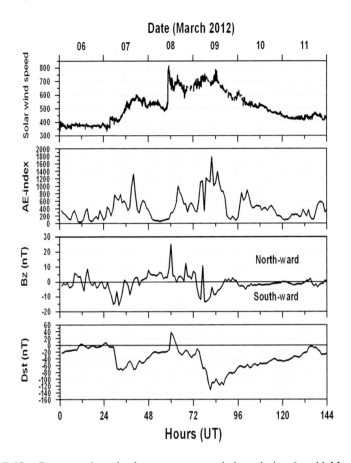

FIGURE 7.13 Geomagnetic and solar parameters variations during 6 to 11 Marsh 2012.

the significant presence of Equatorial Electrojet (EEJ) with moderately enhanced AE-index values (7th March 2012) and (2) under the presence of Counter Electrojet (CEJ) with enhanced AE-index (9th March 2012), as shown in Figure 7.14.

After evaluation of observed results for the first event (7th March 2012), it is clear that the storm time ionization distribution over the equatorial and low-latitude region has been severely affected by the storm time modulated strength of EEJ. The results reveal that a slight depression in EEJ causes a limit to the normal latitudinal expansion of EIA, which results in a positive ionospheric response at equatorial and near-equatorial regions along with negative impacts over a higher-lower latitude region. This is because the suppressed EEJ strength causes a weaker E × B drift velocity, which is unable to lift the plasma at higher altitudes, as also suggested by Mala S. Bagiya et al. (2009) and Sharma et al. (2011). Thus, the EIA crest is formed to the near-low latitudes where the geomagnetic field lines are mapped with equatorial altitudes at which the plasma is lifted due to the weaker upward E × B drift phenomenon. However, the simultaneous enhancement in AE-index suggested the possibility of modification of meridional wind direction, i.e. from poleward to

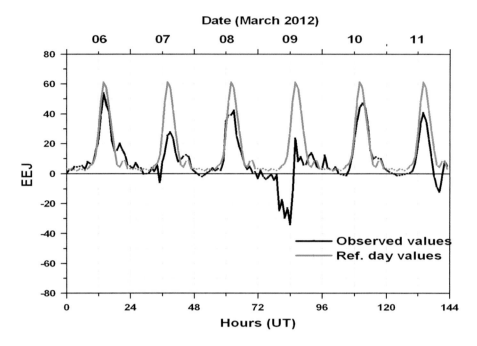

FIGURE 7.14   Equatorial electrojet variations during 6 to 11 March 2012.

equator-ward. The modified meridional wind direction not only responsible for plasma transportation from a high-latitude to lower-latitude region, but also responsible for limiting the poleward expansion of the EIA crest region. The combined effect of weaker EEJ and moderate AE-index leads to becoming favorable for equator-ward transportation of plasma, which not only limits the latitudinal expansion of EIA but is also responsible for the increase in ITEC at the mid-latitude region, as can be seen in Figure 7.12(d).

The enhancement in ITEC over the mid-latitude is manifested due to the modified equator-ward meridional wind, which effectively drives the transport of plasma from a higher latitude to a lower-latitude region, along with the sloppy geomagnetic field lines. It is well known that the impulsive high-energy inputs at high latitudes lead to trigger the traveling atmospheric disturbances (TADs), which travel equator-ward with high velocities (Bauske and Prolss, 1997) and drag the ionization along the inclined magnetic field lines from higher to lower latitudes. A similar result has been simulated by Lin et al. (2005), which shows that the storm-generated equator-ward neutral winds play an important role in producing the TEC enhancement at low- and mid-latitudes. The previously mentioned high-energy input at a high-latitude region, especially during the main phase of the storm, causes a positive storm effect and later decreases due to the convection processes caused by joule's effect, as can be seen in Figure 7.12(e).

However, a different spatial impact due to a severely affected equatorial electrodynamic process was observed 9th March 2012. The intense presence of ring current causes the development of a Counter Electrojet (CEJ) system in place of a normal EEJ, along with a simultaneous increase in AE-index values. The above combination provided a unique opportunity to study the response of the latitudinal distribution of ITEC under such circumstances. The enhanced ITEC over the equatorial region with simultaneous depressed ITEC at low- and low-mid-latitude regions is evidence that the CEJ causes a negative force at the equatorial region. The negative force is responsible for zero or negative upward drag of ionization at an equatorial region, which ultimately prevents the formation of EIA. The effect has been observed in terms of a positive storm effect at an equatorial region along with a negative impact over low- and low-mid-latitude regions. The above observation is not new and has already been reported by Balan et al. (1995) and Veenadhari et al. (2012). However, storm-induced equator-ward meridional winds caused a steep ionization enhancement at high-mid-latitudes and a negative response over the low-mid and mid-latitude regions. The depressed mid-latitude ITEC under the presence of CEJ and high AE-index claims the trans-equatorial plasma transportation due to a modified strong equator-ward meridional wind. The physical process is explained well by Abdu et al. (2007) and Balan et al. (2013). The results reported by Balan et al. (2013) suggested that the upwelling effect of the winds, together with the inward EBX drift (CEJ), causes severe negative ionospheric storms at higher latitudes, which makes the positive storms at equatorial latitudes. Abdu et al. (2007) explained that the effects of the long-lasting wind dynamo electric field during an intense storm dominates the low latitudes. These disturbance winds are due to the heating caused by the high energy triggering the convection process, which is responsible for equator-ward transportation of ionized particles. The observed enhanced ITEC, followed by depression (see Figure 7.12e) over the high-latitude region, supports the theory that particle precipitation initiated the high-latitude heating process followed by convection of ionization, which is propagated more towards the equator-ward direction.

Apart from the previously explained phenomenon, the South Atlantic Magnetic Anomaly also has a significant impact on ionospheric variability, mainly over the three considered stations (Station codes: SSAI, IMBT, and VBCA as detailed in Table 7.2), which comes under the said anomaly zone. Gledhill (1976) suggested that the particles having energy less than 2 keV will lose most of their energy in the F-region, those with energies between 2 and 40 keV in the E-region, and those with energies greater than 40 keV in the D-region. Also, Gledhill and Hoffman (1981) have reported 0.2–26 keV electron flux precipitating in the SAMA region capable of producing detectable effects in the D and E regions of the ionosphere during the nighttime. The mentioned impact has not been explored in this paper due to the lack of such low-energy flux and ground-based experimental data (especially Ionosonde, VLF, and Rio-meter) and the need for future investigation.

ACKNOWLEDGMENTS

The authors would like to thank the National Centre for Polar and Ocean Research, Ministry of Earth Sciences, Goa, India, for providing all required logistical support to conduct the experiments. Also, the authors would like to thank the Indian Institute of Geomagnetism, Mumbai, India, for providing H-component magnetic field data. The authors would also like to acknowledge the model developers of the global solar wind-magnetosphere-ionosphere coupling model "GUMICS" of CCMC (Community Coordinated Modeling Center). The authors also wish to acknowledge those who are involved in the maintenance of the data-providing site www.sonel.org/gps-html?lang=cn and providing the required data set for the above study. The authors also thank the Japanese World Data Center for providing the required geomagnetic data set.

REFERENCES

Abdu, M. A., Maruyama, T., Batista, I. S., Saito, S., and Nakamura, M. (2007). Ionospheric responses to the October 2003 superstorm: Longitude/local time effects over equatorial low and middle latitudes. *J. Geophys Res: Space Phys.*, 112, A10306. https://doi.org/10.1029/2006ja012228.

Akasofu, S. I. (1968). *Polar and Magnetospheric Substorms*, D. Reidel, Norwell, MA.

Bagiya, M. S., Joshi, H. P., Iyer, K. N., Aggarwal, M., Ravindran, S., and Pathan, B. M. (2009). TEC variations during low solar activity period (2005–2007) near the Equatorial Ionospheric Anomaly Crest region in India, *Ann. Geophys.*, 27, 1047–1057.

Baker, D. N., and McPherron, R. L. (1990). Extreme energetic particle decreases near the geostationary orbit: A manifestation of current diversion within the inner plasma sheet, *Geophys. J. Res.*, 95, 6591.

Baker, D. N., and Pulkkinen, T. I. (1991). The earthward edge of the plasma sheet in magnetospheric substorms, in *Magnetospheric Substorms*, Geophysical Monograph Series, Vol. 64, edited by J. R. Kan et al., AGU, Washington, DC, pp. 147–160.

Baker, D. N., Pulkkinen, T. I., Hones Jr., E. W., Belian, R. D., McPherron, R. L., and Angelopoulos, V. (1994a). Signatures of the substorm recovery phase at high-altitude spacecraft, *Geophys. J. Res.*, 99, 10967–10980.

Baker, D. N., Pulkkinen, T. I., Angelopoulos, V., Baumjohann, W., and McPherron, R. L. (1996). Neutral line model of substorms: Past results and present view, *Geophys. J, Res.*, 101(A6), 12975–13010.

Balan, N., and Bailey, G. J. (1995). Equatorial plasma fountain and its effects: Possibility of an additional layer, *J. Geophys. Res.*, 100, 21421.

Balan, N., Liu, J. Y., Otsuka, Y., Liu, H., and Lühr, H. (2011). New aspects of thermospheric and ionospheric storms revealed by CHAMP, *J. Geophys. Res.*, 116, A07305, https://doi.org/10.1029/2010JA016399.

Balan, N., Otsuka, Y., Nishioka, M., Liu, J. Y., and Bailey, G. J. (2013). Physical mechanisms of the ionospheric storms at equatorial and higher latitudes during the recovery phase of geomagnetic storms, *J. Geophys. Res. Space Physics*, 118, 2660–2669, https://doi.org/10.1002/jgra.50275.

Balan, N., Shiokawa, K., Otsuka, Y., Kikuchi, T., Vijaya Lekshmi, D., Kawamura, K., Yamamoto, M., and Bailey, G. J. (2010). A physical mechanism of positive ionospheric storms at low latitudes and midlatitudes, *J. Geophys. Res.*, 115, A02304, https://doi.org/10.1029/2009JA014515.

Balan, N., Shiokawa, K., Otsuka, Y., Watanabe, S., and Bailey, G. J. (2009). Super plasma fountain and equatorial ionization anomaly during penetration electric field, *J. Geophys. Res.*, 114, A03310, https://doi.org/10.1029/2008JA013768.

Bauske, R. and Prolss, G. W. (1997). Modelling the ionospheric response to traveling atmospheric disturbances, *J. Geophys. Res.*, 102, 14555–14562.

Behera, J. K., Sinha, A. K., Singh, A. K., et al. (2015). Substorm-related CNA near the equatorward boundary of the auroral oval in relation to interplanetary conditions, *Adv. Space Res*, 56, 28–37.

Burns, A., Killeen, T., Carignan, G., and Roble, R. (1995). Large enhancements in the O/N2 ratio in the evening sector of the winter hemisphere during geomagnetic storms, *J. Geophys. Res. Space Phys.*, 100, 14661–14671.

Burns, A., Killeen, T., and Roble, R. (1991). R.A theoretical study of thermospheric composition perturbations during an impulsive geomagnetic storm. *J. Geophys. Res. Space Phys.*, 96, 14153–14167.

DeForest, S. E., and McIlwain, C. E. (1971). Plasma clouds in the magnetosphere, *Geophys. J. Res.*, 76, 3587.

Dungey, J. W. (1961). Interplanetary magnetic field and the auroral zones, *Phys. Rev. Lett.*, 6, 47.

Eather, R. H., Mende, S. B., and Judge, R. J. R. (1976). Plasma injection at synchronous orbit and spatial and temporal auroral morphology, *Geophys. J. Res.*, 81, 2805.

Elphinstone, R. D., Heam, D., Murphree, J. S., and Cogger, L. L. (1990). Mapping using Tsyganenko long magnetospheric model and its relationship to Viking auroral images, *J. Geophys. Res. Space Phy*, 96, 1467–1480.

Fuller-Rowell, T., Codrescu, M., Moffett, R., and Quegan, S. (1994). Response of the thermosphere and ionosphere to geomagnetic storms, *J. Geophys. Res. Space Phys.*, 99, 3893–3914.

Galperin, Yu. I., and Feldstein, Ya. I. (1989). Diffuse auroral zone, X. Diffuse auroral zone, oval of discrete auroral forms, and diffuse luminosity polewards from the oval at the nightside as the projections of the magnetospheric tail plasma domains, *Cosmic Res., Engl. Transl.*, 27, 890.

Gledhill, J. A. (1976). Aeronomic effects of the South Atlantic anomaly, *Rev. Geophys.*, 14(2), 173–187, https://doi.org/10.1029/RG014i002p00173.

Gledhill, J. A., and Hoffman, R. A. (1981). Nighttime observations of 0.2- to 26-keV electrons in the South Atlantic Anomaly made by Atmosphere Explorer C, *J. Geophys. Res.*, 86(A8), 6739–6744, https://doi.org/10.1029/JA086iA08p06739.

Gonzalez, W. D., Joselyn, J. A., and Kamide, Y. (1994). What is a geomagnetic storm, *Geophys. J. Res.*, 99(A4), 5771–5792.

Heppner, J. P., Sugiura, M., Skillman, T. L., Ledley, B. G., and Campbell, M. (1967). OGO A magnetic field observations, *Geophys. J. Res.*, 724, 5417.

Kamide, Y., Yasuhara, F., and Akasofu, S.-I. (1974). On the cause of northward magnetic field along the negative X-axis during magnetospheric substorms, *Planetary and Space Science*, 22, 1219–1229.

Kelley, M. C., Vlasov, M. N., Foster, J. C., and Coster, A. J. (2004). A quantitative explanation for the phenomenon is known as storm-enhanced density, *Geophys. Res. Lett.*, 31, L19809, https://doi.org/10.1029/2004GL020875.

Kikuchi, T., Araki, T., Maeda, H., and Maekawa, K. (1978). Transmission of polar electric fields to the equator, *Nature*, 273, 650–651.

Knudsen, W. C. (1974). Magnetospheric convection and the high-latitude F-2 ionosphere, *Geophys. J. Res.*, 79, 1046–1055.

Lin, C. H., Richmond, A. D., Heelis, R. A., Bailey, G. J., Lu, G., Liu, J. Y., Yeh, H. C., and Su, S. Y. (2005). Theoretical study of the low- and mid-latitude ionospheric electron density enhancement during the October 2003 storm: Relative importance of the neutral wind and the electric field, *J. Geophys. Res.*, 110, A12312, https://doi.org/10.1029/2005JA011304.

Lopez, R. E., and Lui, A. T. Y. (1990). A multi-satellite case study of the expansion of a substorm current wedge in the near-Earth magnetotail, *Geophys. J. Res.*, 95, 8009–8017.

Lui, A. T. Y. (1991). Extended consideration of a synthesis model for magnetospheric substorms, in *Magnetospheric Substorm*, Geophysical Monograph Series, Vol. 64, edited by J. R. Kan, T. A. Potemra, S. Kokubun, and T. Ijima, AGU, Washington D.C., pp. 43–60.

Lui, A. T. Y., and Burrows, J. R. (1978). On the location of auroral arcs near substorm onsets, *Geophys. J. Res.*, 83, 3342–3348.

Lui, A. T. Y., Venkatesan, D., Anger, C. D., Akasofu, S. I., Heikkila, W. J., Winningham, J. D., and Burrows, J. R. (1977). Simultaneous observations of particle precipitations and auroral emissions by the Isis 2 satellite in the 19-24 MLT sector, *Geophys. J. Res.*, 82, 2210.

Mannucci, A. J., Tsurutani, B. T., Iijima, B. A., Komjathy, A., Saito, A., Gonzalez, W. D., Guarnieri, F. L., Kozyra, J. U., and Skoug, R. (2005). Dayside global ionospheric response to the major interplanetary events of October 29–30. 2003 "Halloween Storms", *Geophys. Res. Lett.*, 32, L12S02, https://doi.org/10.1029/2004GL021467.

Mayr, H., and Harris, I. (1978). Some characteristics of electric field momentum coupling with the neutral atmosphere. *J. Geophys. Res. Space Phys*, 83, 3327–3336.

McIlwain, C. E. (1974). Substorm injection boundaries, in *Magnetospheric Physics*, edited by B. M. McCormac, D. Reidel, Dordrecht, Netherlands, pp. 143–154.

Prölss, G. W., and Jung, M. J. (1978). Traveling atmospheric disturbances as a possible explanation for daytime positive storm effects of moderate duration at middle latitudes, *J. Atmos. Terr. Phys.*, 40, 1351–1354.

Rothwell, P. L., Block, L. P., Silevitch, M. B., Fälthammar, C -G et al.(1988). A new model for substorm onsets: The pre-breakup and triggering regimes, *Geophys. J. Res. Left*, 15, 1279–1282.

Ruohoniemi, J., and Greenwald, R. (1998). The response of high-latitude convection to a sudden southward IMF turning. *Geophys. Res. Lett.*, 25, 2913–2916.

Sato, T., and Rourke, G. (1964). F-region enhancements in the Antarctic, *Geophys. J. Res.*, 69, 4591–4607.

Sauvaud, J.-A., and Winckler, J. R. (1980). Dynamics of plasma, energetic particles, and fields near synchronous orbit in the nighttime sector during magnetospheric substorms, *Geophys. J. Res.*, 85, 2043–2056.

Sergeev, V. A., Pulkkinen, T. I., and Pellinen, R. J. (1996b). Coupled-mode scenario for the magnetospheric dynamics, *Geophys. J. Res.*, 101, 13047.

Sharma, S., Galav, P., Dashora, N., Alex, S., Dabas, R. S., and Pandey, R. (2011). Response of low-latitude ionospheric total electron content to the geomagnetic storm of 24 August 2005, *J. Geophys. Res.*, 116, A05317, https://doi.org/10.1029/2010JA016368.

Singh, A. K., Saini, S. and Das, R. M. (2019). Impact of geomagnetic variation over the sub-auroral ionospheric region during high solar activity year 2014, *Adv. Space Res.*, https://doi.org/10.1016/j.asr.2019.01.050.

Siscoe, G. L., and Cummings, W. D. (1969). On the cause of geomagnetic bays, *Planet. Space Sci.*, 17, 1795–1802.

Tanskanen, P., Kangas, J., Block, L., Kremser, G., Korth, A., Woch, J., Iversen, I. B., Torkar, K. M., Riedler, W., Ullaland, S., Stadnes, J., and Glassmeier, K.-H. (1987). Different phases of a magnetospheric substorm on June 23, 1979, *J. Geophys. Res.*, 92, 7443–7457.

Tanskanen, E., Pulkkinen, T. I., and Koskinen, H. E. J. (2002). Substorm energy budget during low and high solar activity: 1997 and 1999 compared, *J. Geophys. Res.*, 107(A6), 1086, https://doi.org/10.1029/2001JA900153.

Veenadhari B., Kumar, S., Tulasi Ram, S., Singh, R., and Alex, S. (2012). Corotating interaction region (CIR) induced magnetic storms during solar minimum and their effects on low-latitude geomagnetic field and ionosphere, *Indian J Radio & Space Phys*, 41, 302–315.

Vijaya Lekshmi, D., Balan, N., TulasiRam, S., and Liu, J. Y. (2011). Statistics of geo-magnetic storms, and ionospheric storms at low and mid-latitudes in two solar cycles, *J. Geophys. Res.*, 116, A11328, https://doi.org/10.1029/2011JA017042.

8 Intriguing Relationship between Antarctic Sea Ice, ENSO, and Indian Summer Monsoon

Amita Prabhu, Sujata K. Mandke, R.H. Kripalani and G. Pandithurai
Indian Institute of Tropical Meteorology, Pune, India

CONTENTS

8.1 INTRODUCTION

Indian summer monsoon rainfall (ISMR) from June through September (JJAS) accounts for a major portion (nearly 80%) of its annual rainfall. The economic strength of the country largely depends on the fluctuations of monsoons (Mooley & Parthasarathy, 1982) and on agriculture (Parthasarathy et al., 1992; Gadgil, 2003). Hence, the summer monsoon rainfall over India is treated as an important phenomenon, wherein researchers are actively involved in understanding the potential role of external factors like the cryosphere, ocean, and atmosphere that are

DOI: 10.1201/9781003203742-8

intricately linked with each other (Shukla, 1998; Rajeevan et al., 2007; Prabhu et al., 2018; among many others). Some of the dominant external factors over the tropical region influencing ISMR include two important tropical modes: Indian Ocean Dipole Mode (IODM: Saji et al., 1999; Webster et al., 1999) and El Niño Southern Oscillation (ENSO: Walker, 1923). Recent studies have examined the combined influence of IODM and ENSO on the ISMR (Ashok et al., 2004; Krishnan et al., 2011; Rajagopalan & Molnar, 2012; Krishnaswamy et al., 2014; among many others). In addition, Eurasian and Himalayan snow cover, as suggested by Blanford a century ago over the extra-tropical belt in the Northern Hemisphere, play a crucial role in the Indian monsoon dynamics (Kripalani & Kulkarni, 1999; Bamzai and Shukla, 1999; Dash et al., 2005; Peings and Douville, 2010; and several others). While Prabhu et al. (2016) connected delayed impact of February–March Southern Annular Mode (SAM) on the Indian summer monsoon through the central Pacific Ocean, Liu et al. (2018) suggested the May SAM impact on the South China Sea summer monsoon was transmitted through simultaneously occurring South Pacific Ocean sea surface temperature (SST) variability. Further, they assessed South-North Pacific teleconnection pattern through energy dispersion of a stationary Rossby wave located at the southern tip of this teleconnection that has wave ray trajectories mainly traveling northward into the Southern Hemispheric subtropical Pacific region, which then turn toward the northwest Asian monsoon domain. However, an understanding of such external driving factors from polar regions influencing summer monsoon rainfall over India are limited and still need considerable exploration.

Recent studies have focused on the potential linkages between Arctic sea-ice variability with that of Asian summer monsoon components (Prabhu et al., 2012; Prabhu et al., 2018). Prabhu et al. (2018) revealed two channels linking Greenland's sea ice with summer monsoon rainfall over South Asia (in particular, India) and East Asia (in particular, South Korea). While the channel relaying the Greenland signal towards the Indian monsoon was observed to be routed through the equatorial central Pacific Ocean, the channel linking the former with the Korean monsoon was seen traversing through the Eastern Eurasian snow region. Additionally, a recent review (Yuan et al., 2018) summarizes the advances on tropical-polar inter-annual tele-connections elucidating mainly the role of ENSO phenomenon. However, this review misses an important component of the tropics, the Asian summer monsoon. Hence, the present study seeks to systematically assess the role of the Antarctic sea-ice variability in concurrence with ENSO towards modulation of ISMR.

Interaction of Antarctic sea-ice variability with regional and global processes on different time scales is demonstrated in earlier studies (Yuan & Martinson, 2000; Carleton, 2003; Nuncio & Yuan, 2015; Li et al., 2015; Ludescher et al., 2015; Yuan et al., 2018). Important tropical phenomena, such as ENSO and IODM, have been associated with sea-ice processes in the Southern Ocean surrounding the Antarctic continent (Ledley & Huang, 1997; Liu et al., 2002; Morioka et al., 2017), along with the climate variability over Indo-Pacific region (Ummenhofer et al., 2013; Izumo et al., 2014). Previous studies, based on both observational evidence and numerical experiments, have indicated the role of Antarctic climate in tropical summer monsoon systems (Yang and Huang, 1992; Chen et al., 1996; Xue et al.,

2003; Dugam & Kakade, 2004; Prabhu et al., 2009; Chen et al., 2016; Oza et al., 2017). An eastward propagating planetary wave surrounding the Southern Ocean, known as the Antarctic circumpolar wave, is evident in the oceanic and atmospheric anomalies such as SST, sea level pressure (SLP), and zonal winds (White & Peterson, 1996). A previous study by Peterson and White (1998) indicates a linkage of ENSO with this high-latitude variability. Antarctic sea-ice variations play a vital role in the Southern Hemispheric high-latitude modes (Raphael et al., 2011). In particular, Antarctic sea-ice extent has been linked with the Indian monsoon through the Antarctic circumpolar wave (Prabhu et al., 2010). This study indicated an association between sea-ice variability over the Bellingshausen and Amundsen Seas (BAS) sector during Austral summer (October-December) with that of ensuing Indian summer monsoon. The tele-connection has been channelled through the Pacific longitudes of the Southern Ocean during the boreal autumn that prevails up to the following spring, wherein the signals were relayed through SST and meridional transport of heat variations. Nonetheless, physical mechanisms associated with these linkages are still unclear.

In light of the scientific background, the main objectives of this study are as follows:

i. To examine lead-lag as well as simultaneous relationship between various Antarctic sea ice and ENSO indices with ISMR;

ii. To investigate regional aspects of rainfall over India with respect to its relation with sea ice over BAS and WPO sectors along with ENSO on a monthly as well as a seasonal timescale;

iii. Further, to explore whether the role of central (western) equatorial Pacific SST on relaying the BAS (WPO) sea-ice signal towards the regional aspects of summer monsoon rainfall over India is independent or combined.

iv. Finally, a possible physical mechanism connecting Antarctic sea ice–ENSO–Indian summer monsoon would be assessed.

This chapter is comprised of the following sections: A list of data sets used and methodologies employed are mentioned in Section 2. Trends and tele-connections of Antarctic sea ice with ISMR and ENSO are shown in Section 3. Section 4 elucidates the mechanism for connection between southern polar sea ice and ENSO. Further, Section 5 describes physical approach to Antarctic sea ice–ENSO–ISMR link. Furthermore, regional perspectives of summer monsoon rainfall over India, sea ice, and the ENSO connection are discussed in Section 6. Finally, the last section provides major conclusions and new insights on these tele-connections.

8.2   DATA AND METHODOLOGY

The high-resolution gridded reanalyses together with observed data sets are utilized in this study as listed below:

i. The sea ice area (SIA), over five different sectors of Antarctica (Figure 8.1) and over the total Antarctic (sum of SIA over all these

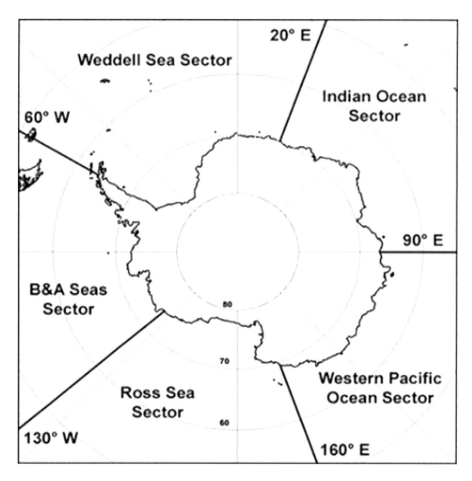

FIGURE 8.1 Location map of five sectors of Antarctica with longitudinal boundaries of the sectors shown in dark solid lines as obtained from NSIDC.

sectors), is obtained from passive microwave sensors on board Nimbus-7 and Defense Meteorological Satellite Program (DMSP) series of satellites (Cavalieri et al., 1999). This data set has been archived from the National Snow and Ice Data Center (NSIDC) for the period 1983–2015.

ii. The monthly ($1° \times 1°$) SST data set (Reynolds et al., 2002) has been obtained from the National Oceanic and Atmospheric Administration.

iii. The ISMR series is constructed from 306 rain gauge stations that are uniformly spread across the country from June to September (JJAS). Area-weighted rainfall series is an appropriate measure depicting strength of the monsoon. This data set, which is obtained from Indian Institute of Tropical Meteorology's data repository (www.tropmet.res.in), is available since 1871 and updated regularly. The above-mentioned rainfall series has been widely used to understand the summer monsoon variability by the international scientific community. In the present study,

ISMR data has been used for the period 1983–2015. Further, for this 33-year period, the ISMR series has a mean and standard deviation of 827 mm and 80 mm, respectively.

iv. Further, a spatially gridded (25 km × 25 km) rainfall data set is developed from a finely distributed set-up of 6,955 rain gauges spread over India. This data set has been acquired from the India Meteorological Department's National Climate Centre (IMD-NCC) (Pai et al., 2014) for the period 1983–2015. Furthermore, for the current study, IMD-NCC rainfall data has been extensively used due to its fine capability to demonstrate the spatial distribution of summer monsoon rainfall over the Indian region during the extreme phases of Antarctic sectors.

v. The changes in the atmospheric circulations during SIA extremes are observed using the National Centers for Environmental Prediction (NCEP) and the National Center for Atmospheric Research (NCAR) monthly mean reanalysis data sets (Kalnay et al., 1996).

vi. The analysis conducted in this chapter is based on the period from 1983 to 2015, which is a common available data length across multiple data sets used. An average over the months June to September (JJAS) is hereafter referred to as the summer monsoon season.

vii. To investigate the polar-tropical linkages, we have widely used statistical tools like correlation and composite analysis. The relationship between the variables is likely to be magnified by significant trends in data series. Hence, before carrying out the relationship between any two entities, trends of each of the inter-annual time series are tested for the period 1983–2015 using a standard F-test statistic that examines the null hypothesis of zero slope (Kendall & Stuart, 1979). Additionally, the correlation coefficients (cc) that are computed to examine the relationship between the two series are tested for significance at a 95% confidence level using a student's t-test statistic (Kendall & Stuart, 1979). The statistical method of composites or means has been employed to demonstrate key patterns in ocean and atmosphere variations both qualitatively and quantitatively. Also, the composite and correlation analyses have been shown to be effective tools in revealing spatial signatures of climate associated with sea-ice variability at high latitudes (Welhouse & Lazzara, 2016).

8.3   ANTARCTIC SEA ICE

Changes in polar sea ice have large-scale effects on the global climate (Baines & Cai, 2000). In recent decades, strong climatic changes occurred over the west Antarctic region, including the Antarctic Peninsula warming, sea-ice redistribution, and accelerated land-ice melting (Li et al., 2015 and references therein). In this study, SIA over five different sectors of the southern polar region (Figure 8.1) are used for examining its possible lead-lag as well as concurrent relation with the summer monsoon rainfall (JJAS) over India and ENSO. Sector-wise definitions for developing SIA time-series for the period 1983–2015 have been adopted following Cavalieri et al. (1999). SIA is estimated for Total Antarctica (TA) and its five sectors: (1) Indian Ocean (IO), (2) Western Pacific Ocean (WPO), (3) Ross Sea (RS), (4) Bellingshausen and Amundsen Seas (BAS), and (5) Weddell Sea (WS).

8.3.1 TREND AND ANNUAL CYCLE

Monitoring of long-term sea-ice trends have become a possibility with availability of passive microwave radiometer observations for the last three decades. Plots of monthly averaged annual cycle and trends of JJAS SIA for the Antarctic sectors based on a 33-year period from 1983 to 2015 are shown in Figures 8.2 and 8.3, respectively. All the sectors including the total Antarctic display SIA maxima in the month of September and minima in the month of February. It is interesting to observe that BAS, WPO, and RS sectors have the least rate of growth of SIA during the months of June through October (Figure 8.2). All the sectors, including the total Antarctic, show an increasing trend during the period 1983–2015 (Figure 8.3). However, except for the total Antarctic and the RS sector, none of the other sectors indicate a significant trend.

8.3.2 TELECONNECTIONS

The microwave satellite data sets have been instrumental for carrying out research on polar teleconnection (Gloersen & Campbell, 1991; Parkinson, 2004). Next, the Southern Hemispheric sea-ice linkage with ISMR, a Northern Hemispheric tropical phenomenon, is investigated.

8.3.2.1   Relation: Sea Ice and ISMR

Association between monthly Antarctic sea-ice variability over all its sectors with that of the Indian summer monsoon is examined. A simple linear correlation

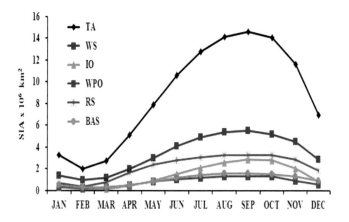

FIGURE 8.2   Annual cycle of SIA based on the period 1983–2015 over Total Antarctic (TA) along with its five sectors: Weddell Sea (WS), Indian Ocean (IO), Western Pacific Ocean (WPO), Ross Sea (RS), and Bellingshausen and Amundsen Seas (BAS) *(Source: Fig. 8.2 of Prabhu et al. 2021)*.

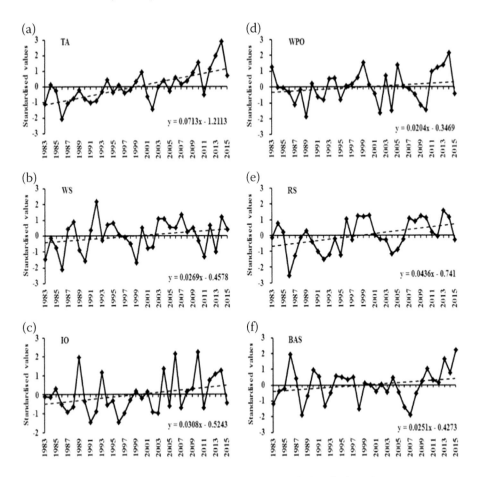

FIGURE 8.3 Inter-annual variability of normalized SIA during the boreal summer monsoon (JJAS) for the period 1983–2015 over (a) Total Antarctic (TA), (b) Weddell Sea (WS), (c) Indian Ocean (IO), (d) Western Pacific Ocean (WPO), (e) Ross Sea (RS), and (f) Bellingshausen and Amundsen Seas (BAS). Dashed lines indicate linear trends *(Source: Fig. 8.3 of Prabhu et al. 2021).*

analysis is carried out between the normalized time-series of raw SIA and ISMR using a monthly mean of SIA for total Antarctic and its five sectors during the period 1983–2015 (Figure 8.4a). The correlation coefficient (cc) ≥ 0.35 is statistically significant at a 95% confidence level for a sample of size 30. This analysis reveals significant correlation of opposite sign for WPO and BAS sectors (Figure 8.4a), which has a maximum for the month of July. A robust positive (negative) relationship with cc = 0.44 (cc = −0.39) is observed between July SIA over WPO (BAS) and ISMR. Correlation between the JJAS averaged SIA over these two sectors and ISMR are also opposite in sign but weaker in magnitude compared to the corresponding July counterpart (Figure 8.4a). Further, correlation between the total Antarctic SIA with that of ISMR is insignificant, perhaps due to a contrasting nature of highly significant SIA-ISMR correlation for the two sectors, as

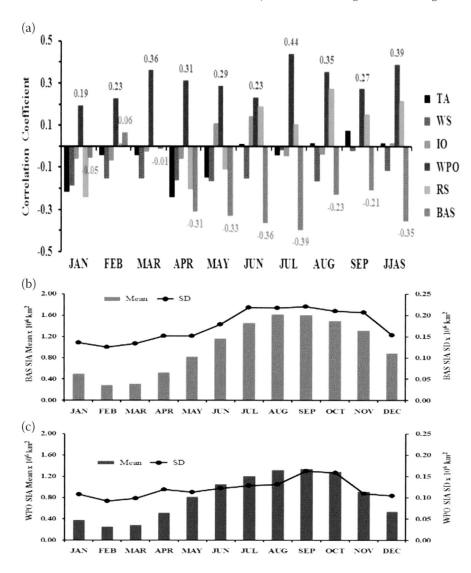

FIGURE 8.4 (a) Correlation coefficient of monthly mean SIA over Total Antarctic along with its five sectors illustrated separately, from January through September and also for JJAS season, with ISMR based on the period 1983–2015 *(Source: Fig. 4 of Prabhu et al. 2021)*; (b) mean and standard deviation (SD) of BAS SIA; (c) same as (b), except for WPO SIA *(Source: Fig. 5 of Prabhu et al. 2021)*.

mentioned previously. The raw time series each of SIA over BAS and WPO sectors along with ISMR is used for analysis hereafter due to their insignificant trends during the period considered. Results do not change even though the trend is removed from the relevant series. Thus, on a large scale, the SIA anomalies over WPO and BAS appear to be organized in a coherent pattern and assume opposite polarity with respect to ISMR.

As noted in the annual cycles (Figure 8.2), the monthly mean of SIA for both of these sectors, which are significantly correlated with ISMR, indicate the least growth. Hence, we need to examine if there exists any inter-annual variability. This is determined by computing the standard deviation and the coefficient of variation (CV = standard deviation/mean) of the SIA for each month based on a 33-year period. The mean (histograms) and standard deviation (solid lines) of SIA are diagrammatically displayed in Figure 8.4b for the BAS sector and in Figure 8.4c for the WPO sector. SIA variability (CV × 100) for each month over both of these sectors range from about 10 to 30%. It can be inferred that SIA variability during the JJAS is about 12.5% over both of these sectors.

In the Southern Hemisphere, WPO and BAS sea-ice sectors are aligned meridionally along the western and central Pacific Ocean, respectively. The ENSO phenomenon has been linked to the climate over the Antarctica Peninsula recently (Yuan et al., 2018 and references therein). Antarctic sea-ice fields linearly covary with the tropical Pacific ENSO phenomenon (Yuan & Martinson, 2000). The southern oscillation has shown a strong relation with sea-ice anomalies over the BAS and WS sectors of the Southern Ocean (Kwok & Comiso, 2002). On the other hand, different flavors of ENSO and its subsequent impact on the Indian summer monsoon have been the focus of active research since it was first observed by Sir Gilbert Walker. In general, an above (below) normal rainfall is observed over India during La Niña (El Niño) events (Pant & Parthasarathy, 1981; Mooley & Parthasarathy, 1983; and several others). El Niño Modoki (Ashok et al., 2007) or the dateline El Niños (Larkin & Harrison, 2005), which are associated with the central equatorial Pacific warming, occurred frequently after the 1980s, has a significant effect on the summer monsoon activity over the Indian continent (Prabhu et al., 2016).

The results of the correlation analysis presented previously leads to two pertinent questions: (1) Does SIA variability over BAS and WPO, which have demonstrated robust connection with ISMR are also linked with ENSO, which is widely known as a major driver of ISMR? (2) What is the underlying physical mechanism connecting all of the above factors involved in modulating the Indian summer monsoon?

8.3.2.2   Relation: Sea Ice and ENSO

To answer the previous questions, we first investigate the relationship between SIA over two important sectors of Antarctica, namely BAS and WPO, with that of ENSO. As noted previously, SIA displays appreciable variability (about 12.5%) over these two sectors. Post-1980s, anomalous warming of SSTs over the equatorial central Pacific has become a common feature (Ashok et al., 2007; Kug et al., 2010) compared to SST warming over the eastern Pacific (Wallace et al., 1998). An increase in atmospheric greenhouse gases is identified as a cause for change in the centers of action from eastern to central Pacific (Yeh et al., 2009). Prabhu et al. (2016) demonstrated a significant inverse relationship between February–March Southern Annular Mode (SAM) with that of ensuing El Niño Modoki (JJAS) for a recent warming period from 1983 to 2013. SAM, which is a high-latitude mode marked by a sub-polar low (~70°S) and mid-latitude high (~40°S), constitutes a dominant mode of atmospheric variability in the Southern Hemisphere (Thompson

& Wallace, 2000). Earlier studies have shown the influence of coupling between various phases of SAM and ENSO on the climate of the Antarctic Peninsula (Goodwin & Thompson, 2016; Clem et al., 2016). The entity El Niño Modoki, which depicts central Pacific anomalous warming flanked by anomalously cool SSTs over the western and eastern Pacific, is linked to the Australian high, the eastern South Pacific subtropical high and annular modes in the Southern Hemisphere (Weng et al., 2007; Prabhu et al., 2016). The El Niño Modoki Index (EMI) is constructed following Ashok et al. (2007), which is a mathematical equation showing anomalous equatorial central Pacific warming associated with cooling over the eastern and western ends of equatorial Pacific Ocean basin. Since the variability of El Niño Modoki entity suits both regions of interest over the Pacific, namely the central and western Pacific, it is correlated with the longitudinally collocated sectors of Antarctica, namely, SIA over the BAS and WPO sectors, respectively, as both of these sectors have a significant relationship with ISMR (Figure 8.4a).

The lead-lag correlation between standardized EMI averaged during JJAS (EMI-JJAS) with monthly standardized SIA over BAS and WPO sea-ice sectors from January of the previous year through to December of the subsequent year for the period 1983–2015 is shown in Figure 8.5a. The most striking result of Figure 8.5a is a highly significant direct (inverse) relationship of July SIA over BAS (WPO) with JJAS EMI of the concurrent year. A positive (negative) relation of July SIA over BAS (WPO) with JJAS EMI of the same year (Figure 8.5a) could be due to the Rossby wave train emanating from the tropics (Yuan, 2004; Cai et al., 2011). The maximum relation of July SIA with JJAS EMI suggests a robust linkage between the tropics and the polar regions. A similar lead-lag correlation between standardized JJAS SIA over BAS and WPO sectors separately with monthly standardized EMI from January of the previous year through to December of the following year is also delineated (Figure 8.5b). The lead-lag correlation of monthly EMI with JJAS WPO SIA tends to be negative continuously from the previous year to the ensuing year, although it is significant only with the EMI of the concurrent June. The lead-lag relation of the monthly EMI with JJAS BAS SIA changes the sign from negative in the previous year to positive in the concurrent year and again negative in the following year (Figure 8.5b), but cc is significant only with June EMI of previous year and October–November of concurrent year. The lead-lag relation of BAS and WPO SIA with EMI (Figures 8.5a, b) probably suggests a two-way interaction between the tropics and polar regions. Furthermore, it is likely that EMI and SIA could influence ISMR independently or through a combined effect of the two. This feature is ascertained through partial correlation analysis (Table 8.1). The main inferences drawn from this table are as follows:

a. For the SIA (JJAS)-ISMR connection, the partial correlation between BAS SIA (JJAS) and ISMR (–0.35) becomes insignificant (–0.26) on removing the EMI (JJAS) effect (Table 8.1, Set No. I). However, for the WPO sector, partial correlation between WPO SIA (JJAS) and ISMR (0.39) weakens (0.34) on removing EMI (JJAS) effect (Table 8.1, Set No. II), implying the BAS (WPO) SIA relation with ISMR to be completely

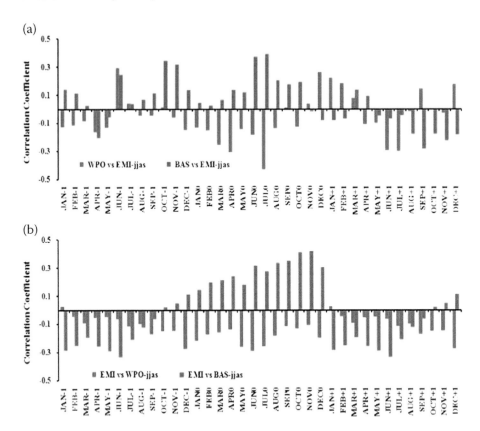

FIGURE 8.5 (a) Lead-lag correlations between standardized EMI of JJAS (EMI-JJAS) and monthly standardized SIA averaged over BAS and WPO sectors, for 36 months from Jan-1 through to Dec+1 based on the period 1983–2015; (b) same as (a), but for EMI and SIA during JJAS season over BAS (BAS-JJAS) and WPO (WPO-JJAS); notation "0" as used in x-axis label denotes the reference year, while -1 and +1 refer to the year before and after the reference year.

    (moderately) dependent on EMI (JJAS). Likewise, for the EMI (JJAS)-ISMR relation, eliminating the respective influences of SIA (JJAS) over BAS (WPO) (Table 8.1, Set No. I, SET No. II) indicate complete (moderate) dependence of EMI on SIA over BAS (WPO). In other words, sea-ice variability over BAS and WPO sectors modulate summer monsoon rainfall over India in association with equatorial Pacific SSTs.

  b. Further, the partial correlations of both BAS SIA (July) with ISMR and WPO SIA (July) with ISMR decreases on eliminating EMI (July) effect (Table 8.1, Set Nos. III and IV). These results clearly suggest that the relationship of SIA (July) over both BAS and WPO with ISMR strengthen in combination with equatorial Pacific SST for the month of July.

  c. The partial correlation of EMI (July) with ISMR becomes insignificant on removing the respective influences of SIA (July) over both BAS and WPO

TABLE 8.1

Correlation and Partial Correlation Analysis Carried Out for the Period 1983–2015

Set No.	Sr. No.	Correlation Between	Correlation Coefficient	Statistical Significance
I	1.	BAS SIA (JJAS) and ISMR	−0.35	95% CL
	2.	BAS SIA (JJAS) and EMI (JJAS)	0.33	90% CL
	3.	EMI (JJAS) and ISMR	−0.38	95% CL
	4.	BAS SIA (JJAS) and ISMR removing EMI (JJAS) effect	−0.26	Insignificant
	5.	EMI (JJAS) and ISMR removing BAS SIA (JJAS) effect	−0.28	Insignificant
II	1.	WPO SIA (JJAS) and ISMR	0.39	95% CL
	2.	WPO SIA (JJAS) and EMI (JJAS)	−0.30	90% CL
	3.	EMI (JJAS) and ISMR	−0.38	95% CL
	4.	WPO SIA (JJAS) and ISMR removing EMI (JJAS) effect	0.34	95% CL
	5.	EMI (JJAS) and ISMR removing WPO SIA (JJAS) effect	−0.31	90% CL
III	1.	BAS SIA (July) and ISMR	−0.39	95% CL
	2.	BAS SIA (July) and EMI (July)	0.35	95% CL
	3.	EMI (July) and ISMR	−0.37	95% CL
	4.	BAS SIA (July) and ISMR removing EMI (July) effect	−0.30	90% CL
	5.	EMI (July) and ISMR removing BAS SIA (July) effect	−0.27	Insignificant
IV	1.	WPO SIA (July) and ISMR	0.44	95% CL
	2.	WPO SIA (July) and EMI (July)	−0.48	95% CL
	3.	EMI (July) and ISMR	−0.37	95% CL
	4.	WPO SIA (July) and ISMR removing EMI (July) effect	0.32	90% CL
	5.	EMI (July) and ISMR removing WPO SIA (July) effect	−0.21	Insignificant

sectors, suggesting a dominant impact of July SIA on ISMR with the support of equatorial Pacific SSTs (Table 8.1, Set Nos. III and IV).

d. Additional details of the partial correlation coefficients can be inferred from Table 8.1.

In short, the main summary of this partial correlation analysis clearly suggests a possible two-way interaction between Antarctic (BAS and WPO sectors) SIA and equatorial Pacific SSTs. Such two-way interactions have been reported earlier. A tropical-polar teleconnection (Flatau & Kim, 2013) on an intra-seasonal timescale is revealed, where the Madden Julian Oscillation (MJO) appears to force annular modes in both hemispheres. However, they inferred based on longer timescales that a persistent Arctic oscillation/Antarctic oscillation anomaly appears to influence the convection in the tropical belt and impact the distribution of MJO preferred phases. Further, they noted that in the Southern Hemisphere, the SST anomalies are to some extent also related to a persistent Antarctic oscillation pattern. Besides, a negative mode of SAM resembles a tropically forced Rossby wave train over the longitudes of the Pacific Ocean sector, having significant correlations with central Pacific SSTs during the boreal summer (Ding et al., 2012).

Next, the time-evolution of cc between EMI from January through September and the ISMR is made (Figure 8.6a). The correlation between the two normalized series for the period 1983–2015 builds up from the month of June and persists for the entire boreal summer with a significant inverse relationship (Figure 8.6a). The spatial characteristics of correlation between the summer monsoon (JJAS) averaged rainfall at each grid point over Indian land and monthly EMI separately for June to September along with the JJAS season are illustrated (Figure 8.6b–f). The spatial pattern demonstrates a significant inverse correlation over parts of north India and

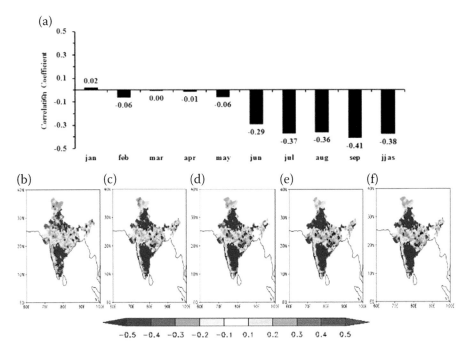

FIGURE 8.6 (a) Evolution of correlation coefficient between standardized EMI from January through September and also for JJAS season, with standardized ISMR for the period 1983–2015 (*Source: Fig. 6(c) of Prabhu et al. 2021*). (b–f) Concurrent correlation coefficient (shaded) between standardized EMI and standardized IMD rainfall at each grid point over the Indian region during June, July, August, and September and also for JJAS season, respectively. Black solid (dotted) contours represent significance at 95% level of confidence for positive (negative) correlations.

peninsular region for EMI from June through September (Figure 8.6b–f). The spatial plots also highlight building up of an inverse correlation pattern between the two from June through September. Having examined the tele-connection between Antarctic sea ice, ENSO, and ISMR, a possible physical mechanism associated with this tele-connection is subsequently investigated.

8.4   PHYSICAL MECHANISM: SEA ICE AND ENSO CONNECTION

To assess an imprint of sea-ice variability on the surface atmospheric condition, composite analysis is performed by identifying the ten strongest cases of positive and negative SIA extremes over both BAS and WPO regions. The composite analysis is conducted only for WPO and BAS sectors as SIA over these two sectors are significantly correlated with ISMR (Figure 8.4a). A composite technique is an effective tool for analyzing and inferring possible impacts. The years with a standardized SIA above (below) zero are categorized as positive (negative) SIA events. Accordingly, for the period 1983–2015, the ten strongest positive and negative SIA extremes for BAS as well as WPO sectors are displayed in Table 8.2 (set I and II of upper and lower panels). The years, which match the criteria as mentioned previously, have been arranged in a descending order of standardized SIA magnitude averaged over the respective regions of BAS and WPO.

As evident from the results of partial correlation analysis discussed in the previous section, the presence of co-occurring SIA and equatorial Pacific SST extremes indicate that both of these entities may be linked, thereby influencing large-scale circulations. To comprehend the impact of central ($170°W–120°W$, $5°S$-$5°N$) and western ($100°E$-$150°E$, $5°S$-$5°N$) equatorial Pacific SST variability on the surface atmospheric condition using composite analysis, the ten strongest cases of positive and negative SST extremes averaged over both of these regions are considered. The years with standardized values of SST above (below) zero are categorized as positive (negative) SST events. Accordingly, for the period 1983–2015, the ten strongest episodes each of positive and negative equatorial central Pacific SST as well as equatorial western Pacific SST are listed in Table 8.2 (upper and lower panels of set III and upper and lower panels of set IV, respectively). The years chosen as per the defined criteria have been arranged in a descending order of SST magnitude averaged over the respective regions of central and western equatorial Pacific (Table 8.2).

The characteristics of spatial distribution estimated from composites of (positive-negative) extremes can provide useful insights about the possible linkages over remote regions. For this purpose, we depict (Figure 8.7) spatial distribution of SLP over the Southern Hemisphere computed using ten (positive-negative) extremes of both equatorial Pacific SST and Antarctic sea-ice episodes separately (as mentioned in Table 8.2). Significant signatures of sub-polar high and mid-latitude low locked over the central Pacific longitudes (~$170°W$-$120°W$, Figure 8.7a) are discernible based on the composites of equatorial central Pacific SST extremes, while the SLP pattern of opposite signatures are noticeable over the western Pacific longitudes (~$100°E$-$150°E$, Figure 8.7a). A similar pattern of sub-polar high and mid-latitude low is observed over the western Pacific longitudes, along with a sign of opposite

TABLE 8.2
List of Years Corresponding to Strongest Ten Positive and Negative Extremes of SIA Averaged over BAS and WPO Sectors Based on Standardized SIA over the Respective Regions. Also, Strongest Ten Positive and Negative Extremes of SST Averaged over Central and Western Equatorial Pacific Based on Standardized SST over the Respective Regions

Sr. No.	Set I: Years of Positive Extremes for BAS SIA (JJAS)	Set II: Years of Positive Extremes for WPO SIA (JJAS)	Set III: Years of Positive Extremes for Central Pacific SST (JJAS)	Set IV: Years of Positive Extremes for Western Pacific SST (JJAS)
1	2015	2014	2015	2010
2	1986	1999	1997	1998
3	2013	2013	1987	1996
4	2010	2005	2002	2013
5	1990	1983	2009	2014
6	2014	2012	1991	2007
7	1994	2011	2004	1995
8	1991	2003	2012	2001
9	1997	1998	2006	2000
10	1995	1994	2014	1989

Sr. No.	Set I: Years of Negative Extremes for BAS SIA (JJAS)	Set II: Years of Negative Extremes for WPO SIA (JJAS)	Set III: Years of Negative Extremes for Central Pacific SST (JJAS)	Set IV: Years of Negative Extremes for Western Pacific SST (JJAS)
1	2007	1989	1988	1993
2	1988	2002	2010	1987
3	1998	2004	1998	1997
4	2006	2010	1999	1994
5	1992	2009	1989	1991
6	1983	1987	2007	1983
7	1989	1992	1985	1992
8	2008	1995	1984	2004
9	1993	1991	2000	2015
10	2003	2008	2011	1984

polarity over the central Pacific longitudes is noticed, for the composite of ten equatorial western Pacific SST episodes (Figure 8.7b). Next, analyses of atmospheric condition over the Southern Hemisphere using SLP patterns corresponding to SIA extremes over the BAS and WPO sectors of Antarctica are carried out. Figures 8.7(c) and (d) highlight spatial distribution of SLP for ten (positive-negative) extremes of BAS and WPO SIA, respectively. The prevalence of an anomalous sub-polar high (~65°S) and mid-latitude low (~40°S) is observed mainly

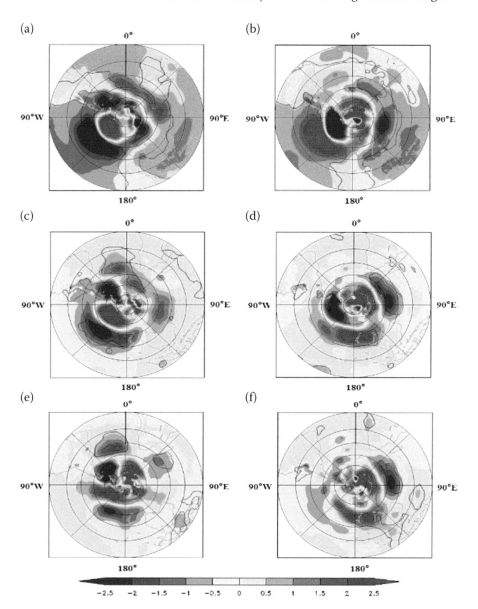

FIGURE 8.7   (a) Spatial distribution of SLP (hPa) during JJAS based on composite of ten strongest (positive - negative) central equatorial Pacific SST extremes for the period 1983–2015. (b) Same as (a) but for western equatorial Pacific SST extremes. (c) and (d) Same as (a) and (b), respectively, except for co-occurring SIA extremes over BAS and WPO sectors, respectively. (e) and (f) Same as (c) and (d), respectively, except for pure BAS and WPO SIA extremes. Solid (dotted) contour indicates significant difference in SLP at a 95% level of confidence.

over the sectors of BAS and Ross Sea from 170°E–90°W during the composites of BAS SIA (positive-negative) extremes (Figure 8.7c), which resemble the SLP patterns over the central Pacific (as seen in Figure 8.7a). A similar pattern (as seen in Figure 8.7b) is noticed over WPO and Indian Ocean sectors from 20°–135°E for the composites of WPO SIA (positive-negative) extremes (Figure 8.7d). Such a scenario, marked by a sub-polar high and mid-latitude low over a particular region in the Southern Ocean, may lead to a weakened regional meridional Ferrell cell, resulting in a tropical-polar interaction (Ciasto et al., 2015; Sen Gupta and England, 2006; Ciasto and Thompson, 2008). Further, a negative mode of SAM, marked by sub-polar high and mid-latitude low, resembles a tropically forced Rossby wave train over the longitudes of the Pacific Ocean sector, with a significant correlation with central Pacific SSTs during the boreal summer (Ding et al., 2012).

However, it is imperative to note from Table 8.2 that some of the SIA extremes have co-occurred during equatorial Pacific SST extremes. Henceforth, an event with absence (presence) of co-occurring equatorial Pacific SST extreme year is referred to as a pure (co-occurring) SIA extreme event. A comparison of positive extremes for the BAS SIA and equatorial central Pacific SST (Set I and III, upper panel of Table 8.2) reveals four co-occurring episodes (2015, 2014, 1991, and 1997). Likewise, three co-occurring positive extremes of WPO SIA and equatorial western Pacific SST (Set II and Set IV, upper panel of Table 8.2) are 2014, 2013, and 1998. Correspondingly, four co-occurring episodes each are also observed for negative SIA extremes of BAS SIA that co-occur with equatorial central Pacific SST (2007, 1988, 1998, and 1989) and WPO SIA, which co-occur with equatorial western Pacific SST (2004, 1987, 1992, and 1991) sectors (lower panels of Table 8.2). The interplay of equatorial Pacific SST and SIA over BAS and WPO may have different implications for ISMR, depending on the strengths and phases of the two as compared to their independent impact. Hence, to determine an independent effect of SIA on the large-scale atmospheric circulations, SLP patterns for the composite (positive-negative) extremes are recomputed for the pure SIA extremes. Analogous SLP patterns of sub-polar high (~65°S) and mid-latitude low over the WPO and Indian Ocean sectors from 20°–135°E are also evident considering the pure WPO SIA events compared to that of co-occurring WPO SIA and equatorial western Pacific SST episodes (comparison of Figures 8.7d and 8.7f). SLP patterns of pure and co-occurring SIA episodes over BAS and WPO sectors (comparison of Figure 8.7c with Figure 8.7e and Figure 8.7d with Figure 8.7f) share common characteristics of anomalous sub-polar high and a mid-latitude low. Thus, SLP patterns prevailing over the high and mid-latitudes of the Southern Hemisphere (Figure 8.7) may not be entirely governed by the equatorial Pacific SST variability but partially influenced by SIA variability over WPO and BAS sectors. Prabhu et al. (2016) reported large-scale interactions between the regions of sub-polar and mid-latitudes during a negative mode of SAM that is associated with equatorial central Pacific warming. It is intriguing to observe that an excessive SIA over the respective regions of BAS and WPO manifest signatures similar to that of negative mode of SAM, which are longitudinally locked around these two sectors with a sub-polar high and mid-latitude low, in agreement with an earlier study by Raphael et al. (2011). Further, the impact of SIA variability on the upper atmosphere during the summer monsoon season is also assessed by analyzing

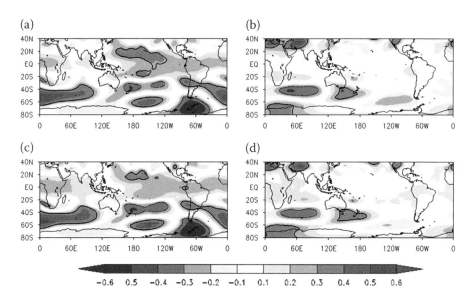

FIGURE 8.8   (a) Concurrent correlation between geopotential height (200 hPa) at each grid point and BAS SIA during JJAS for the period 1983–2015. (b) Same as (a), except for WPO SIA. (c) and (d) Same as (a) and (b), respectively, except for geopotential height at 300 hPa (*Source: Fig. 7(c,d) of Prabhu et al. 2021*). Solid (dotted) contour in black indicates significant positive (negative) relationship at a 95% level of confidence.

the correlation between geopotential height (200 hPa) at each grid point with BAS SIA (Figure 8.8a) and WPO SIA (Figure 8.8b). A similar correlation between geopotential height (300 hPa) at each grid point with BAS SIA (Figure 8.8c) and WPO SIA (Figure 8.8d) is also elucidated. The spatial pattern of correlation for geopotential height at 200 hPa broadly resembles the features at 300 hPa, except for minor differences. A meridional wavelike pattern is visible connecting tropical-polar belt along the BAS longitudes. In the case of WPO, a similar wavelike structure is observed over the western Indian Ocean longitudes. These atmospheric patterns connected to Antarctic sea-ice variability may have linkages with southern annular mode (Sen Gupta and England, 2006).

Next, the meridional circulations related to SIA anomalies over the BAS and WPO sectors are examined. In the subsequent analyses, the combined effect of equatorial Pacific SST and sea ice is addressed by considering all ten SIA and equatorial Pacific SST co-occurring episodes. Nonetheless, emphasis would also be given to analysis based on pure SIA extreme years to understand the relative importance of Antarctic SIA variability against dominant equatorial Pacific SST variability in connection to the Indian summer monsoon.

Figure 8.9 depicts the latitude-pressure section of meridional circulation, shown as (v,ω) wind vector, from southern high latitude to northern tropics, wherein Figures 8.9a and 8.9b display the summer monsoon season climatology averaged over the longitudes of BAS and WPO sectors, respectively. The climatology clearly

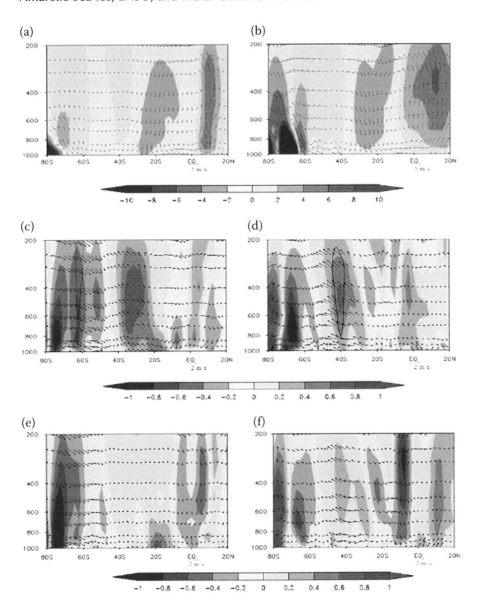

FIGURE 8.9   Latitude-pressure section of meridional circulation during JJAS based on the period 1983–2015, averaged over (a) BAS (130°-60°W) climatology (b) WPO (90°E-160°E) climatology. (c) and (d) Same as (a) and (b), respectively, except for composite of ten strongest (positive - negative) co-occurring BAS and WPO SIA extremes. (e) and (f) Same as (c) and (d), respectively, except for pure BAS and WPO SIA extremes. Wind-vector (v, ω) whereby v-wind (m/s) and vertical velocity ω (Pa/s, scaled by −0.01), overlaid by vertical velocity (shaded). Solid (dotted) contour in black indicates significant difference in vertical velocity at a 95% level of confidence.

reveals an ascending motion over the equatorial region of the Northern Hemisphere (Eq-20°N) and a descending motion over the Southern Hemisphere (10°–30°S). The climatological meridional circulations appear stronger for the WPO sector (Figure 8.9b) than for the BAS sector (Figure 8.9a). During the summer monsoon season, meridional circulation prepared from composites of ten (positive–negative) co-occurring BAS SIA extremes (Figure 8.9c), averaged over the longitudinal belt of 130°-60°W, and also for ten co-occurring WPO SIA extremes (Figure 8.9d), averaged over the longitudinal belt of 90°-160°E, show distinctly a descending motion corresponding to sub-polar high (~65°S) and an ascending motion corresponding to mid-latitude low (~40°S), further having an ascending motion over the equatorial region (Eq-10°N). As pointed out earlier, such circulation patterns are accompanied by a negative mode of SAM and exhibit wavelike structures, which are often meridional in nature and are associated with the atmospheric and oceanic anomalies over the equatorial Pacific (Prabhu et al., 2016). Likewise, composites for the pure SIA extremes (Figures 8.9e and 8.9f) also reveal analogous features, but of relatively weaker magnitude. The weakening of circulation could be attributed to the fact that the excluded years of equatorial Pacific SST extremes also happen to be co-occurring strong SIA extreme years. This result is suggestive of the mutual interaction between the two. Anomalous atmospheric circulations triggered by an excessive SIA over both sectors display a negative mode of SAM-like structure with a sub-polar high (~50–70°S) and mid-latitude low (~30–40°S) that is accompanied by an ascending branch over the equatorial Pacific. It is noteworthy that above (below) normal SIA variability over the BAS (WPO) sector is in concurrence with equatorial central (western) Pacific warming. The characteristics of the meridional circulation based on composites of co-occurring and pure SIA extremes for the WPO sector (Figures 8.9d and 8.9f) convey similar inferences. However, some differences are conspicuous in meridional circulation made from the extremes of co-occurring and pure SIA over the BAS sector, especially over the equatorial region (Figures 8.9c and 8.9e). Further, meridional circulations illustrated in Figure 8.9 highlight the potential impact of SIA extremes on atmospheric circulations besides a strong influence played by the equatorial Pacific SST episodes. In the following section, an implication of equatorial central and western Pacific SST extremes on large-scale circulation patterns over the Indian longitudes during the boreal summer season is further explored.

8.5   TELE-CONNECTION MECHANISM: SEA ICE, ENSO, AND ISMR

After the pioneering work of Sir Gilbert Walker (Walker, 1923), several studies revealed a tendency for deficient (excess) ISMR during a traditional El Niño (La Niña) occurrence associated with eastern (western) Pacific warming (Pant & Parthasarathy, 1981; Mooley & Parthasarathy, 1982; Khandekar & Neralla, 1984; and several others). However, recently, central Pacific warming based SST variability has been gaining vast importance, which further has potential consequences on the Asian summer monsoon variability (Gadgil et al. 2004; Kug et al., 2010; Prabhu et al., 2016). Thus, to comprehend the influence of central and western equatorial Pacific SST variability on the Indian region through anomalous zonal circulation, ten cases of positive and negative SST extremes for both of these regions are considered. The list of years corresponding to ten positive and negative SST extremes of central and western equatorial Pacific are provided in Table 8.2 and the procedure adopted for their identification is described earlier in Section 4.

The oceanic signatures prevalent over the Pacific Ocean during the dominant modes of SST variability over both central and western equatorial Pacific regions are demonstrated using composite difference of positive and negative SST extremes (Figure 8.10). The spatial distribution of SST over the Pacific Ocean based on the composite of ten (positive-negative) pure SST extremes of central and western equatorial Pacific are illustrated in Figures 8.10a and 8.10b, respectively. The spatial structure of SST in pure SST episodes (Figures 8.10a and 8.10b) is quite consistent with that of co-occurring SIA composite episodes of BAS and WPO sectors (Figures 8.10c and 8.10d), although the spatial extent of positive (negative) SST over central (western) Pacific in pure SST episodes is much more extensive than in respective co-occurring SIA episodes. Furthermore, the spatial distribution of SST in pure SIA extreme composite of BAS and WPO (Figures 8.10e and 8.10f) moderately differ from that of corresponding co-occurring SIA episodes (Figures 8.10c and 8.10d). The amplitude of composite (positive-negative) SST in pure SIA extremes of the two sectors (Figures 8.10e and 8.10f) tends to be weaker than that of respective co-occurring SIA episodes (Figures 8.10c and 8.10d) due to removal of SST extreme years. Therefore, a combined effect of both equatorial Pacific SST and Antarctic SIA variability, as illustrated separately in Figures 8.10c and 8.10d, is likely to trigger different responses of atmospheric circulation on either side of the Pacific Ocean, which may consequently facilitate or inhibit the monsoon activity over the Indian sub-continent. In the final step, we shall evaluate this aspect through zonal circulation, represented as longitude-pressure section of (u,ω) wind vector, over the Indo-Pacific region averaged over the latitudes spanning 5°S–30°N (Figure 8.11). Climatology of zonal circulation distinctly illustrates ascending motion over the Indian-west Pacific domain (60°–160°E) and a descending motion over the central Pacific sector, east of 150°W (Figure 8.11a). Zonal circulation prepared from the composite of ten (positive-negative) SST extremes of the equatorial central Pacific shows a strong ascending motion over this region (Figure 8.11b), which coincides with statistically significant anomalously warm equatorial belt spread along the longitudes 150°E–120°W (Figure 8.10a). An ascending branch over the central Pacific is flanked by a descending branch over the Indian longitudes (60°-90°E), resulting in reduced convection and subdued rainfall. In the case of zonal circulation based on the composite of ten (positive-negative) SST extremes over the equatorial western Pacific, a descending motion over the region (170°E-120°W) is accompanied by an ascending motion over the Indian sector (60°–80°E), causing strong convection and an enhanced rainfall activity (Figure 8.11c). Analogous zonal circulation patterns prevail with respect to the ten co-occurring SIA extremes of BAS and WPO sectors (comparison of Figure 8.11b with Figure 8.11d and Figure 8.11c with Figure 8.11e). However, a comparison of SST distribution with the corresponding zonal circulation patterns based on the composites of ten (positive-negative) pure SIA extremes (comparison of Figure 8.10e with Figure 8.11f and Figure 8.10f with Figure 8.11g) appear a little

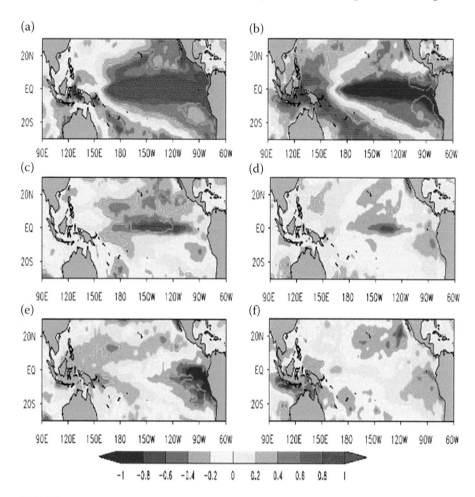

FIGURE 8.10 (a) Spatial distribution of SST (in °C) during JJAS based on composite of ten strongest (positive–negative) central equatorial Pacific SST extremes for the period 1983–2015. (b) Same as (a), but for western equatorial Pacific SST extremes. (c) and (d) Same as (a) and (b), respectively, except for co-occurring BAS and WPO SIA extremes. (e) and (f) Same as (c) and (d), respectively, except for pure BAS and WPO SIA extremes. Solid contour in green indicates significant difference in SST at a 95% level of confidence.

intriguing. A descending (ascending) motion along 50°–60°E (60°–90°E), as depicted in Figure 8.11f (Figure 8.11g), is interesting. The sign of the SST anomalies over the eastern Indian Ocean are typically similar to that over the adjoining western Pacific through the Indonesian through flow (Clarke et al., 1998). A careful examination of Figure 8.10e suggests likelihood of positive SST anomalies over the eastern Indian Ocean off the Sumatra coast. Positive SST anomalies over the eastern equatorial Indian Ocean prevail during the negative phase of IODM (Saji et al., 1999; Webster et al., 1999). The summer monsoon rainfall over India weakens

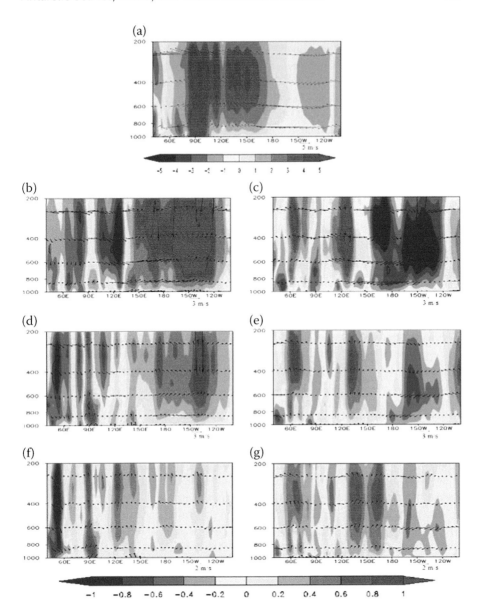

FIGURE 8.11 Longitude-pressure section of zonal circulation during JJAS averaged over the latitudes (5°S – 30°N) based on the period 1983–2015 for (a) climatology over Indo-Pacific domain. (b) and (c) Same as (a), except for composite of ten strongest (positive–negative) central and western equatorial Pacific SST extremes, respectively. (d) and (e) Same as (b) and (c), respectively, except for co-occurring BAS and WPO SIA extremes. (f) and (g) Same as (d) and (e), respectively, except for pure BAS and WPO SIA extremes. Wind vector (u,ω), where u-wind (m/s) and vertical velocity ω (Pa/s, scaled by −0.01), overlaid by vertical velocity (shaded). Solid (dotted) contour in black indicates significant difference in vertical velocity at a 95% level of confidence.

during a negative phase of IODM (Ashok et al., 2001), which could be ascertained from a descending motion observed over the Indian longitudes (~50°–60°E) (Figure 8.11f). On the other hand, negative SST anomalies over the Sumatra coast (Figure 8.10f) are characterized by a positive phase of IODM, which favors the Indian summer monsoon (Ashok et al., 2001), as implied by the upward motion over the Indian longitudes (60°–90°E) (Figure 8.11g). However, the connection between Antarctic sea ice and IODM needs a careful and detailed investigation, which is beyond the scope of the current study.

Rainfall averaged over India as a whole is considered in this study so far. The summer monsoon rainfall over northeast India and neighborhood is out of phase with rainfall over the rest of the country (Parthasarathy et al., 1993; Prabhu et al. 2017), indicating that regional nature of summer monsoon rainfall over India is inhomogeneous. In view of this, the relationship between spatial distribution of rainfall over various regions of India with ENSO and sea ice are examined in the subsequent section.

8.6 REGIONAL PERSPECTIVES: INDIAN SUMMER MONSOON, SEA ICE, AND ENSO LINKAGES

To evaluate the Antarctic sea ice and ENSO connection with rainfall on a regional scale over India, spatial distribution of summer monsoon rainfall is examined based on the composites of (positive-negative) pure SST and co-occurring SIA and pure SIA extremes, respectively (Figure 8.12). Significant negative rainfall anomalies are observed to be spread across large areas of northern and southern parts of India, for the composites of anomalously warm central equatorial Pacific SSTs (Figure 8.12a). Analogous spatial distribution of rainfall anomalies, but with an opposite sign, prevail for composites of anomalously warm western equatorial Pacific SSTs (Figure 8.12b), in agreement with anomalous Walker circulation and its consequent implication on summer monsoon rainfall over India (Figures 8.11b and 8.11c). Besides, small pockets of negative rainfall anomalies scattered over central India are also noticed (Figure 8.12b). A comparison of rainfall patterns constructed from pure central Pacific SST composite and corresponding co-occurring BAS SIA composite reveal a large reduction in the spatial extent of negative rainfall anomalies over northern parts and displacement over southern India (compare Figure 8.12a with Figure 8.12c). Large differences are also apparent in the spatial structure of significant positive rainfall anomalies in the pure western pacific SST composite (Figure 8.12b) as compared to the co-occurring WPO SIA composite (Figure 8.12d). Finally, significant negative rainfall anomalies in ten co-occurring BAS SIA composites (Figure 8.12c) are observed to entirely disappear in seven pure BAS SIA composites (Figure 8.12e), implying that the connection of BAS SIA with summer monsoon rainfall over India is completely dependent on central Pacific SST. On the contrary, the spatial features of Figures 8.12(d) and 8.12(f) display striking resemblances except for weakened magnitude, suggesting that the relationship of the WPO SIA with the summer monsoon rainfall over India is moderately dependent on the western Pacific SST. The significant negative (positive) rainfall anomalies over the Indian sub-continent are apparent for composites of

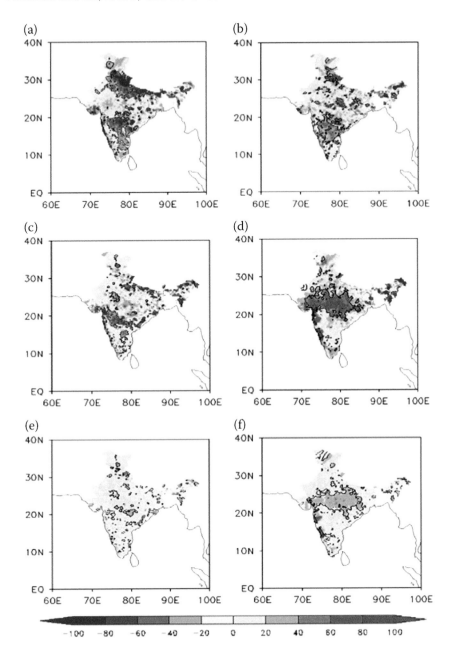

FIGURE 8.12 (a) Rainfall anomalies (in mm/day) during JJAS based on composite of ten strongest (positive-negative) central equatorial Pacific SST extremes based on the period 1983–2015. (b) Same as (a), but for western equatorial Pacific SST extremes. (c) and (d) Same as (a) and (b), respectively, except for co-occurring BAS and WPO SIA extremes. (e) and (f) Same as (c) and (d), respectively, except for pure BAS and WPO SIA extremes. Dashed (dotted) contour in black indicates significant difference in rainfall at a 95% level of confidence.

both the pure central (western) equatorial Pacific SST, as shown in Figure 8.12a (Figure 8.12b), along with co-occurring BAS (WPO) SIA, as depicted in Figure 8.12c (Figure 8.12d), clearly illustrating the combined influence of central (western) equatorial Pacific SST and BAS (WPO) sea ice on spatial distribution of rainfall adversely (favorably).

The impact of the Pacific SST perturbation was evaluated by examining the zonal east-west Walker cell (Figure 8.11), in the previous Section 8.5. However, anomalous rainfall distribution patterns depicted in Figure 8.12, suggest probing the vertical meridional north-south Hadley cell as well (Figure 8.13). The climatology of the Hadley cell averaged over the longitudes covering central India (75–85°E) clearly depicts a strong rising motion over the Indian subcontinent (Figure 8.13a: 25°–35°N) associated with rainfall over central parts of India. Further, the anomalous north-south (or Hadley) circulation patterns for the ten strongest (positive–negative) pure SST extremes of central Pacific (Figure 8.13b) and western Pacific (Figure 8.13c) are evaluated. Negative rainfall anomalies north of 30°N and south of 20°N (Figure 8.12a) are clearly associated with a co-located descending branch of Hadley circulation with indications of an ascending motion in between (Figure 8.13b). Again, a Hadley cell comparison of ten pure SST composite with that of ten co-occurring SIA composite patterns (compare Figure 8.13b with Figure 8.13d and Figure 8.13c with Figure 8.13e) convey similar inferences. In addition, seven pure SIA composite patterns (Figure 8.13f and Figure 8.13g) also convey analogous implications. Thus, it appears that ISMR is associated with Antarctic sea ice through the east-west Walker circulation over the wide Indo-Pacific basin, which in turn induces a north-south Hadley cell over the Indian sub-continent, resulting in rainfall distribution and vertical motions, as displayed in Figure 8.12 and Figure 8.13, respectively. This is an alternate channel suggested in this study.

Finally, Figure 8.14 represents spatial correlation maps of rainfall at each grid point over the Indian land separately for the months of July, August, and September, and also for the averaged July through September (JAS) season, with that of July SIA averaged over BAS (upper panel) and WPO (lower panel) sectors. Major features of regional correlation pattern of July SIA with rainfall (Figures 8.14a and 8.14e) resemble its counterpart for the JAS season (Figures 8.14d and 8.14h) for both BAS and WPO sectors. As such, the July SIA index could be used to foreshadow spatial distribution of rainfall during the sub-sequent months. However, the most intriguing feature is a considerable decline in correlation pattern during the month of August (Figures 8.14b and 8.14f). Based on the 50-year (1901–1950) period, rainfall data has been obtained from several well-distributed stations over India. Ananthakrishnan (1970) noted that the peak in rainfall amount was attained in July, declining to a minimum value in mid-August; thereafter, an increase in rainfall during early September and its final demise is observed by the end of September. He further observed that such a minimum in August was noted for a 50-year period, which was more or less repetitive from year to year, and attributed this pattern as a unique feature of the monsoon. It is of interest to inquire whether there is independent evidence for any abnormality in the monsoon circulation during the middle of August. Based on an 80-year period

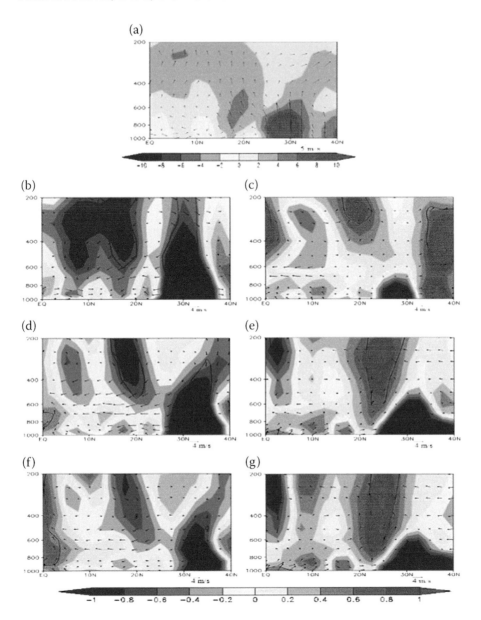

FIGURE 8.13 Latitude-pressure section of meridional circulation averaged over longitudes of central India (75°E-85°E) during JJAS based on the period 1983–2015. (a) Climatology over Indian domain. (b) and (c) Same as (a), except based on composite of ten strongest (positive–negative) central equatorial Pacific SST extremes and western equatorial Pacific SST extremes, respectively. (d) and (e) Same as (b) and (c), respectively, except for co-occurring BAS and WPO SIA extremes. (f) and (g) Same as (d) and (e), respectively, except for composite of pure BAS and WPO SIA extremes. Wind-vector (v, ω) whereby v-wind (m/s) and vertical velocity ω (Pa/s, scaled by -0.01), overlaid by vertical velocity (shaded). Dashed (dotted) contour in black indicates significant difference in vertical velocity at a 95% level of confidence.

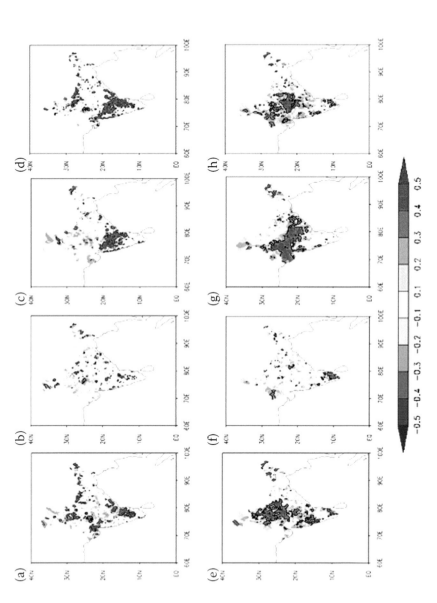

FIGURE 8.14 Correlation between standardized BAS SIA (July) with that of standardized IMD rainfall at each grid point for the period 1983–2015, during the months of (a) July, (b) August, and (c) September, and (d) averaged over the months of July, August, and September. (e), (f), (g), and (h) Same as (a), (b), (c), and (d), respectively, except for WPO SIA (July). Dashed (dotted) contour in black indicates significant relationship at a 95% level of confidence.

(1888–1967), Ramamurthi (1969) inferred that the mid-August period is characterized by a maximum frequency of occurrences of breaks over India, which is marked by a considerable decrease of rainfall over large areas of the country. During breaks, the monsoon trough moves close to the foot of the Himalayas, leading to a decrease in rainfall over major parts of central India. Interestingly, a normal position of the monsoon trough is found to have moved westwards by about three degrees of longitude during recent decades (Preethi et al., 2017). They attributed this feature as a possible reason for the declining trend of the ISMR during the last four to five decades. Thus, the minima in August rainfall and the recent westward shift of the monsoon trough explain the correlation drop observed for August (Figures 8.14b and 8.14f). A possible reason for such breaks over the Indian subcontinent has been attributed to an interaction between the easterly and the subtropical jet streams (Ramaswamy, 1962).

8.7 SUMMARY

Antarctic sea ice fluctuations have an immense capability to impose large-scale variations on the global climate. Physical processes underlying the polar-tropical linkages are complex in nature. However, this study identifies a sector-based Antarctic sea-ice relationship with the Indian summer monsoon that may serve as a crucial diagnostic tool in climate forecast system models (Pattanaik & Kumar, 2009), in addition to enhancing our understanding of essential mechanisms driving these tele-connections. The composite fields presented could be used as indicators of general conditions prevalent over the identified Antarctic sectors and tropics during the sea-ice extremes. To address the objectives of this study, passive microwave satellite data has been utilized for the period 1983–2015, leading to the following insights.

Out of the five sea-ice sectors considered along with total Antarctic, sea-ice variability over BAS and WPO is observed to have robust concurrent linkages with the Indian summer monsoon rainfall. During the boreal summer, an in-phase significant relationship is observed between the WPO sea ice and ISMR, while for the same period, an out-of-phase relationship is observed between the BAS sea ice and ISMR. Next, a significant inverse relationship between the ENSO and ISMR is observed to gradually strengthen from the month of June that persists for the entire boreal summer monsoon season. Further, lead-lag correlations demonstrate a significant direct (inverse) relationship of July SIA over BAS (WPO) with JJAS EMI of the concurrent year. A detailed discussion of these results has been documented recently (Prabhu et al., 2021).

Furthermore, regional aspects of rainfall over India with respect to the relation with sea ice over BAS and WPO along with ENSO are also investigated. A spatial correlation of rainfall over India with ENSO and BAS SIA during JJAS shows a significant negative relation over northern and southern parts of India. On the other hand, the WPO SIA is positively correlated with rainfall over parts of central and northwest India. The chief features of a regional correlation pattern for concurrent July SIA with rainfall over India are similar to its counterpart during the JAS season for both BAS and WPO sea-ice sectors. As such, the July SIA index could be used

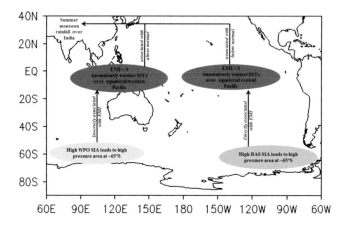

FIGURE 8.15 Schematic representation of SIA-ENSO-ISMR connection based on the period 1983–2015 for 'High' episodes of SIA variability over BAS and WPO sectors of Antarctica. Reverse is truefor 'Low' episodes of respective SIA sectors (*Source: Fig. 11 of Prabhu et al. 2021*).

to foreshadow the spatial distribution of rainfall during the subsequent months. The BAS (WPO) sea-ice relation with spatial distribution of rainfall during JJAS is completely (moderately) dependent on central (western) equatorial Pacific SST.

Thus, the physical mechanism proposed in communicating the simultaneous southern polar variability signal towards the Indian sub-continent during June through September is through the Pacific Ocean channel, wherein the co-occurrence of SST warming over central (western) equatorial Pacific in addition to excessive sea-ice variability over the BAS (WPO) sector is instrumental in influencing the summer monsoon rainfall adversely (favorably). Anomalous large-scale atmospheric circulations pertaining to the BAS (WPO) SIA and central (western) equatorial Pacific SST variability have a simultaneous bearing on the summer monsoon rainfall with opposite polarity, as mentioned earlier. In conclusion, the ISMR variability may be connected with Antarctic sea ice through the Walker cell as well as the Hadley cell over the Indo-Pacific, which are some new insights proposed in this study. A schematic representation highlighting a physical mechanism governing the SIA–ENSO–ISMR linkage is shown in Figure 8.15.

In order to further augment our understanding regarding the observed complex sea ice–ENSO–monsoon connection, a dedicated model sensitivity study considering both independent and combined Antarctic sea ice and Pacific SST impacts on the Indian summer monsoon needs to be carried out in the future.

ACKNOWLEDGMENTS

We thank Director, Indian Institute of Tropical Meteorology (IITM), Pune for all the support to carry out this work. IITM is funded by the Ministry of Earth Sciences, Government of India, New Delhi. The authors are thankful to NCEP and NCAR for the reanalysis data sets. The authors are also grateful to NSIDC, University of Colorado, Boulder, U.S.A., for the sea-ice data. The authors acknowledge other data sets such as the gridded SST product from NOAA, spatial rainfall data from IMD, and the rain gauge–derived ISMR series from IITM. We gratefully acknowledge the Center for Ocean Land Atmosphere, U.S.A., for the Grid Analysis and Display System (GrADS) software, which has been used to plot figures.

REFERENCES

Ananthakrishnan, R. (1970). Some salient features of the space-time variation of rainfall over India and neighbourhood. *Curr. Sci.*, 29(5), 101–105.

Ashok, K., Toshio, Z.G., and Yamagata, T. (2001). Impact of Indian Ocean dipole on the relationship between the Indian Monsoon Rainfall and ENSO. *Geophys. Res. Lett.*, 28(3). doi: 10.1029/2001GL013294.

Ashok, K., Guan, Z., Saji N. H., and Yamagata, T. (2004). Individual and combined influences of ENSO and the Indian Ocean dipole on the Indian summer monsoon. *J Clim.*, 17, 3141–3154.

Ashok, K., Behera, S. K., Rao, S. A., Weng, H., and Yamagata T. (2007). El Niño Modoki and its possible teleconnection. *J. Geophys. Res.*, 1 and 2, C11007. doi: 10.1029/2006JC003798.

Baines, P. G., and Cai, W. J. (2000). Analysis of an interactive instability mechanism for the Antarctic circumpolar wave. *J. Clim.*, 13, 1831–1844. doi: 10.1175/1520-0442(2000)013.

Bamzai, A. S., and Shukla, J. (1999). Relation between Eurasian snow cover, snow depth and the Indian monsoon: An observational study. *J. Clim.*, 12, 3117–3132.

Cai, W., Rensch, P. V., Cowan, T., and Hendon, H. (2011). Teleconnection pathways of ENSO and the IOD and the mechanism for impact on the Australian rainfall. *J. Clim.*, 24, 3910–3923.

Carleton, A. M. (2003). Atmospheric teleconnections involving the Southern Ocean. *J. Geophys. Res.*, 108 (C4), 8080. doi: 10.1029/2000JC000379.

Cavalieri, D. J., Parkinson, C. L., Gloersen, P., Comiso, J. C., Zwally, H. J. (1999). Deriving long-term time series of Sea ice cover from satellite passive-microwave multisensor data sets. *J. Geophys. Res.*, 104(C7), 15803–15814.

Chen B., Smith S. R., and Brownwich D. H. (1996) Evolution of the tropospheric split jet over the Southern Pacific Ocean during the 1986–89 ENSO cycle. *Mon Wea Rev.* 124, 1711–1731.

Chen, S., Wang, Y., Cheng, H., Edwards, R. L., Wang, X., Kong, X., and Liu, D. (2016). Strong coupling of Asian Monsoon and Antarctic climates on sub-orbital timescales. *Sci. Rep.*, 6. doi: 10.1038/srep32995.

Ciasto, L. M., and Thompson, D. W. J. (2008). Observations of large-Scale Ocean-Atmosphere interaction in the Southern Hemisphere. *J Clim.*, 21, 1244–1259. doi: 10.1175/2007JCLI1809.1.

Ciasto, L. M., Simpkins, G. R., and England, M. H. (2015). Teleconnections between tropical Pacific SST anomalies and extratropical Southern Hemisphere Climate. *J. Clim.*, 28, 56–65. doi: 10.1175/JCLI-D-14-00438.1.

Clarke, A.J., Liu, X., and Van Gorder, S. (1998). Dynamics of the biennial oscillation in the Equatorial Indian and Far Western Pacific Oceans, *J. Clim.*, 11, 987–1001.

Clem, K. R., Renwick, J. A., McGregor, J., and Fogt, R. L. (2016). The relative influence of ENSO and SAM on Antarctic Peninsula climate. *J. Geophys. Res. (Atmos)*, 121, 9324–9341. doi: 10.1002/2016JD025305.

Dash, S. K., Singh, G. P., Shekhar, M. S., Vernekar, A. D. (2005). Response of the ISM circulation and rainfall to seasonal snow depth anomaly over Eurasia. *Clim. Dyn.*, 24, 1–10.

Ding, Q., Steig, E. J., Battisti, D. S., and Wallace J. M. (2012). Influence of the tropics on the Southern Annular Mode. *J. Clim.*, 25, 6330–6348.

Dugam, S. S., and Kakade, S. B. (2004). Antarctica Sea ice and monsoon variability. *Indian J. Radio and Space Phys.*, 33, 306–309.

Flatau, M., and Kim, Y-J. (2013). Interaction between the MJO and Polar circulation. *J. Clim.*, 26, 3562–3574.

Gadgil, S. (2003). The Indian monsoon and its variability. *Annual Rev. Earth Planetary Sci.*, 31, 429–467.

Gadgil, S., Vinayachandran, P. N., Francis P. A., and Gadgil, S. (2004). Extremes of the Indian summer monsoon rainfall, ENSO and equatorial Indian Ocean oscillation. *Geophys. Res. Lett.*, 31, L12213.

Gloersen, P., and Campbell, W. J. (1991). Recent variations in Arctic and Antarctic Sea ice covers. *Nature*, 352, 33–36.

Goodwin, B. P., and Thompson, E. M. (2016). Accumulation variability in the Antarctic Peninsula: The role of large-scale aatmospheric oscillations and their interactions. *J. Clim.*, 29, 2579–2596.

Izumo, T., Lengaigne, M., Vialard, J., Luo, J.-J., Yamagata, T., and Madec, G. (2014). Influence of Indian Ocean Dipole and Pacific recharge on following year's El Niño: Interdecadal robustness. *Clim. Dyn.*, 42, 291–310.

Kalnay, E., Kanamitsu, M., and Kistler, R., et al. (1996). The NCEP/NCAR 40-year re-analysis project. *Bull. Am. Met. Soc.*, 77, 437–472.

Kendall, M. G., and Stuart, A. (1979). *The advanced theory of statistics, volume 2: inference and relationship, griffin*, 4th edn, Hodder Arnold publisher, London, pp. 758 (ISBN: 0852642555).

Khandekar, M. L., and Neralla, V. R. (1984). On the relationship between the sea surface temperatures in the equatorial Pacific and the Indian monsoon rainfall. *Geophys. Res. Lett.*, 11, 1137–1140.

Kripalani, R. H., and Kulkarni, A. (1999). Climatology and variability of historical Soviet snow depth data: Some new perspectives in snow-Indian monsoon tele-connection. *Clim. Dyn.*, 15: 475–489.

Krishnan, R., Sundaram, S., Panickal, S., and Mujumdar, M. (2011). The crucial role of ocean–atmosphere coupling on the Indian monsoon anomalous response during dipole events. *Clim. Dyn.*, 37, 1–17.

Krishnaswamy, J., Vaidyanathan, S., Rajagopalan, B., Bonell, M., Sankaran, M., Bhalla, R., and Badiger, S. (2014). Non-stationary and non-linear influence of ENSO and Indian Ocean Dipole on the variability of Indian monsoon rainfall and extreme rain events. *Clim. Dyn.*, 45, 175–184.

Kug, J.-S., Choi, J., An, S.-I., Jin, F.-F., and Wittenberg, A. T. (2010). Warm pool and cold tongue El Niño events as simulated by the GFDL 2.1 coupled GCM. *J. Clim.*, 23, 1226–1239. doi:10.1175/2009JCLI3293.1.

Kwok, R., and Comiso, J. C. (2002). Southern Ocean climate and Sea ice anomalies associated with the Southern Oscillation. *J. Clim.*, 15, 487–501.

Larkin, N. K., and Harrison, D. E. (2005). On the definition of El Niño and associated seasonal average U.S. weather anomalies. *Geophys. Res. Lett.*, 32, L13705. 10.1029/2005GL022738.

Ledley, T. S., and Huang, Z. (1997). A possible ENSO signal in the Ross Sea. *Geophys. Res. Lett.* 24, 3253–3256.

Li, X., Holland, D. M., Gerber E. P., and Yoo, C. (2015). Rossby waves mediate impacts of tropical oceans on west Antarctic atmospheric circulation in Austral Winter. *J Clim* 28: 8151–8164.

Liu, J., Yuan, X., Rind, D., and Martinson, D. G. (2002). Mechanism study of the ENSO and southern high latitude climate teleconections. *Geophys. Res. Lett.*, 29. doi: 1029/2002 GL015143.

Liu, T., Li, J., Li, Y., Zhao, S., Zheng, F., Zheng, J., and Yao, Z. (2018). Influence of the May Southern Annular mode on the South China Sea summer monsoon. *Clim. Dyn.*, 51, 4095–4107.

Ludescher, J., Bunde, A., Franzke, C. E., and Schellnhuber, H. J. (2015). Long-term persistence enhances uncertainty about anthropogenic warming of Antarctica. *Clim. Dyn.*, 457, 459–462.

Mooley, D. A., and Parthasarathy, B. (1982). Fluctuations in the deficiency of summer monsoon over India and their effect on economy. *Arch. Für Met. Geoph. md Bioklim.*, B30, 383–398.

Mooley, D.A., and Parthasarathy, B. (1983). Indian summer monsoon and El Nino. *Pure Appl. Geophys.*, 121, 339–352.

Morioka, Y, Engelbrecht, F., and Behera, S. K. (2017). Role of Weddell Sea ice in South Atlantic atmospheric variability. *Clim. Res.*, 74, 171–184.

Nuncio, M., and Yuan, X. (2015). The influence of the Indian Ocean Dipole on Antarctic Sea ice. *J. Clim.*, 28, 2682–2690. doi: 10.1175/JCLI-D-14-00390.1.

Oza, S. R., Rajak, D. R., Dash, M. K., Bahuguna, I. M., and Kumar, R. (2017). Advances in Antarctic Sea ice Studies in India. *Proceed Indian Nat. Sci. Acad.*, 83(2), 427–435. doi: 10.16943/ptinsa/2017/48947.

Pai, D. S., Sridhar, L., Rajeevan, M., Sreejith, O. P., Satbhai, N. S., and Mukhopadhyay, B. (2014). Development of a new high spatial resolution (0.25 × 0.25 degree) long period (1901–2010) daily gridded rainfall data set over India and its comparison with existing data sets over the region. *Mausam*, 65(1), 1–18.

Pant, G. B., and Parthasarathy, B. (1981). Some aspects of an association between the southern oscillation and Indian summer monsoon. *Arch Für Met Geoph md Bioklim*, B29, 245–252.

Parthasarathy, B., Rupa Kumar, K., and Munot, A. A. (1992). Forecast of rainy season foodgrain production based on monsoon rainfall. *Indian J. Agric. Sci.*, 62, 1–8.

Parthasarathy, B., Rupa Kumar, K., and Munot, A. A. (1993). Homogeneous Indian monsoon rainfall: Variability and prediction. *Proc. Indian Acad. Sci. (Earth Planet Sci)* 102, 121–155.

Parkinson, C. L. (2004). Southern Ocean Sea ice and its wider linkages: Insights revealed from models and observations. *Antarctic Sci.*, 16(4), 387–400. doi:10.1017/S0954102 004002214

Pattanaik, D. R., and Kumar, A. (2009). Prediction of summer monsoon rainfall over India using the NCEP climate forecast system. *Clim. Dyn.*, 34, 557–572.

Peings, Y, and Douville, H. (2010). Influence of the Eurasian snow cover on the Indian summer monsoon variability in observed climatologies and CMIP3 simulations. *Clim. Dyn.*, 34, 643–660.

Peterson, R. G., and White, W. B. (1998). Slow oceanic teleconnections linking the Antarctic Circumpolar Wave with the tropical El Niño-Southern Oscillation. *J. Geophys. Res.*, 103(C11), 24573–24583. doi:10.1029/98JC01947.

Prabhu, A., Mahajan, P., Khaladkar R, and Bawiskar, S. (2009). Connection between Antarctic Sea ice extent and Indian summer monsoon rainfall. *Int. J. Rem Sens.*, 30, 3485–3494.

Prabhu, A., Mahajan, P. N., Khaladkar, R. M., and Chipade, M. D. (2010). Role of Antarctic circumpolar wave in modulating the extremes of Indian summer monsoon rainfall. *Geophys. Res. Lett.*, 37, 1–5. doi:10.1029/2010GL043760.

Prabhu, A., Mahajan, P. N., and Khaladkar, R. M. (2012). Association of the Indian summer monsoon rainfall variability with the geophysical parameters over the Arctic region. *Int. J. Climatol.*, 32, 2042–2050.

Prabhu, A., Kripalani, R. H., Preethi, B., and Pandithurai, G. (2016). Potential role of the February–March Southern Annular Mode on the Indian summer monsoon rainfall: A new perspective. *Clim. Dyn.*, 47: 1161–1179. doi:10.1007/s00382-015-2894-5

Prabhu, A., Oh, J., Kim, I.-W. , Kripalani, R. H., Mitra, A. K., and Pandithurai, G. (2017). Summer monsoon rainfall variability over North East regions of India and its association with Eurasian snow, Atlantic Sea Surface temperature and Arctic Oscillation. *Clim. Dyn.* 49, 2545–2556. doi:10.1007/s00382-016-3445-4.

Prabhu, A., Oh, J., Kim I.-W., Kripalani R. H., and Pandithurai, G. (2018). SMMR-SSM/I derived Greenland Sea ice variability: Links with Indian and Korean Monsoons. *Clim. Dyn.*, 50, 1023–1043. doi:10.1007/s00382-017-3659-0.

Prabhu, A., Mandke, S. K., Kripalani, R. H., and Pandithurai, G. (2021). Association between Antarctic Sea ice, Pacific SST and the Indian summer monsoon: An observational study. *Polar Sci.*, doi.org/10.1016/j.polar.2021.100746.

Preethi, B., Mujumdar, M., Kripalani, R. H., Prabhu, A., and Krishnan, R. (2017). Recent trends and tele-connections among South and East Asian summer monsoon in a warming environment. *Clim. Dyn.*, 48, 2489–2505.

Rajeevan, M., Pai D. S., and Anil Kumar, R. (2007). New statistical models for long range forecasting of southwest monsoon rainfall over India. *Clim. Dyn.*, 28(7), 813–828.

Rajagopalan, B., and Molnar, P. (2012). Pacific Ocean sea-surface temperature variability and predictability of rainfall in the early and late parts of the Indian summer monsoon season. *Clim. Dyn.*, 39(6), 1543–1557.

Ramamurthi, K. (1969). Some aspects of the breaks in the Indian southwest monsoon in July and August. India Met Department Forecasting Manual, Rep No. iv-18-3.

Ramaswamy, C. (1962). Breaks in the Indian summer monsoon as phenomena of interaction between the easterly and the subtropical westerly jet streams. *Tellus*, 14A, 337–349.

Raphael, M. N., Hobbs, W., and Wainer, I. (2011). The effect of Antarctic Sea ice on the Southern Hemisphere atmosphere during the southern summer. *Clim. Dyn.*, 36, 1403–1417.

Reynolds, R. W., Rayner, N. A., Smith, T. M., Stokes, D. C., and Wang W. (2002). An improved in situ and satellite SST analysis for climate.*J. Clim.*, 15, 1609–1625.

Saji, N. H., Goswami, B. N., Vinayachandran, P. N., and Yamagata, T. (1999). A dipole mode in the tropical Indian Ocean. *Nature*, 401, 360–363

Sen Gupta, A, and England, M. (2006). Coupled ocean–atmosphere feedback in the Southern Annular Mode. *J. Clim.*, 20, 3677–3692.

Shukla, J. (1998). Predictability in the midst of chaos: A scientific basis for climate forecasting. *Science*, 282, 728–731.

Thompson, D. W. J., and Wallace, J. M. (2000). Annular modes in the extratropical circulation. Part I: Month-to-month variability. *J. Clim.*, 13, 1000–1016.

Ummenhofer, C. C., D'Arrigo, R. D., Anchukaitis, K. J., Buckley, B. M., and Cook, E. R. (2013). Links between Indo-Pacific climate variability and drought in the Monsoon Asia Drought Atlas. *Clim. Dyn.*, 40, 1319–1334.

Walker, G. T. (1923). Correlation in seasonal variation in Weather, VIII. *Memoirs of the India Met Department*, 24, 75–131.

Wallace, J. M., Rasmusson, E.M., Mitchell, T. P., Kousky, V. E., Sarachik, E. S., and von Storch, H. (1998). On the structure and evolution of ENSO related climate variability in the tropical Pacific: Lessons from TOGA. *J. Geophys. Res.*, 103, 14241–14260.

Welhouse, L. J., and Lazzara, M. A. (2016). Composite analysis of the effects of ENSO events on Antarctic. *J. Clim.*, 29, 1797–1808.

Webster, P. J., Moore, A. M., Loschnigg, J. P., and Leban, R. R. (1999). Coupled ocean-atmosphere dynamics in the India Ocean during 1997–98. *Nature*, 401, 356–360.

Weng, H., Ashok, K., Behera, S. K., Rao, S. A., and Yamagata, T. (2007). Impacts of recent El Niño Modoki on dry/wet conditions in the Pacific rim during boreal summer. *Clim. Dyn.*, 29, 113–129.

White, W. B., and Peterson, R. G. (1996). An Antarctic circumpolar wave in surface pressure, wind, temperature and sea-ice extent. *Nature*, 380, 699–702. doi:10.1038/380699a0.

Xue, F, Guo, P., and Yu, Z. (2003). Influence of interannual variability of Antarctic Sea ice on summer rainfall in eastern China. *Adv. Atmos. Sci.*, 20, 97–102.

Yang, X. Q., and Huang, S. S. (1992). A numerical experiment of climate effect of Antarctic Sea ice during the Northern Hemisphere summer. *Chin. J Atmos Sci.*, 16, 80–89.

Yeh, S. W., Kug, J.S., Dewitte, B., Kwon, M. H., Kirtman, B. P., and Jin, F. F. (2009). El Niño in a changing climate. *Nature*, 461, 511–514.

Yuan X (2004) ENSO-related impacts on Antarctic Sea ice: A synthesis of phenomena and mechanism. *Ant. Sci.*, 16(4), 415–425.

Yuan, X., and Martinson, D. G. (2000). Antarctic Sea ice extent variability and its global connectivity. *J. Clim.*, 13, 1697–1717.

Yuan, X., Kaplan, M., and Cane, M. (2018). The interconnected global climate system – A review of tropical-polar teleconnections. *J. Clim.*, 31, 5765–5792.

9 Quantifying the Predictability of Southern Indian Ocean Sea-Ice Concentration in a Changing Climate Scenario

Suneet Dwivedi and Lokesh Kumar Pandey
K Banerjee Centre of Atmospheric and Ocean Studies and
M N Saha Centre of Space Studies, University of Allahabad,
Allahabad, India

CONTENTS

9.1 INTRODUCTION

The Southern Ocean sea ice influences the global climate system by modulating the Earth's energy budget and general circulation of the atmosphere and ocean (Steele et al., 1997; Massom and Stammerjohn, 2010; Bouchat and Tremblay, 2014). The biogeochemical cycles in the Southern Ocean are also influenced by the sea-ice variability (Thomas et al., 2009). The formation, evolution, and melting of the Southern Ocean sea ice, on the other hand, critically depends on the seasonal changes in the air-sea fluxes (Hall and Visbeck, 2002; Raphael, 2007; Massom and Stammerjohn, 2010; Holland and Kwok, 2012; Rae et al., 2013; Kumar et al. 2018a). Climate change significantly influences the sea-ice properties of the Southern Ocean (Comiso et al., 2017; Parkinson & Cavalieri, 2012). By analyzing the Antarctic sea-ice variability and trends during the years 1979–2010, Parkinson and Cavalieri (2012) found a positive trend in the Antarctic sea-ice cover since the late 1970s. Parkinson (2019) has also shown that the Antarctic sea-ice extent has been slightly increasing during the years 1979–2018. On the other hand, studies have also reported a rapid loss in the sea ice in the western sector of the Antarctic (Holland, 2014).

The realistic simulation of the Southern Ocean sea ice by the state-of-the-art climate models has remained a challenging problem (Turner et al., 2013; Shu et al., 2015; Roach et al., 2018; Holmes et al., 2019). The Coupled Model Intercomparison Project (CMIP) models are useful for assessing the performance (even though with 'low confidence') of Antarctic sea-ice simulations against the available observations (Turner et al., 2013; Shu et al., 2015; Roach et al., 2020; Shu et al., 2020). It has been shown that the CMIP5 (Taylor et al., 2012) and CMIP6 (Eyring et al. 2016) models can adequately reproduce the seasonal cycle of the Antarctic sea ice extent, though with a large intermodel spread (Shu et al., 2015; Roach et al., 2020; Shu et al., 2020). The importance of employing CMIP models for this purpose lies in the fact that these models not only help us in knowing the historical and present changes in the sea ice but they can also help us in investigating the possible changes in the future sea-ice extent in a warming environment, especially under the different greenhouse gas (GHG) emission scenarios.

It has been found that the ice sheet in the Indian Ocean sector of the Antarctic (e.g. East Antarctic) is more stable compared to West Antarctica (Naish et al., 2009; Pollard and DeConto, 2009), but it has become more dynamic in recent years and is receding faster as a result of climate change (Pollard et al., 2015; Aitken et al., 2016). Even though the coupled ocean sea-ice modeling studies giving realistic simulation of the sea ice have been carried out in the southern Indian Ocean region of the Antarctic with special emphasis over the Indian Antarctic Stations, Maitri [11.7°E; 70.7°S] and Bharati [76.1°E; 69.4°S] (Kumar et al., 2017, 2018a, 2018b), however, to the best of our knowledge, the examples of research utilizing CMIP output for gauging the performance of sea-ice simulations specifically over this region are not there. This study aims to evaluate the performance of CMIP6 sea-ice simulations against the satellite observation in the eastern Antarctic sector, specifically in the southern Indian Ocean region of the Antarctic covering both the Indian Antarctic stations.

The nonlinear dynamical techniques such as the fractal and multi-fractal dimension analysis have been widely used on the time series data of the meteorological (atmosphere, ocean, land, sea ice) and other geophysical processes for the determination of their dimensionality, multi-fractality, variability, as well as predictability (Rangarajan and Sant, 2004; Dwivedi, 2012; Valle et al., 2013; Baranowski et al., 2015; Hou et al., 2018; Kumar et al., 2018b; Laib et al., 2018; Lopenz Lambrano et al., 2018; Agbazo et al., 2019; Cadenas et al., 2019; Chandrasekaran et al., 2019; Kimothi et al., 2019). In particular, researchers have also utilized these nonlinear time series analysis techniques on the sea-ice data of the Arctic and Antarctic (Chmel et al., 2005; Agarwal et al., 2012; Moon et al., 2019; Rampal et al., 2019 and references therein). However, the predictability (and chaotic nature) of East Antarctic sea-ice fraction in the context of climate change has not been quantified in any scientific study so far. The estimation of predictability of the southern Indian Ocean sea ice around the Maitri and Bharati regions is of vital importance in the Indian context. These stations are situated at very different locations in the southern Indian Ocean and their sea-ice properties are also very different from each other (Kumar et al., 2017).

One of the aims of this chapter is to estimate the sea ice variability and predictability of the southern Indian Ocean and compare the predictability of the Maitri and Bharati Indian Antarctic stations using the historical data of CMIP6 model output. The effect of climate change on the sea-ice predictability of these regions shall also be investigated in terms of the generalized Hurst exponent (Di Matteo et al., 2003) and Climate Predictability Index (Rangarajan and Sant, 2004) as quantifiers. The future projection data of the SSP5–8.5 GHG emission scenario of CMIP6 models shall be utilized for this purpose. We organize the chapter as follows. Section 2 briefly describes the methodology and CMIP6 models used. The results are given in Section 3. Conclusions are in Section 4.

9.2 DATA AND METHOD

We analyze a total of 33 CMIP6 models for which historical data of the sea-ice concentration (SIC) was available. The details of these models are summarized in Table 9.1. These CMIP6 models are interpolated to a $0.5° \times 0.5°$ regular grid in the southern Indian Ocean region around [10E–100E; 55S–75S]. The SIC of these models is compared against the observed satellite data, namely, NOAA/NSIDC Climate Data Record of Passive Microwave Sea Ice Concentration Version 3 (Peng et al., 2013). The observed data is available from 1979 to 2014. To investigate the effect of climate change on the predictability of the sea ice, we use the SIC of the SSP5–8.5 projection scenario of the CMIP6 models. For the SSP5–8.5 scenario, the SIC data is available only in 15 CMIP6 models (details given in Table 9.1).

The fractal dimension analysis helps us to determine whether the SIC variability of the southern Indian Ocean is deterministic or it is random (Peitgen et al., 2004). It also helps us to quantify the predictability of SIC. For quantifying the fractal dimension of the SIC time series data, we use the generalized Hurst exponent computation given in Di Matteo et al. (2003). The Hurst exponent (H) (Di Matteo et al., 2003; Peitgen et al., 2004; Rangarajan and Sant, 2004) is a measure of the persistence, i.e. long-term memory in a data set. The fractal dimension (D) of a time series is related to the Hurst exponent (H) as $H = 2 - D$. A time series with a fractal dimension of 1.5 (i.e. $H = 0.5$) is representative of Brownian motion in which the present state does not correlate with past or future state. In such a time series, it will not be possible to determine any trend in amplitude, and, therefore, such processes will be unpredictable. The time series in which $0.5 < H \leq 1$ (i.e. $1 \leq D < 1.5$) will exhibit 'persistence' and increase in predictability where future values are likely to vary according to a known trend in the data. Similarly, the time series in which $0 \leq H < 0.5$ (i.e. $1.5 < D \leq 2$) will exhibit 'anti-persistence'. The predictability of such data also increases since a decrease in the amplitude of the process is more likely to lead to an increase in the future (Rangarajan and Sant, 2004).

TABLE 9.1

CMIP6 Models Used in the Study. The Models Are Classified as Good/Poor according to Their Performance in Realistically Simulating the SIC Based on the Taylor Diagram Metric. The Third Column Represents Those Good Models for Which SSP5–8.5 Future Projection Data Was Available

S. No.	CMIP6 Model	Good/Poor	Good Models with SSP5–8.5 Data
1	ACCESS-CM2	Good	Available
2	ACCESS-ESM1–5	Good	Available
3	AWI-ESM-1-1-LR	Good	Not Available
4	BCC-CSM2-MR	Poor	–
5	BCC-ESM1	Poor	–
6	CAMS-CSM1-0	Poor	–
7	CAS-ESM2-0	Good	Not Available
8	CESM2-FV2	Good	Not Available
9	CESM2-WACCM-FV2	Good	Not Available
10	CESM2-WACCM	Good	Available
11	CESM2	Good	Available
12	CIESM	Good	Available
13	CMCC-CM2-SR5	Good	Available
14	CanESM5	Good	Available
15	E3SM-1-0	Good	Not Available
16	E3SM-1-1-ECA	Good	Not Available
17	E3SM-1-1	Good	Not Available
18	EC-Earth3	Good	Available
19	FGOALS-f3-L	Good	Available
20	FIO-ESM-2-0	Good	Available
21	GFDL-CM4	Good	Available
22	GFDL-ESM4	Good	Available
23	GISS-E2-1-H	Poor	–
24	INM-CM5-0	Poor	–
25	IPSL-CM6A-LR	Good	Available
26	MIROC6	Poor	–
27	MPI-ESM1–2-HR	Poor	–
28	MRI-ESM2-0	Good	Available
29	NESM3	Good	Available
30	NorCPM1	Good	Not Available
31	NorESM2-LM	Poor	–
32	SAM0-UNICON	Good	Not Available
33	TaiESM1	Good	Not Available

The Predictability Index (PI) is defined as (Rangarajan and Sant, 2004):

$$PI = 2|D-1.5| = 2|0.5-H| \qquad (9.1)$$

where | | denotes the absolute value. Equation (9.1) suggests that predictability increases in both cases, when H is greater than 0.5 (persistence) and when it becomes less than 0.5 (anti-persistence).

9.3   RESULTS AND DISCUSSION

We begin by showing the performance of all the CMIP6 models in simulating the mean SIC of the southern Indian Ocean region [10E–100E; 55S–75S] (Figure 9.1). This region covers both the Indian Antarctic stations Maitri and Bharati. We see from the figure that the EC-Earth3 model shows the poorest performance in simulating the mean SIC. The mean SIC of this model is even less than half of the observed SIC. Other than this model, ACCESS-CM2, CIESM, MIROC6, and NESM3 models also simulate significantly less mean SIC than the observed value. The mean SIC of all other 28 models is reasonably well simulated when compared with observation. We then evaluate the overall fidelity of individual CMIP6 models in simulating the SIC variability using the Taylor diagram (Taylor, 2001). The Taylor diagrams are well-accepted performance metrics for climate models that provide a brief statistical outline of how well spatiotemporal patterns of a variable match against the observation in terms of their correlation coefficients (CC), their root-mean-square error (RMSE), and the simulated to an observed ratio of their variances (Sharmila et al., 2015). The Taylor diagram of SIC averaged over the southern Indian Ocean region [10E–100E; 55S–75S] corresponding to all 33 CMIP6 models is shown in Figure 9.2. The models for which CC with observed SIC is high, the standard deviation is close to the observation, and the RMSE value is small are considered as good models. By applying this criterion, we find that out of 33 models, only 25 models can simulate the SIC variability correctly. The eight models, namely, BCC-CSM2-MR, BCC-ESM1, CAMS-CSM1-0, GISS-E2-1-H, INM-CM5-0, MIROC6, MPI-ESM1-2-HR, and NorESM2-LM, are not able to pass through the Taylor diagram statistical performance metric. We, therefore, use the 25 CMIP6 models (Table 9.1) by excluding these eight models in further analysis of SIC data carried out in this chapter. We call these 25 models as good CMIP5 models for this study.

We show in Figure 9.3 the SIC annual cycle of these 25 good CMIP6 models. The Multi-Model Mean (MMM) annual cycle of the SIC, as well as the annual cycle of observed data, is also shown in Figure 9.3. We notice a large inter-model spread in the annual cycle of the models. However, it is interesting to note that the SIC annual cycle of the MMM matches very well with the observed annual cycle, thus giving confidence in the use of CMIP6 data of the SIC. The annual cycle curve suggests that the SIC of the southern Indian Ocean region remains lowest in the February–March months (melting season), whereas it remains highest in the September–October months (freezing season) when almost 70% of the entire region remains occupied with the sea ice.

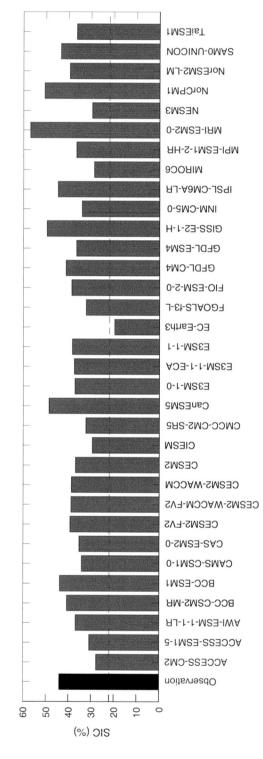

FIGURE 9.1 Mean SIC of the southern Indian Ocean region area averaged over [10E–100E; 55S–75S] for a total of 33 CMIP6 models (red). The mean SIC of observation is also shown (black). The dashed horizontal line represents half of the mean SIC of observation.

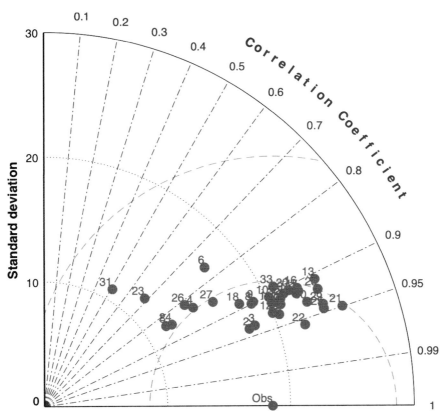

FIGURE 9.2 Taylor diagram of area-averaged SIC over the southern Indian Ocean region corresponding to 33 CMIP6 models (red). The CC value is shown in blue, standard deviation in black, and RMSE in green color.

To carry out the fractal analysis for computing the predictability index of the SIC of the southern Indian Ocean region, we generate a long time series of the monthly SIC data area-averaged over this region from 1900–2014 (115 years) using the historical data of good CMIP6 models. To quantify the effect of climate change on the predictability of the sea ice, we also carry out fractal analysis on CMIP6 SSP5–8.5 (high global warming) future scenario data for the years 2015–2100 (86 years). The MMM of historical and SSP5–8.5 data are made for those good 15 models for which historical, as well as projection data of SIC, are available. The CMIP6 MMM SIC time series of historical and SSP5–8.5 data are shown in Figures 9.4(a) and (b), respectively. We see from the figure that both SIC time series show a decreasing trend. However, it is found that SIC will decrease at an alarming rate of 0.016% per month (~0.2% per year) in the SSP5–8.5 scenario. We also compare the annual cycle of the historical and SSP5–8.5 SIC data (Figure 9.4c). We find that even though the seasonal variability of the SIC will remain intact in the SSP5–8.5 global warming scenario, however, the magnitude of

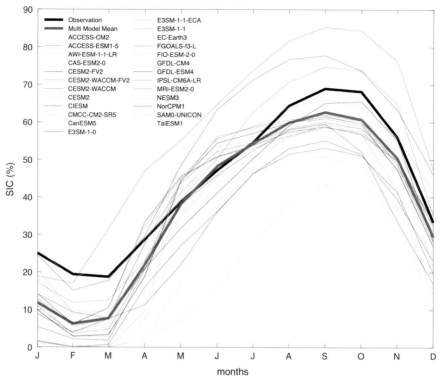

FIGURE 9.3 Annual cycle of area-averaged SIC corresponding to good CMIP6 models. The Multi-Model Mean (MMM) annual cycle of the SIC (thick red curve), as well as the annual cycle of observed data (thick black curve), is also shown for comparison.

SIC will decrease in the SSP5–8.5 scenario in all the months, with the maximum decrease being in the melting season of August–September–October.

We remove the long-term linear trend from the historical as well as SSP5–8.5 data of the SIC. The detrended time series is used for computing the generalized Hurst exponent of the SIC data in the southern Indian Ocean region. Using the method described in Di Matteo et al. (2003), we obtained the generalized Hurst exponent H(q) of the SIC time series for q-order moments of 1–5. The results for historical as well as SSP5–8.5 projection scenarios are shown in Figure 9.5(a). It is clear from the figure that H(q) is a nonlinear function of q for both the historical as well as SSP5–8.5 SIC data, suggesting multi-fractality of the SIC time series. In other words, the SIC of the southern Indian Ocean exhibits multi-fractal nature both in the historical period as well as in a high GHG global warming scenario. The Hurst exponent H(1) corresponding to q = 1 describes the scaling behavior of the absolute value of the increments and is commonly used to assign a single fractal dimension to a variable. The H(2) is associated with the scaling of the autocorrelation function and is related to the power spectrum (Di Matteo et al., 2003). In this paper, we focus on the Hurst exponent H(q = 1) for the determination of predictability of the southern Indian Ocean SIC. We also see from the figure that the Hurst exponent of SSP5–8.5 data remains

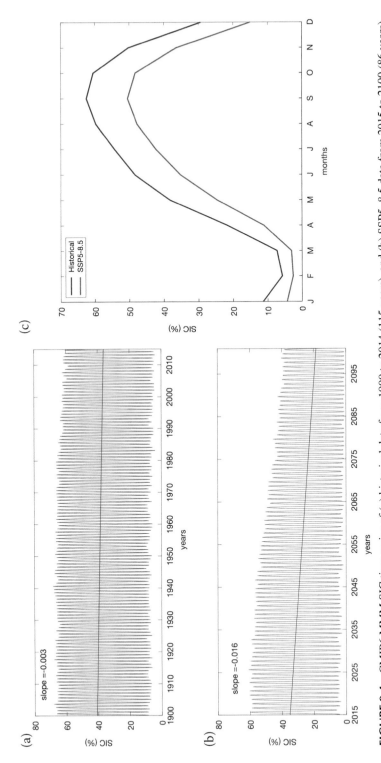

FIGURE 9.4 CMIP6 MMM SIC time series of (a) historical data from 1900 to 2014 (115 years), and (b) SSP5–8.5 data data from 2015 to 2100 (86 years). The solid black line in (a) and (b) represents linear trend; (c) annual cycle of the historical (blue) and SSP5–8.5 (red) SIC data.

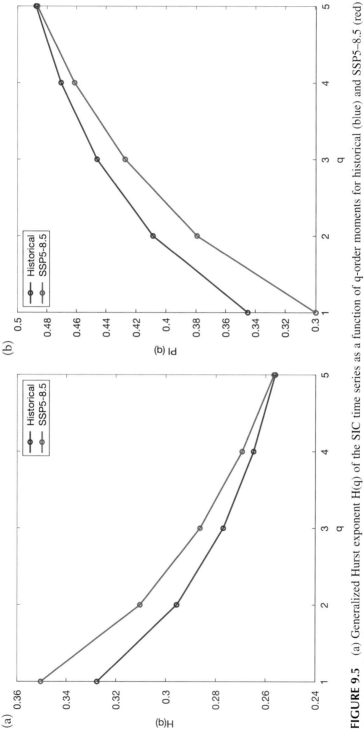

FIGURE 9.5   (a) Generalized Hurst exponent H(q) of the SIC time series as a function of q-order moments for historical (blue) and SSP5–8.5 (red) data; (b) predictability Index (PI) of the SIC time series as a function of q for historical (blue) and SSP5–8.5 (red) data.

higher compared to the corresponding historical data. For higher q, H(q) of both time series approach each other. We find that H(1) of historical and SSP5–8.5 data is 0.33 (i.e. fractal dimension D = 1.67) and 0.35 (i.e. fractal dimension D = 1.65), respectively. These H values are less than 0.5. Thus, the SIC variability of the southern Indian Ocean is different from a random unpredictable Brownian motion process and shows 'anti-persistence' and predictable nature.

Motivated by this observation, we compute the Predictability Index (PI) using Equation (9.1). The PI values for historical as well as SSP5–8.5 data are shown in Figure 9.5(b). It is to be emphasized here that the PI values of SIC are related to the interannual variability of the SIC, and not to the magnitude of SIC. The PI values are especially useful in determining how predictable a process is (Rangarajan and Sant, 2004). By definition (Eq. 9.1), PI varies between 0 (unpredictable) and 1 (highly predictable). The higher the value of PI, the more predictable will be the process. We see from Figure 9.5(b) that the PI of historical data remains higher compared to SSP5–8.5 data. Thus, the predictability of the southern Indian Ocean SIC will be decreasing in a warming environment. As the GHG concentration will increase, the southern Indian Ocean SIC and its predictability will decrease.

The quantification of predictability of the SIC of Maitri and Bharati Indian Antarctic stations in a changing climate is very important from India's point of view. We generate the area-averaged monthly SIC time series over the region [10E–14E; 65S–69S] around Maitri and over the region [65E–77E; 65S–69S] around the Bharati station for historical as well as SSP5–8.5 CMIP6 data. We compute the Predictability Index (PI) of these time series using the method described previously. The Hurst exponent, fractal dimension, and PI values of SIC around the Maitri and Bharti regions are described in Table 9.2 for historical as well as SSP5–8.5 data. We see from the table that the SIC of the Maitri region (PI = 0.34) is more predictable compared to the Bharati region (PI = 0.31) during the historical period of 1900–2014. Further, with an increase in GHG concentration, the predictability of SIC of these regions shall decrease in the SSP5–8.5 scenario during the years 2015–2100.

TABLE 9.2
Hurst Exponent, Fractal Dimension and Predictability Index of the SIC Time Series Data around the Maitri and Bharati Indian Antarctic Stations for Historical (1900–2014) and SSP5–8.5 (2015–2100) Period

Stations	Hurst Exponent		Fractal Dimension		Predictability Index	
	Historical	SSP5–8.5	Historical	SSP5–8.5	Historical	SSP5–8.5
Maitri	0.33	0.36	1.67	1.64	0.34	0.29
Bharati	0.34	0.37	1.66	1.63	0.31	0.26

9.4 CONCLUSIONS

In this chapter, we evaluated the performance of a total of 33 CMIP6 models in simulating the sea-ice concentration (SIC) of the southern Indian Ocean region against the corresponding satellite observation. For this purpose, the CMIP6 data of the SIC is taken over the region [10E–100E; 55S–75S]. The region covers both Indian Antarctic Stations Maitri and Bharati. The statistical analysis using the Taylor diagram metric suggests that only 25 out of 33 models can be considered good models for SIC simulation in our region of interest. We find large inter-model spread in the annual cycle of SIC over the southern Indian Ocean. Interestingly, the multi-model mean annual cycle of the SIC matches well with the corresponding observed satellite data. We find that the SIC decreases at a slower rate of nearly 0.04% per year during 1900–2014, whereas it will decrease at an alarming rate of nearly 0.2% per year in the SSP5–8.5 scenario during 2015–2100. We investigate the effect of global warming and the increase in greenhouse gas concentration on the predictability of southern Indian Ocean SIC. The predictability of the SIC is quantified in terms of the Predictability Index (PI) and generalized Hurst exponent. The first moment Hurst exponent of historical and SSP5–8.5 SIC data of the region is 0.33 and 0.35, respectively. This suggests that the SIC variability of the region is predictable. We find that the PI value obtained from the analysis of historical SIC data is greater than the corresponding value obtained from SSP5–8.5 SIC data. Thus, the predictability of southern Indian Ocean SIC will decrease in a warming environment with high GHG forcing. The analysis of the SIC time series of Maitri and Bharati Indian Antarctic stations reveals that the SIC around Maitri is more predictable compared to the Bharati region. We also find that the predictability of SIC around both Indian Antarctic stations will decrease in the SSP5–8.5 future projection scenario as a result of anthropogenic climate change.

ACKNOWLEDGMENTS

SD thanks DST-CCP for providing financial assistance in the form of research projects. LKP is thankful to the DST-CCP for providing research fellowship. The CMIP6 data of the SIC are downloaded from https://esgf-node.llnl.gov/search/cmip6/.

REFERENCES

Agarwal, S., Moon, W., and Wettlaufer, J. S. (2012). Trends, noise and re-entrant long-term persistence in Arctic sea ice. *Proc. R. Soc. A*, 468, 2416–2432.

Agbazo, M., Koto N'gobi, G., Alamou, E., Kounouhewa, B., Afouda, A., and Kounkonnou, N. (2019). Multifractal behaviors of daily temperature time series observed over Benin Synoptic Stations (West Africa). *Earth Sci. Res. J.*, 23(4), 365–370.

Aitken, A. R. A., Roberts, J. L., Van Ommen, T. D., Young, D. A., Golledge, N. R., Greenbaum, J. S., Blankenship, D. D., and Siegert, M. J. (2016). Repeated large-scale retreat and advance of Totten Glacier indicated by inland bed erosion. *Nature*, 533(7603), 385–389.

Baranowski, P., Krzyszczak, J., Slawinski, C., Hoffmann, H., Kozyra, J., Nieróbca, A., Siwek, K., and Gluza, A., (2015). Multifractal analysis of meteorological time series to assess climate impacts. *Clim. Res.*, 65, 39–52.

Bouchat, A., and Tremblay, B. (2014). Energy dissipation in viscous-plastic sea-ice models. *J. Geophys. Res.*, 119, 976–994.

Cadenas, E., Campos-Amezcua, R., Rivera, W., et al. (2019). Wind speed variability study based on the Hurst coefficient and fractal dimensional analysis. *Energy. Sci. Eng.*, 7, 361–378.

Chandrasekaran, S., Poomalai, S., Saminathan, B., Suthanthiravel, S., Sundaram, K., and Abdul Hakkim, F. F. (2019). An investigation on the relationship between the Hurst exponent and the predictability of a rainfall time series. *Meteorol. Appl.*, 26, 511–519.

Chmel, A., Smirnov, V. N., and Astakhov, A. P. (2005). The Arctic sea-ice cover: Fractal space-time domain. *Physica A*, 357, 556–564.

Comiso, J. C., Gersten, R., Stock, L. V., Turner, J., Perez, G. J., and Cho, K. (2017). A positive trend in the Antarctic Sea ice cover and associated changes in surface temperature. *J. Clim.*, 30(6), 2251–2267.

Di Matteo, T., Aste, T., and Dacorogna, M. M. (2003). Scaling behaviours in differently developed markets. *Phys. A: Stat. Mech. Appl.*, 324, 183–188.

Dwivedi, S. (2012). Quantifying predictability of Indian summer monsoon intraseasonal oscillations using nonlinear time series analysis. *Meteorol. Z.*, 21(4), 413–419.

Eyring, V., Bony, S., Meehl, G. A., Senior, C. A., Stevens, B., Stouffer, R. J., and Taylor, K. E. (2016). Overview of the Coupled Model Intercomparison Project Phase 6 (CMIP6) experimental design and organization. *Geosci. Model Dev.*, 9, 1937–1958, https://doi.org/10.5194/gmd-9-1937-2016

Hall, A., and Visbeck, M. (2002). Synchronous variability in the Southern Hemisphere atmosphere, sea ice, and ocean resulting from the Annular Mode. *J. Clim.*, 15, 3043–3057.

Holland, P. R. (2014). The seasonality of Antarctic sea ice trends. *Geophys. Res. Lett.*, 41, 4230–4237.

Holland, P. R., and Kwok, R. (2012). Wind-driven trends in Antarctic sea-ice drift. *Nat. Geosci*, 5(12), 872–875.

Holmes, C. R., Holland, P. R., and Bracegirdle, T. J. (2019). Compensating biases and noteworthy success in the CMIP5 representation of Antarctic sea ice processes. *Geophys. Res. Lett.*, 46, 4299–4307.

Hou, W., Feng, G., Yan, P., and Li, S. (2018). Multifractal analysis of the drought area in seven large regions of China from 1961 to 2012. *Meteorol. Atmos. Phys.*, 130, 459–471.

Kimothi, S., Kumar, A., Thapliyal, A., Ojha, N., Soni, V. K., and Singh, N. (2019). Climate predictability in the Himalayan foothills using fractals. *Mausam*, 70, 357–362.

Kumar, A., Dwivedi, S., and Pandey, A. C. (2018b). Quantifying predictability of sea ice around the Indian Antarctic stations using coupled ocean sea ice model with shelf ice. *Pol. Sci.*, 18, 83–93.

Kumar, A., Dwivedi, S., and Rajak, D. R. (2017). Ocean sea-ice modelling in the Southern Ocean around Indian Antarctic stations. *J. Earth. Syst. Sci.*, 126(5), 70. https://doi.org/10.1007/s12040-017-0848-5.

Kumar, A., Dwivedi, S., Rajak, D. R., and Pandey, A. C. (2018a). Impact of air-sea forcings on the Southern Ocean sea ice variability around the Indian Antarctic stations. *Pol. Sci.*, 18, 197–212.

Laib, M., Golay, J., Telesca, L., and Kanevski, M. (2018). Multifractal analysis of the time series of daily means of wind speed in complex regions. *Chaos, Solitons and Fractals*, 109, 118–127.

Lopenz Lambrano, A. A. et al. (2018). Spatial and temporal Hurst exponent variability of rainfall series based on the climatological distribution in a semiarid region in Mexico. *Atmósfera*, 31(3), 199–219.

Massom, R. A., and Stammerjohn, S. E. (2010). Antarctic sea ice change and variability-Physical and ecological implications. *Pol. Sci.*, 4, 149–186.

Moon, W., Nandan, V., Scharien, R. K., Wilkinson, J., Yackel, J. J., Barrett, A., Lawrence, I., Segal, R. A., Stroeve, J., Mahmud, M., and Duke, P. J. (2019). Physical length scales of wind-blown snow redistribution and accumulation on relatively smooth Arctic first-year sea ice. *Environ. Res. Lett.*, 14, 104003, https://doi.org/10.1088/1748-9326/ab3b8d.

Naish, T., Powell, R., Levy, R., Wilson, G., Scherer, R., Talarico, F., Krissek, L., Niessen, F., Pompillo, M., Wilson, T. (2009). Obliquity-paced Pliocene West Antarctic ice sheet oscillations. *Nature*, 458, 322–328.

Parkinson, C. L. (2019). A 40-y record reveals gradual Antarctic sea ice increases followed by decreases at rates far exceeding the rates seen in the Arctic. *Proc. Natl. Acad. Sci. U.S.A.*, 116(29), 14414–14423.

Parkinson, C. L., and Cavalieri, D. J. (2012). Antarctic sea ice variability and trends, 1979-2010. *Cryosphere*, 6, 871–880.

Peitgen, H.-O., Jurgens, H., and Saupe, D. (2004). *Chaos and Fractals: New Frontiers of Science*, Springer-Verlag, New York, pp. 1–864.

Peng, G., Meier, W. N., Scott, D., and Savoie, M. (2013). A long-term and reproducible passive microwave sea ice concentration data record for climate studies and monitoring. *Earth Syst. Sci. Data*, 5, 311–318. https://doi.org/10.5194/essd-5-311-2013.

Pollard, D., and DeConto, R. M. (2009). Modelling West Antarctica ice sheet growth and collapse through the past five million years. *Nature*, 458, 329–332.

Pollard, D., DeConto, R. M., and Alley, R. B. (2015). Potential Antarctic ice sheet retreat is driven by hydrofracturing and ice cliff failure. *Earth Planet Sci. Lett.*, 412, 112–121.

Rae, J., Hewitt, H., Keen, A., Ridley, J., Edwards, J., and Harris, C. (2013). A sensitivity study of the sea ice simulation in the global coupled climate model, HadGEM3. *Ocean Model*, 74, 60–76.

Rampal, P., Dansereau, V., Olason, E., Bouillon, S., Williams, T., Korosov, A., and Samaké, A. (2019). On the multi-fractal scaling properties of sea ice deformation. *The Cryosphere*, 13, 2457–2474.

Rangarajan, G., and Sant, D. A. (2004). Fractal dimensional analysis of Indian climatic dynamics. *Chaos Solit. Fractals*, 19, 285–291.

Raphael, M. N. (2007). The influence of atmospheric zonal wave three on Antarctic sea ice variability. *J. Geophys. Res.*, 112(D12). https://doi.org/10.1029/2006JD007852.

Roach, L. A., Dean, S. M., and Renwick, J. A. (2018). Consistent biases in Antarctic sea ice concentration simulated by climate models. *The Cryosphere*, 12, 365–383.

Roach, L. A., Dörr, J., Holmes, C. R., Massonnet, F., Blockley, E. W., Notz, D., et al. (2020). Antarctic sea ice area in CMIP6. *Geophys. Res. Lett.*, 47, e2019GL086729. https://doi.org/10.1029/2019GL086729.

Sharmila, S., Joseph, S., Sahai, A. K., Abhilash, S., and Chattopadhyay, R. (2015). Future projection of Indian summer monsoon variability under climate change scenario: An assessment from CMIP5 climate models. *Glob. Planet. Change*, 124, 62–78.

Shu, Q., Song, Z., and Qiao, F. (2015). Assessment of sea ice simulations in the CMIP5 models. *The Cryosphere*, 9(1), 399–409.

Shu, Q., Wang, Q., Song, Z., Qiao, F., Zhao, J., Chu, M., and Li, X. (2020). Assessment of sea ice extent in CMIP6 with comparison to observations and CMIP5. *Geophys. Res. Lett.*, 47, e2020GL087965. https://doi.org/10.1029/2020GL087965.

Steele, M., Zhang, J., Rothrock, D., and Stern, H. (1997). The force balance of sea ice in a numerical model of the Arctic Ocean. *J. Geophys. Res.: Oceans*, 102(C9), 21061–21079.

Taylor, K. E. (2001). Summarizing multiple aspects of model performance in a single diagram. *J. Geophys. Res.*, 106 (D7), 7183–7192.

Taylor, K. E., Stouffer, R. J., and Meehl, G. A. (2012). An overview of CMIP5 and the experiment design. *Bull. Amer. Meteor. Soc.*, 93, 485–498.

Thomas, H., Schiettecatte, L. S., Suykens, K., Kone, Y. J. M., Shadwick, E. H., Prowe, A. E. F., Bozec, Y., de Baar, H. J. W., and Borges, A. V. (2009). Enhanced ocean carbon storage from anaerobic alkalinity generation in coastal sediments. *Biogeosciences*, 6, 267–274.

Turner, J., Bracegirdle, T. J., Phillips, T., Marshall, G. J., and Hosking, J. S. (2013). An initial assessment of Antarctic sea ice extent in the CMIP5 models. *J. Clim.*, 26(5), 1473–1484.

Valle, M. A. V., García, G. M., Cohen, I. S., Klaudia Oleschko, L., Ruiz Corral, J. A., and Korvin, G. (2013). Spatial Variability of the Hurst Exponent for the Daily Scale Rainfall Series in the State of Zacatecas. *Mexico. J. Appl. Meteor. Climatol.*, 52, 2771–2780.

10 Antarctic Decadal Sea-Ice Variability

Alvarinho J. Luis

Earth System Science Organization – National Centre for Polar and Ocean Research, Ministry of Earth Sciences, Goa, India

CONTENTS

10.1 INTRODUCTION

Sea ice is a layer of ice over the ocean, which exists as far as 38°N in the Northern Hemisphere, and extends as far as 54°N in the Southern Hemisphere (SH) (Parkinson, 2019). The bright surface reflects 80% of the solar radiation into space, and inhibits the exchange of heat and momentum fluxes between ocean and atmosphere, and maintaining lower temperatures. When the sea ice completely melts, this radiation is absorbed by the darker ocean surface, creating positive feedback (Screen and Simmonds, 2010), wherein the ocean releases the heat back to the atmosphere in autumn. When sea ice forms, most of the salty and dense water sinks and flows along the ocean bottom toward the equator, while the warm water from mid-depth to the surface flows from the equator to the poles; this contributes to the ocean's global conveyor-belt circulation (Rahmstorf, 2003). The sea ice also plays a crucial role in the biogeochemical cycles of the Southern Ocean (SO) (Deppeler and Davidson, 2017), wherein it contributes to variability in primary production by serving as a habitat for algal biomass and growth; affecting nutrient distribution, ocean stratification, and light availability; and promoting widespread phytoplankton blooms upon its melt in Austral spring and summer (Arrigo et al., 2009). Sea ice is known to regulate the SO biological pump and the sequestration of atmospheric

CO_2 (Tréguer and Pondaven, 2002), as well as ocean upwelling and outgassing and ocean acidification (Barker and Ridgwell, 2012).

The Antarctic sea-ice extent (SIE) increases from about 3×10^6 km^2 in Austral summer (January–March) to 19×10^6 km^2 in Austral winter (July–September). At the north, the Arctic SIE spans ~6×10^6 km^2 in summer (July–September) to 15×10^6 km^2 in winter (January–March). This difference is a result of geography, wherein the expansion of the sea ice formed over the North Pole is restricted by the landmass of Eurasia, North America, and Greenland. Only 15% of Antarctic sea ice survives in the summer because it forms at lower, warmer latitudes, compared to the 40% of the Arctic sea ice remaining at the end of summer over the North Pole. Overall, the average Antarctic is thinner compared to the multi-year 3–4 m thick Arctic ice. The Antarctic sea ice shows less variability in Austral summer, and more variability in Austral winter because the sea ice can expand and drift freely across the SO with no land boundaries to the north. Weather phenomena like unusual air pressure patterns give rise to the wind shifts and exert a greater influence on Arctic minimum and the Antarctic maximum sea ice.

The polar sea ice has received considerable attention largely because of significant decreases in the Arctic SIE (Stroeve and Notz, 2018), and increases in the Antarctic SIE since the late 1970s (Zwally et al., 2002; Parkinson and Cavalieri, 2012). From November 1978 through 2019, though the Arctic sea-ice extent declined in all months, Antarctic SIE showed an overall slight positive trend in all months except November (-0.1 ± 0.8%/decade), with the greatest increase in March (1.9 ± 3.7%/decade) (www.nsidc.org). It is noted that despite overall increases in the Antarctic sea ice (Parkinson and Cavalieri, 2012), the SIE has decreased considerably in the Bellingshausen and Amundsen Seas (BAS), immediately to the west of the Antarctic Peninsula (Stammerjohn et al., 2008). The record maximum Antarctic SIE of 19.72×10^6 km^2 was attained on 24th September 2012 (Turner et al., 2013a) and a second record maximum cover of 19.58×10^6 km^2 was reached on 30th September 2013, according to the National Snow and Ice Data Center Sea Ice Index (Fetterer et al., 2017). However, during the Austral spring of 2016, the Antarctic sea-ice cover decreased at a record rate of 2.25×10^6 km^2, after which the extent stayed at low levels and culminated on 1st March 2017 to a record low of 2.07×10^6 km^2 (Turner et al., 2017), since a continuous record of multi-frequency passive microwave satellite observations began in 1979 (Parkinson and Cavalieri, 2012).

A decreasing trend in the Arctic sea ice (Parkinson, 2014) is in line with Arctic warming (Walsh, 2013) in response to rising atmospheric greenhouse gases, but the positive trend in Antarctic sea ice (Parkinson and Cavalieri, 2012), in an era of marked global warming, is inscrutable. Although many mechanisms have been proposed, it is still unclear whether the increasing trend is anthropogenically driven or only caused by internal natural variability. Intuitive explanations put forth for the variability are the influence of the large-scale atmospheric patterns associated with Southern Annular Mode (SAM), El Niño Southern Oscillation (ENSO), and ozone depletion (Harangozo, 2006; Holland and Kwok, 2012). Under positive SAM (resulting in strengthened SH westerlies) and neutral ENSO conditions, sea ice tends to advance earlier in the western Ross Sea (RS) sector but later in the Bellingshausen Sea sector (Stammerjohn et al., 2008; Turner et al., 2009; Simpkins et al., 2012).

The deepening of the Amundsen Sea Low (ASL) is consistent with negative trends in the sea ice, in particular the loss of ice and warming of the Bellingshausen Sea, and a positive trend in the eastern RS (Turner et al., 2013b). ASL has an annual cycle, with the low being found immediately west of the Antarctic Peninsula from December to February and moving westward to the RS during June–August (Fogt et al., 2012). Using model experiments, Turner et al. (2009) pointed out that the increase of SIE in the RS and the deepening of the ASL in Austral autumn are related largely to the decrease of stratospheric ozone. Models studies also suggest that ozone depletion is accompanied by a decrease in SIE all through the year and that the recently observed positive trends in SIE are likely linked to the large internal variability (Sigmond and Fyfe, 2014). The study of Holland and Kwok (2012) using concurrent sea-ice drift vectors and near-surface wind fields showed that the sea-ice anomalies off West Antarctica are primarily a result of wind-driven changes in sea-ice advection, while wind-driven thermodynamics play a greater role elsewhere.

Other mechanisms suggested for the sea-ice increase are ocean-ice feedback processes (Zhang, 2007), ocean change (Jacobs and Comiso, 1997), and Atlantic Multidecadal Oscillation (AMO) (Li et al., 2014). There was an exceptional increase in the sea ice from 2000 to 2014 ($0.57 \pm 0.33 \times 10^6$ km^2/decade) by a factor of five compared to the increase during 1979–1999 has been attributed to the negative phase of Interdecadal Pacific Oscillation (IPO). Another potential mechanism is that an increased mass loss of the Antarctic ice sheet due to increasing ocean warming and sub-surface ice-shelf melting causes the uppermost ocean layer to become fresher and thus lighter than the underlying warmer water and this decreases upward ocean heat transfer. It is argued that the freshwater layer can cool more rapidly in the autumn and early winter and contribute to sea-ice expansion (Bintanja et al., 2013). The vast majority of Coupled Model Intercomparison Project Phase 5 (CMIP) multi-model ensemble mean trends for 1979–1999 exhibit a statistically significant decrease in all seasons with larger decreasing trends in July–August and September–November (Yu et al., 2017), which has been attributed to the underestimated changes in wind-induced ocean circulation in the models (Purich et al., 2016).

The present work intends to consolidate the previous research scattered in literature to evolve a general picture of the drivers that cause the Antarctic sea-ice expansion. We examine the trend using monthly mean extent from the U.S. National Snow and Ice Data Center (Fetterer et al., 2017). We highlight the trends over the four decades, 1979–1988, 1989–1998, 1999–2008, and 2009–2018, and discuss the possible roles of local drivers and climate indices on the observed trends. In the discussion, we consolidate the research that explains the interconnection between local and remote drivers for SIE variability. It is intended that this chapter would cater to the knowledge of future scientists on the present-day scenario in consolidated form.

10.2   DATA

The SIE was obtained from the brightness temperature measurements done by Scanning Multichannel Microwave Radiometer (SMMR) on the National Aeronautics and Space Administration's *Nimbus 7* satellite; the Special Sensor

Microwave Imager (SSM/I) on the *F8, F11,* and *F13* satellites of the U.S. Department of Defense's Defense Meteorological Satellite Program (DMSP); and the SSM/I Sounder (SSM/IS) on the DMSP *F17* satellite. These satellites measure emitted radiation from Earth's surface in microwave wavelengths which discriminates liquid water from sea ice in all weather and day and night conditions and monitor full polar sea-ice cover every 1 or 2 days. The passive-microwave data have undergone rigorous inter-calibration between the SSM/I and SSM/IS sensors (Cavalieri et al., 1999; Cavalieri et al., 2012) to create a homogeneous data set for long-term trend studies.

The sea-ice product G02135 (Fetterer et al., 2017) was processed through the NASA Team algorithm (Cavalieri et al., 1984) and mapped onto rectangular grids overlaid on polar stereographic projections with a grid size of approximately 25 × 25 km. The data are distributed as a consistent time series covering 1978 to date by the National Snow and Ice Data Center (NSIDC). The advantage of the NASA team algorithm is that it is produced in near-real-time by the NSIDC (Fetterer et al., 2017) and has been extensively evaluated in earlier studies (e.g. Steffen and Schweiger, 1991; Emery et al., 1994; Agnew and Howell, 2003).

SIE was computed as the total area of all satellite pixels where the sea-ice cover exceeded 15%. Because of the sensitivity of microwave emissivity to melt, most studies focus on SIE instead of the sea-ice area to examine the temporal evolution of sea ice. Following the literature (cf. Parkinson, 2019), the region around Antarctica is partitioned into five sectors: the Weddell Sea (WS), Indian Ocean (IO), western Pacific Ocean (PO), RS, and BAS (Figure 10.1). Figure 10.2 shows the time series of the seasonal cycle of sea ice for different sectors. Monthly anomalies were generated by the removing long-term monthly-mean from that month, i.e. January mean for 1979–2018 was removed from the January SIE of each year, and likewise for other months. Using the anomalies, we fit a linear regression line to estimate the slope (or trend). The SIE trend is also expressed in percentage, which was computed relative to the 1981–2010 mean SIE.

Normally, the relationship between sea ice and the atmosphere is explained mostly in terms of local drivers through the atmosphere and ocean dynamics and thermodynamics (Liu et al., 2004; Zhang, 2007). To infer the role of atmosphere, we use 10-m wind speed (WS), mean sea level pressure (MSLP), heat flux (latent plus sensible heat flux), and sea surface temperature (SST), which were obtained from the ECMWF atmospheric reanalysis of the global climate (ERA5) reanalysis, which is regarded as the most reliable reanalysis over Antarctica (Bromwich et al., 2011; Bracegirdle and Marshall, 2012). ERA5 reanalyses model output is available at a horizontal resolution of 31 km and 137 vertical levels from the surface to 0.01 hPa which represents retrospective forms of numerical weather prediction, using a fixed prediction model and data assimilation system to provide global estimates of atmospheric variables from 1979 onwards. Air temperature (TA) was derived from the global air temperature data set HadCRUT4 (http://climexp.knmi.nl/daily2longer.cgi). HadCRUT4 consists of gridded temperature anomalies across the world as well as averages for 75°S–60°S representing the SH (Scafetta and Mazzarella, 2015).

Previous studies have found associations between SIE and low-frequency global climate indices such as the Pacific Decadal Oscillation (PDO), Atlantic Meridional

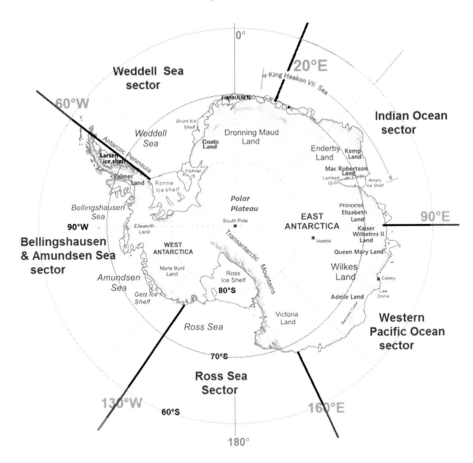

FIGURE 10.1   Study area. Different ocean sectors considered in this study are shown.

Oscillation (AMO), SOI, and SAM. A dominant year-round pattern of monthly North Pacific sea surface temperature, PDO is akin to a long-lived El Niño-like pattern of the Pacific (Zhang et al., 1997). We used updated standardized values for the PDO index, derived from the leading principal component of monthly SST anomalies in the North Pacific Ocean, poleward of 20°N (Mantua et al., 1997). We used unsmoothed AMO series, which was computed using detrended Kaplan SST, with the area-weighted average over the North Atlantic (0–70°N) (https://psl.noaa.gov/data/timeseries/AMO/). The Southern Oscillation Index (SOI) was downloaded from http://www.cpc.ncep.noaa.gov/. We used the station-based index of the SAM, which is based on the zonal pressure difference between the latitudes of 40°S and 65°S (Marshall, 2003).

The correlation coefficient was used to assess the degree of association between the time series of the reanalysis/climate indices and SIE, whose significance was judged at 95% and 99% significance levels using a two-tail student test with the null hypothesis of a zero trend, with 38 degrees of freedom.

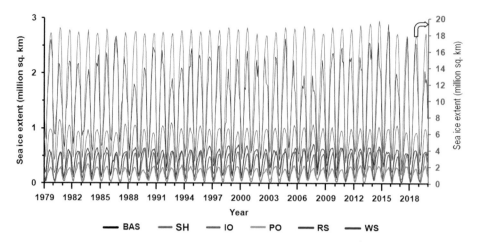

FIGURE 10.2   Season cycle of sea-ice extent variation for different sectors in the Southern Hemisphere.

10.3 RESULTS

10.3.1 SEA-ICE TRENDS

The trends are presented based on yearly-mean and seasonal SIE over the period 1979–2018. In Austral summer, there was a gradual increasing trend in the SIE in the WS from -29535 ± 2505 km^2/yr (-19.2%/decade) in 1979–1988 to 46260 ± 5468 (30.1%/decade) during 1999–2008; however, in the next decade, the trend is reversed to -30975 ± 5689 km^2/yr (-20.2%/decade) (Table 10.1). In Austral autumn, the WS SIE trend exhibited an increasing trend from 1979–1988 from -63011 ± 3068 km^2/yr (-17%/decade) to 44513 ± 5921 km^2/yr (12.3%/decade) in 1999–2008; however, the SIE trends switched to negative (-37774 ± 5776 km^2/yr; -10.4%/decade) during 2009–2018. In Austral winter, we observed negative SIE trends during 1979–1988 and 2009–2018. On the other hand, positive trends were found during 1979–2008, with negative trends in 2009–2018. In brief, we found a non-significant negative (positive) trend during 2009–2018 (1999–2008) across the seasons. The highest sea-ice loss was found to be in summer (-20.2%/decade), and the second-highest during autumn (-10.4%/decade). The trend for yearly-mean suggests a sea-ice loss of -29317 ± 4181 (-6.9%/decade). The trend for yearly-mean SIE for WS showed an increase from -30551 ± 2996 km^2/yr (-7.2%/decade) in 1979–1988 to 50383 ± 3481 km^2/yr (11.9%/decade) in 1999–2008, which was followed by a decline of -29317 ± 4181 km^2/yr (-6.9%/decade) that started in 2015 (Table 10.1). Overall, the SIE in the WS showed an increase of 0.96%/decade for yearly-mean SIE, 8.93%/decade (significant at 99% confidence) in summer, 2.36%/decade in autumn, -0.5%/decade in winter, and -0.52%/decade in spring during 1979–2018. The actual percentage may differ in literature depending upon the version of the SIE data used, but it is a fact that the SIE decrease in winter and

TABLE 10.1

Sea-Ice Trends for the Weddell Sea Sector

Period	Unit	Weddell Sea Sector			
		1979–1988	1989–1998	1999–2008	2009–2018
Summer	km²/yr(%/decade)	−29535±2505(−19.24±1.63)	−13452±3250(−8.76±2.12)	46260±5468(30.14±3.56)	−30975±5689(−20.18±3.71)
Autumn	km²/yr(%/decade)	−63011±3068(−17.38±0.85)	19240±3524(5.31±0.97)	44513±5921(12.28±1.63)	−37774±5776(−10.42±1.59)
Winter	km²/yr(%/decade)	−39050±5766(−6.16±0.91)	14561±4130(2.3±0.65)	50242±3222(7.93±0.51)	−10435±3377(−1.65±0.53)
Spring	km²/yr(%/decade)	9390±4582(1.72±0.84)	23169±4773(4.25±0.87)	60517±4107(11.1±0.75)	−38084±4965(−6.98±0.91)
Yearly	km²/yr(%/decade)	−30551±2996(−7.21±0.71)	10879±3294(2.57±0.78)	50383±3481(11.89±0.82)	−29317±4181(−6.92±0.99)

spring and increases in other seasons concurs with previous works (Turner et al., 2020; Kumar et al., 2021).

For the IO sector, the summer SIE showed a decline from -3571 ± 822 km^2/yr $(-11.2\%/\text{decade})$ to -9914 ± 797 km^2/yr $(-31.1\%/\text{decade})$ during 1979–1988 and 1989–1998, increases in the sea ice during 1999–2008, followed by a loss in sea ice $(-8322 \pm 1553$ km^2/yr) during 2009–2018 (Table 10.2). Likewise, the sea ice decreased in autumn from -17613 ± 1866 km^2/yr $(-13.2\%/\text{decade})$ to -14101 ± 2327 km^2/yr $(-10.6\%/\text{decade})$; a positive trend of 10380 ± 1644 $(7.8\%/\text{decade})$ during 1999–2008, and a significant (at 99% confidence) sea-ice loss of -26776 ± 1293 km^2/yr $(-20.1\%/\text{decade})$ was observed. In winter, the IO sector lost sea ice in all the decades $(-23138 \pm 2880$ km^2/yr was highest during 1989–1998), except for 1999–2008. With positive trends in the spring of 1979–1988 $(129 \pm 2496$ km^2/yr) and 1999–2008 $(11027 \pm 2065$ km^2/yr), the IO sector exhibited a negative trend of -19415 ± 2927 km^2/yr $(-7\%/\text{decade})$ during 1989–1998. The highest sea-ice loss of -62831 ± 4350 km^2/yr $(-22.8\%/\text{decade})$ during 2009–2018 was due to a series of remarkable storms in 2016 during spring, which brought warm air and strong winds from the north that promoted melting of sea ice (Turner et al., 2017). The year-mean SIE trend depicted sea-ice loss during 1979–1988, 1989–1998, and 2009–2018 corresponding to -7589 ± 1083 $(-3.9\%/\text{decade})$, -16642 ± 1841 km^2/yr $(-8.7\%/\text{decade})$, and -38229 ± 2254 km^2/yr $(-20.1\%/\text{ decade})$; a positive trend in SIE was detected $(9123 \pm 1211$ km^2/yr) during 1999–2008. During 1979–2018, the SIE in the IO showed an increase of 1.3%/decade for yearly-mean, 6.8%/decade for summer, 2.8%/decade for autumn, 0.8%/decade for winter, and 0.62%/decade for spring.

The PO sector showed a negative trend during 1979–1988 $(-17802 \pm 1470$ km^2/yr; $-38\%/\text{decade})$, which was followed by positive trends during 1989–1998, 1999–2008, and 2009–2018, with a decreasing sea-ice tendency from 11923 ± 937 km^2/yr $(25.5\%/\text{decade})$ to 7513 ± 1707 km^2/yr $(16\%/\text{decade})$ in summer (Table 10.3). In autumn, the negative trend was observed during 1979–1988 $(-3608 \pm 1441$ km^2/yr), with the highest decline in sea ice $(15.7\%/\text{decade})$ occurring during 1999–2008 $(-17523 \pm 1698$ km^2/yr). The SIE exhibited increases in SIE during the winters of 1979–1988 and 2009–2018, corresponding to 6249 ± 2457 and 8345 ± 2102 km^2/yr, and a significant (at 95% confidence) positive trend in 1989–1998 $(29314 \pm 1886$ km^2/yr). The highest sea-ice loss amounting to -29858 ± 2205 $(-21.6\%/\text{decade})$ occurred in spring during 1979–1988 and -5135 ± 2131 km^2/yr during 2009–2018. The yearly-mean SIE trends depicted a significant decline in sea ice (at 99% confidence) during 1979–1988 $(-11255 \pm 1486$ km^2/yr) by $-9.5\%/\text{decade}$; however, the other decades showed increases in SIE. For the 1979–2018 period, the yearly-mean SIE increased by 2.4%/decade, with 7.8%/decade for summer, 3.5%/decade for autumn, 1%/decade for winter, and 0.73%/decade for spring.

The RS exhibited a gradual change from a positive trend $(81264 \pm 3990$ km^2/yr, significant at 99% confidence) during 1979–1988 and $(8044 \pm 2507$ km^2/yr) during 1989–1998 to a negative trend $(-53246 \pm 4091$ km^2/yr) during 2009–2018 (Table 10.4). Likewise, a gradual shift from increasing trend during 1979 and 1989–1998 to negative trends during 1999–2008 and 2009–2018 was found. In winter, the SIE showed positive trends of 10098 ± 3458 and 25175 ± 3144 km^2/yr

TABLE 10.2

Sea-Ice Trends for the Indian Ocean Sector

Period	Unit	1979–1988	1989–1998	Indian Ocean Sector 1999–2008	2009–2018
Summer	km²/yr(%/decade)	−3571±822(−11.21±2.58)	−9914±797(−31.12±2.5)	11805±1493(37.05±4.69)	−8322±1553(−26.12±4.88)
Autumn	km²/yr(%/decade)	−17613±1866(−13.23±1.40)	−14101±2327(−10. 6±1.75)	10380±1644(7.8±1.23)	**−26776±1293(−20.12±0.97)**
Winter	km²/yr(%/decade)	−9304±980(−2.91±0.31)	−23138±2880(−7.23±0.9)	3280±3045(1.02±0.95)	−54986±3658(−17.18±1.14)
Spring	km²/yr(%/decade)	129±2496(0.05±0.91)	−19415±2927(−7.05±1.06)	11027±2065(4.0±0.75)	−62831±4350(−22.81±1.58)
Yearly	km²/yr(%/decade)	−7589±1083(−3.99±0.57)	−16642±1841(−8.75±0.97)	9123±1211(4.8±0.64)	***−38229±2254(−20.11±1.19)***

TABLE 10.3

Sea-Ice Trends for the Western Pacific Ocean Sector. The Values Shown in Bold Italics Are Significant at 95%, and Those in Bold Are Significant at 99%

Period	Unit	Western Pacific Ocean Sector			
		1979–1988	1989–1998	1999–2008	2009–2018
Summer	km²/yr(%/decade)	−17802±1470(−38.12±3.15)	11923±937(25.53±2.01)	9745±1564(20.87±3.35)	7513±1707(16.09±3.65)
Autumn	km²/yr(%/decade)	−3608±1441(−3.24±1.29)	8634±910(7.75±0.82)	−17523±1698(−15.72±1.52)	4916±1494(4.41±1.34)
Winter	km²/yr(%/decade)	6249±2457(3.49±1.37)	***29314±1886(16.38±1.05)***	−4501±2220(−2.51±1.24)	8345±2102(4.66±1.17)
Spring	km²/yr(%/decade)	−29858±2205(−21.6±1.59)	23445±1981(16.96±1.43)	13200±1176(9.55±0.85)	−5135±2131(−3.71±1.54)
Yearly	km²/yr(%/decade)	**−11255±1486(−9.47±1.25)**	**18329±981(15.42±0.83)**	230±1215(0.19±1.02)	3910±1480(3.29±1.24)

TABLE 10.4

Sea-Ice Trends for the Ross Sea Sector. The Values Shown in Bold Italics Are Significant at 95%, and Those in Bold Are Significant at 99%

Period	Unit	Ross Sea Sector			
		1979–1988	1989–1998	1999–2008	2009–2018
Summer	km²/yr(%/decade)	**81264±3990(74.48±3.66)**	8044±2507(7.37±2.3)	−23006±3642(−21.09±3.34)	−53246±4091(−48.8±3.75)
Autumn	km²/yr(%/decade)	56002±5069(19.8±1.79)	14992±3004(5.3±1.06)	−3030±2467(−1.07±0.87)	−44641±3908−15.78±1.38
Winter	km²/yr(%/decade)	10098±3458(2.54±0.87)	25175±3144(6.33±0.79)	−15265±3134(−3.84±0.79)	***−47269±2957(−11.89±0.74)***
Spring	km²/yr(%/decade)	**74571±4018(21.26±1.15)**	48266±3414(13.76±0.97)	17349±4109(4.95±1.17)	−42670±3043(−12.17±0.87)
Yearly	km²/yr(%/decade)	***55483±3064(19.46±1.07)***	24119±2603(8.46±0.91)	−5988±2464(−2.1±0.86)	***−46957±2871(−16.47±1.01)***

during 1979–1988 and 1989–1998, respectively; on the other hand, we found a decreasing trend in the sea ice during 1999–2008 and a significant trend in 2009–2018. The positive trends gradually showed a declining trend from 1979 to 2008 and a switch to a negative trend (-42670 ± 3043 km^2/yr) during 2009–2018. The yearly-mean SIE trends showed a gradual decline from positive (55483 ± 3064 km^2/yr, significant at 95% confidence) to negative trend (-46957 ± 2871 km^2/yr) during 2009–2018. During 1979–2018, the yearly-mean SIE increased by 1.7%/ decade, by –0.2%/decade in summer, 2.7%/decade (significant at 95% confidence) in autumn, 1.75%/decade (significant at 95% confidence) in winter, and 2.4%/ decade (significant at 95% confidence) in spring.

The BAS sector showed a decreasing trend in the Austral summer ($-13628 \pm$ 1193 km^2/yr or -21.98%/decade) during 1979–1988, which was followed by an increasing trend (5440 ± 1175 km^2/yr or 8.77%/decade) during 1989–1998 (Table 10.5). During 1999–2008, the SIE decreasing trend was 6% higher compared to that during 1979–1988. This was followed by an upward trend of $9183 \pm$ 1611 km^2/yr or 14.81%/decade). The autumn period is dominated by negative trends, except for the 1989–1998 period. In winter, the SIE trends showed a decline in the SIE from -12106 ± 3313 km^2/yr to -26704 ± 2040, with a positive trend during 1989–1998 and 2009–2018. We observed a continuous sea loss in spring. The yearly mean showed SIE decline in all seasons, with maximum loss during 1999–2008 (-27730 ± 1562 km^2/yr, significant at 95% confidence). During the period 1979–2008, the SIE exhibited a trend of –1.9%/decade (significant at 95% confidence) for yearly-mean, –17%/decade (significant at 99%) for summer, –6.1%/ decade (significant at 99% confidence) for autumn, 1.74%/decade for winter, and –0.38%/decade for spring.

The SO as a whole exhibited positive trends during 1979–1988 (17010 ± 3494 km^2/yr), 1989–1998 (1717 ± 4957 km^2/yr), and 1999–2008 (29576 ± 7253 km^2/yr) (Table 10.6). We observed high sea-ice loss in summer (–18.8%/decade). In autumn, the SIE trends were negative during 1979–1988 and 2009–2018. Likewise, the wintertime sea-ice decline was observed during 1979–1988 and 2009–2018. A statistically significant positive trend (at 95% confidence) in the SIE was found during 1989–1998 and 1999–2008 in spring; the decline in the sea ice is evident for 2009–2018. For yearly-mean SIE, a negative trend was found for 1979–1988 and 2009–2018. For the period 1979–2008, the SIE exhibited an increasing trend of 1%/ decade (significant at 95%) for yearly-mean, 2.1%/decade for summer, 1.64%/ decade (significant at 95%) for autumn, 0.68%/decade (significant at 95%) for winter, and 0.5%/decade for spring.

10.3.2 REGIONAL DRIVERS FOR SEA-ICE CHANGE

Regional factors such as wind intensification result in the ridging of sea ice. The zero-lag correlation between SIE and WS, MSLP, HFLUX, TA, and SST is depicted in Figure 10.3. The correlation coefficients exceeding 0.64 and 0.76 were found to be significant at 95% and 99% confidence levels, respectively. Since the winds, which are a significant contributor to the atmospheric circulation, are driven by pressure systems in the vicinity of the continent, we have considered MSLP. The

TABLE 10.5

Sea-Ice Trends for the Bellingshausen and Amundsen Seas. The Values Shown in Bold Italics Are Significant at 95%, and Those in Bold Are Significant at 99%

Period	Unit	1979–1988	1989–1998	1999–2008	2009–2018
			Bellingshausen and Amundsen Sea		
Summer	km²/yr(%/decade)	−13628±1193(−21.98±1.92)	5440±1175(8.77±1.9)	−14448±1876(−23.3±3.02)	9183±1611(14.81±2.6)
Autumn	km²/yr(%/decade)	−10271±2663(−8.42±2.18)	4181±1896(3.43±1.55)	−30973±2798(−25.38±2.29)	−10978±2597(−9±2.13)
Winter	km²/yr(%/decade)	−12106±3313(−5.8±1.59)	4466±2956(2.14±1.42)	−26704±2040(−12.8±0.98)	4036±2662(1.93±1.28)
Spring	km²/yr(%/decade)	−30110±3305(−17.67±1.94)	−31181±3047(−18.3±1.79)	−38796±2975(−22.77±1.75)	−16837±3165(−9.88±1.86)
Yearly	km²/yr(%/decade)	−16529±2087(−11.74±1.48)	−4273±1550(−3.04±1.1)	*−27730±1562(−19.7±1.11)*	−3649±1226(−2.59±0.87)

TABLE 10.6

Sea-Ice Trends for the Southern Hemisphere. The Values Shown in Bold Italics Are Significant at 95%, and Those in Bold Are Significant at 99%

Southern Hemisphere

Period	Unit	1979–1988	1989–1998	1999–2008	2009–2018
Summer	km^2/yr(%/decade)	17010±3494(4.22±0.87)	1717±4957(0.43±1.23)	29576±7253(7.33±1.8)	−75939±10083(−18.83±2.5)
Autumn	km^2/yr(%/decade)	−38707±7542(−3.82±0.74)	32788±3479(3.24±0.34)	3353±6929(0.33±0.68)	−115535±11162(−11.42±1.1)
Winter	km^2/yr(%/decade)	−44283±3286(−2.55±0.19)	50162±2358(2.88±0.14)	6808±4161(0.39±0.24)	−100323±7339(−5.77±0.42)
Spring	km^2/yr(%/decade)	23838.38±6105(1.61±0.41)	***44081±2762.52(2.98±0.19)***	***63172±3684.8(4.27±0.25)***	−165596±10951(−11.19±0.74)
Yearly	km^2/yr(%/decade)	−10535±2442(−0.91±0.21)	32187±2383(2.78±0.21)	25727±3788(2.22±0.33)	−114348±9022(−9.87±0.78)

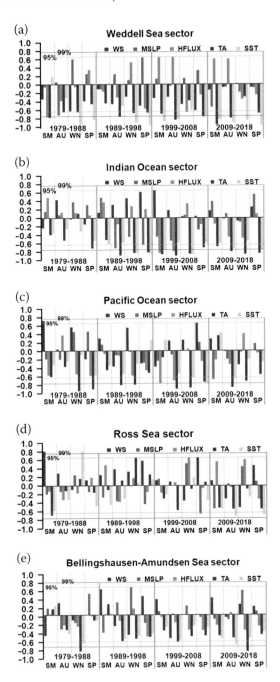

FIGURE 10.3 Correlation between local forcing factors and sea-ice extent. The factors include wind speed (WS), mean sea level pressure (MSLP), heat flux (sensible and latent), air temperature (TA), and sea surface temperature (SST). Abbreviations are SM: summer, AU: autumn, WN: winter, and SP: spring.

net radiation (shortwave minus longwave radiation) in summer (winter) is in the range of 5–26 (–43 to –12) W/m^2, and the yearly net radiation was estimated between 2 and 22 W/m^2 (van den Broeke et al., 2004). Reanalysis data sets in the SO suffer from a large inter-product spread in estimates of the net air-sea heat flux (e.g. Liu et al., 2011), stemming from insufficient observations, particularly in winter. So we have not considered the contribution of net solar and net longwave radiation to HFLUX.

During the 1979–1988 decade for the WS sector, overall we found a negative correlation between the local drivers and SIE. The increase in the heat flux in the surface planetary boundary layer promoted warmer TA during summer which resulted in an SIE decrease by –19.2%/decade. A negative correlation was observed between SIE, and MSLP and TA in autumn, while TA and SST counteracted increases in the SIE during winter. In spring, the SST also played an important role in reducing the SIE. During 1989–1998, a significant and negative correlation between TA and SIE was detected in autumn, while TA and SST acted against increases in the SIE, but favorable gyral circulation induced by MSLP increased SIE in spring (Figure 10.3a). During 1999–2008, TA and SST were negatively correlated with SIE in summer, autumn, and spring. The SIE in summer and autumn was positively correlated with HFLUX. Increased SIE in winter reduced the HFLUX. During 2009–2018, we found the SST and SIE were negatively correlated for all seasons. Likewise, air temperature's influence on SIE was overwhelmed, except for winter. A decrease in SIE in summer and autumn (Table 10.1) facilitated increases in HFLUX.

For the IO sector, TA and SST were found to be some of the significant forcings on SIE during 1979–1988 in spring. During autumn, MSLP, TA, and SST exerted maximum influence on the reduction of SIE by-13.2%/decade (Table 10.2), while TA and SST were drivers for the decrease in SIE by –2.91%/decade in winter during 1989–1998. A positive correlation between MSLP and SIE was also detected in winter and spring. We encountered a negative correlation between TA, SST, and HFLUX in summer and autumn. Likewise, during winter, TA and SST exerted a negative influence on SIE. During 2009–2018, TA and SST exerted a negative influence on SIE in all the seasons.

For the PO sector, a positive correlation between MSLP and SIE during summer, and a negative correlation between TA and SIE in winter and spring were detected during 1979–1988. Warmer atmosphere reduced SIE during winter and spring. Likewise, during 1989–1998 and 1999–2008, we found a negative correlation between TA and SIE in autumn and winter. In spring, the MSLP was positively correlated, while TA and SST exerted a negative influence on SIE. During 2009–2018, TA was negatively correlated to SIE in autumn.

For the RS sector, the WS (TA and SST) is positively (negatively) correlated to SIE in summer, while a highly positive correlation between SST and SIE (significance at 99% confidence) was detected in winter during 1979–1988. Except for the high correlation of TA with SIE (significant at 95% significance), no other variables showed a significant correlation with SIE during 1989–1998. We detected a significant positive correlation between HFLUX (WS) with SIE in winter (spring); however, MSLP exhibited a negative correlation significant at 95% significance with SIE in spring during 1999–2008. It is noted that SST and TA showed an inverse relationship with SIE, which was not significant; however, decreasing SST was highly correlated with SIE (significant at 95% confidence) in autumn and spring during 2009–2018.

For the BAS sector, decreases in TA and SST facilitated increases in SIE in winter during 1979–1988. During 1989–1998, the MSLP was positively correlated to SIE in winter. During 1999–2008, we encountered a negative correlation between TA and SIE in autumn. It was observed that decreasing TA facilitated increases in SIE in winter during 2009–2018.

10.3.3 Role of Remote Forcing on Sea-Ice Change

Literature reveals that the SH sea ice responds to the extra-polar ocean and at-mosphere variability, through thermal and mechanical processes (Lefebvre and Goosse, 2005). On shorter time scales, internal modes of variability, such as ENSO and SAM and the acceleration of westerly winds around Antarctica (Turner et al., 2009), the deepened ASL induced by the tropical Pacific (Meehl et al., 2016) have been identified for having played important roles in influencing seasonal variations of SIE. Positive SAM is the key atmospheric circulation mechanism with AMO and PDO as potential contributors to sea-ice variability (Raphael and Hobbs, 2014). The influence of ENSO on Antarctic sea ice is primarily through the atmospheric Rossby wave trains forced by anomalous tropical convection that accompanies anomalous sea SST in the tropical Pacific.

For the WS, the AMO influences SIE in winter and spring (not significant) during 1979–1988. For 1989–1998, SAM and SIE were found to be negatively correlated to SIE in summer (r = 0.64, significant at 95%), while positive corre-lation, though not significant, between PDO and SIE was also detected in spring and winter during 1999–2008. AMO exerts a negative influence on the SIE in summer during 2009–2018 (0.59).

For the IO sector, AMO exerts a negative and significant influence (r > 0.65) on SIE in autumn and winter during 1979–1988. During 1999–2008, we detected a negative correlation between PDO (SAM) and SIE in summer (winter), while the SOI favored increases in SIE in summer and autumn, which was not significant. For the PO sector, the PDO showed a negative association with SIE in the summer during 1979–1988. AMO showed a positive correlation (r = 0.64) with SIE in winter during 1999–2008, while during 2009–2018 we detected an inverse re-lationship (r = 0.6, not significant) between SOI and SIE in winter.

For the RS sector, AMO was positively correlated to SIE in spring during 1989–1998. During 1999–2008, we encountered a high positive correlation between SOI and SIE in autumn and winter (r > 0.62), and a negative correlation (r < 0.7) between PDO and SIE in spring. During 2009–2018, the AMO exerts a negative influence on SIE in summer and autumn, though the correlation is not significant (r = 0.53). For the BAS, we detected a strong negative correlation between SOI and SIE in winter (r = 0.72) during 1979–1988 and 1989–1998. PDO and SAM forced a decreasing trend in the SIE in spring during 1989–1998. SAM and SIE exhibited a negative correlation in spring (significant at 95% confidence) and summer during 1999–2008. During 2009–2018, we encountered a high positive and significant correlation between SAM and SIE in autumn. Positive correlations (not significant) were observed in summer and autumn for SOI, likewise, negative correlations (r = 0.54) were detected between AMO/ SOI and SIE in autumn and winter.

Another perspective mechanism for the Antarctic sea-ice increase is enhanced mass loss of the Antarctic ice sheet due to increasing ocean warming and subsurface ice-shelf melting, which causes the uppermost ocean layer to become fresher and thus less dense than the underlying warmer water and this decreases upward ocean heat transport. The upper ocean layer can then cool more rapidly in the autumn and early winter, thus leading to sea-ice expansion (Bintanja et al., 2013).

10.4   DISCUSSION

The sea-ice changes around Antarctica are influenced by several large-scale modes of atmospheric circulation. Antarctic sea-ice variability is spatially heterogeneous, and links between the atmospheric circulation modes and the sea-ice variability are not clearly understood. To date, studies have indicated that this diverse behavior is most likely because of the strengthening of the WS due to positive SAM (Marshall, 2003), which leads to ridging in sea ice; changes in the atmospheric circulation due to MSLP changes;HFLUX; and SST changes induced as a part of the feedback due to exposure of the ocean to the atmosphere when the sea ice retreats; surface warming imposed by airflow over orography, thereby creating conditions that lead to air warming to the east of the Antarctic Peninsula (Turner et al., 2021); the dynamic and thermodynamic processes in the ice pack (Holland et al., 2005); the interaction with the SO (Hall and Visbeck, 2002; Swart and Fyfe, 2013; Holland et al., 2005); and the modification of the hydrological cycle (Liu and Curry, 2010).

Depending on wind direction, meridional winds advect warm, moist air from lower latitudes or cold, dry air from the continent, affecting sea-ice production, extent, and melt. The wind is also the major driver of sea-ice motion (Holland and Kwok, 2012). We observed that the WS SIE exhibited an increasing trend in autumn, winter, and spring during 1989–1998 and 1999–2008, and 1999–2008 summertimes. The increasing trend in the SIE is attributed by Kumar et al. (2021) to the intensification of the westerly winds and increase in wind velocity in the band 55°–50°S resulting in higher sea-ice concentration towards the north of 65°S. We note that positive SAM has a significantly positive (negative) relationship with SIE during summer (winter). Hence, the WS showed sea-ice expansion (retreat) in summer (winter) during positive SAM. However, Kumar et al. (2021) point to an increase in sea-ice concentration towards the south of 65°S due to the poleward shift of the westerly winds caused by positive SAM, which strengthened the Weddell gyre. The Weddell Gyre is associated with cooler SST in the gyre and warmer SST to its north.

The IPO had positive phases during 1978–1998, followed by a negative phase, which led to the deepening of the ASL that has contributed to regional circulation changes in the RS region and expansion of sea ice (Clem and Renwick, 2015). Seasonal Antarctic-average sea-ice extent trends before the IPO transition to negative showed slight decreases for the 1979–1999 period during December–February ($-0.04 \pm 0.33 \times 10^6$ km^{2}decade^{-1}) to small increases during the other seasons (0.2×10^6 km^{2}decade^{-1})(Meehl et al., 2016). PDO was found to be significantly negatively correlated (at the 95% confidence level) with the SIE in the King Hakon VII sector during ice advance (winter) and also with the SIE in the Ross-Amundsen sector during ice retreat (summer) (Raphael and Hobbs, 2014).

In the IO sector, strengthening of the westerly wind during the summer cools the upper ocean due to net heat loss from the ocean surface, thereby enhancing the vertical stability (Bintanja et al., 2013). The winds also facilitate northward advection of cool and fresh water through enhanced surface currents between 50°S and 62°S. The combined effect favors sea-ice expansion in the subsequent seasons (Jena et al., 2018). The freshening has been detected in different sector of the SO, which is facilitated by the accelerated freshwater fluxes from the Antarctic glacial melt (Jacob et al., 2002; Paolo et al., 2015), atmospheric fluxes by excess precipitation over evaporation (Durack et al., 2012), and the northward transport of sea ice (Holland and Kwok, 2012; Haumann et al., 2016). The accelerated freshening of the sea surface can lower the SST by weakening convection and vertical mixing through enhanced thermohaline stratification in the water column, which favors the sea-ice expansion (Zhang, 2007).

The Ross Sea is the only region where significant trends in SIE have occurred year-round (Table 10.4; Hobbs et al., 2016). Over the RS continental shelf, the duration of the summer ice-free period has decreased (Parkinson, 2014; Stammerjohn et al., 2008), ice production of the RS polynya has increased (Comiso et al., 2011), and sea surface temperature has decreased (Comiso et al., 2011; Comiso et al., 2017). These changes have been attributed to the increased strength of the cold southerly winds that extend over the Ross ice shelf and the continental shelf (Turner et al., 2009; Holland and Kwok, 2012; Turner et al., 2015), which in turn have been linked to changes in the ozone hole over Antarctica (Turner et al., 2009) and deepening of the ASL in response to variability in the tropical Pacific Ocean (Ding et al., 2011) and the north and tropical Atlantic Ocean (Li et al., 2014).

How the SIE responds to tropical SST anomalies via the ENSO (Turner, 2004, Yuan, 2004; Simpkins et al., 2012) can be explained as follows. Atmospheric anomalies excited by tropical Pacific convection propagate south-eastwards via an atmospheric Rossby wave train to the high latitude southeastern Pacific (Yu et al., 2011), called the Pacific South-American (PSA) mode (Mo and Higgins, 1998; Mo, 2000; Mo and Paegle, 2001). The response is most evident in the eastern RS and BAS sectors (Kwok and Comiso, 2002; Turner, 2004; Yuan, 2004), and peaks during the late Austral winter and spring (Simpkins et al., 2012). During warm ENSO events (or El Niños), a high-pressure anomaly is established centered at about 90°W, 55°S, i.e. a weakened ASL (Turner, 2004; Yuan, 2004; Yuan and Li, 2008). Strong evidence suggests that the response to ENSO is modulated by the background state of the SAM. The ENSO teleconnection is only statistically significant when SAM is in a weak (neutral) phase, or when SAM is in phase with the ENSO event. Nonetheless, the teleconnection with La Niña (El Niño) is significant when the SAM switches to a *positive* (negative) phase (Fogt and Bromwich, 2006; Stammerjohn et al., 2008; Fogt et al., 2010).

It was observed that the SIE trend over RS is highly statistically significant, particularly during 1979–1988 in summer, spring, and during 2009–2018 in winter

(Table 10.4), while the trends of the other sub-regions are well within the bounds of natural variability (Yuan et al., 2017). Accordingly, only the trend in the RS SIE cannot be explained by natural variability alone and external forcing must contribute to this trend. Both global warming and ozone depletion may deepen the ASL, weaken the westerlies near the RS, and strengthen the southerly cold winds from the continent. This way, the sea ice in the RS may become more isolated and increase due to the southerly cold winds.

One of the significant features of the circumpolar atmosphere's circulation is the quasi-stationary atmospheric low-pressure anomaly at approximately 60°–70°S, known as the ASL (Hosking et al., 2013). Its climatological mean zonal location shifts from 250°E in summer to 220°E in winter. It occurs as a result of zonal flow around the continental topography (Baines and Fraedrich, 1989) and therefore is closely related to the SAM (Lefebvre et al., 2004; Turner et al., 2013a). Cyclonic flow around this low-pressure center drives warm poleward winds into the Antarctic Peninsula/Bellingshausen Sea region, and a cold equator-ward wind over the RS, with implications for the dipole in sea-ice trends between these two regions (Lefebvre et al., 2004; Holland and Kwok, 2012; Hosking et al., 2013). It has been shown that the changes in autumn sea-ice concentration in the Ross and Amundsen seas are connected to wind-driven sea-ice motion trends (Holland and Kwok, 2012), while a similar relationship was detected for annual mean trends (Haumann et al., 2014). Hobbs et al. (2016) showed that a cyclonic trend that forms over the Bellingshausen Sea is consistent with a deepening ASL that promotes a warm and moist northerly airflow to the Bellingshausen Sea and decreases the sea-ice concentration. The sea-ice growth in the Bellingshausen Sea is predominately thermodynamic rather than dynamic, making it susceptible to advection of warm air (Kimura and Wakatsuchi, 2011).

In the RS, there is a positive SIC trend at the ice edge over the entire sector that is consistent with ASL-related southerly wind trends. Hobbs et al. (2016) found an ASL-related northerly wind trend over the Bellingshausen Sea, although somewhat reduced, in the RS the sea-ice increase is largely confined to the western RS. They also argued that the sea-ice changes in the Bellingshausen Sea are related to the deepening ASL, but in the western RS, while local wind trends may be responsible for the sea-ice changes (Haumann et al., 2014), the role of the ASL in explaining these wind changes is not so clear. It should be noted that ice-ocean feedbacks may be able to sustain sea-ice trends for some time after an initial atmospheric forcing (Goosse and Zunz, 2014), and that wind trends in the earlier period may have triggered an ongoing sea-ice increase.

Literature provides a link of West Antarctic surface temperature changes to multi-decadal modes of Pacific variability (Schneider and Steig, 2008; Ding et al., 2011; Okumura et al., 2012). A possible link exists between multi-decadal tropical Atlantic variability and West Antarctic sea ice (Li et al., 2014; Simpkins et al., 2014), with a physical mechanism that is similar to the tropical Pacific teleconnection, wherein a Rossby wave train triggered by tropical SSTs modulates the depth of the climatological ASL. In the warming tropical Atlantic conditions during the late 1970s, the ASL tends to deepen, promoting sea-ice changes in Bellingshausen and eastern RS in winter (Li et al., 2014) and spring (Simpkins et al., 2014) (Tables 10.4 and 10.5). Additionally, there is no impact in the warmer seasons when the sea-ice trends are strongest

(Tables 10.4 and 10.5), as would be expected since atmospheric teleconnections to the high latitude of the SH are weak in the Austral summer (Jin and Kirtman, 2009).

Apart from the atmosphere, the ocean plays a key role in sea-ice interannual variability. The large annual cycle of Antarctic sea ice is driven by the interactions between wind speed, SIE, solar radiation, and the ocean mixed layer accelerate the spring melt (Gordon, 1981). Sea ice forms when surface water is cooled below its freezing point, and a near-isothermal surface mixed layer is formed, whose temperature is altered at the surface by incoming solar radiation, and at the base of it by entrainment of warm water; both processes are in turn affected by sea ice. The vertical structure of the near-surface water column is highly seasonal in the sea-ice zone due to ice-related heat and freshwater fluxes.

In his model, Zhang (2007) demonstrated that reduced sea-ice production in the advanced season (winter) decreases brine rejection and increases stratification. This reduces the entrainment of warm, deep water into the winter mixed layer, and the spring melt is consequently reduced even more than the ice growth. Thus, there is negative feedback through the ocean between autumn ice production and the subsequent summer sea ice. Conversely, other model studies suggest that an *increase* in ice coverage can lead to the same increased stratification (Goosse et al., 2009; Goosse and Zunz, 2014). By this mechanism, once the pack ice is established, the surface water tends to be fresher when there is greater SIE. This again increases stratification and reduces melt by reducing the upwelling of warm water. Though these processes were hypothesized based on the model experiments, both feedbacks have some concurrence with observed trends. The RS has a lower trend in autumn SIE (Table 10.4), but this is counteracted by increases in the spring, leading to an overall summer increase in SIE during 1979–1988 and 1989–1998 (Table 10.4), although there is no indication of a surface warming there. However, in the Bellingshausen Sea, observed surface salinification has been linked to a reduction in sea-ice production (Meredith and King, 2005) and consequently showed mostly negative trends (Table 10.5).

The previous inferences have implications for both current and future understanding of Antarctic sea-ice variability as moderated by the large-scale atmospheric circulation. It is feasible that changes in the regional ocean circulation have a role in explaining the trend pattern. Model experiments (Thoma et al., 2008) and observations (Schmidtko et al., 2014) suggest that there has been an increased advection of warm Circumpolar Deep Water (CDW) onto the continental shelf of the Amundsen Sea, forced by wind stress trends ascribed to tropical Pacific variability (Steig et al., 2012), but with considerable interannual variability (Dutrieux et al., 2014). However, observations show that both vertical and horizontal transport of CDW onto the continental shelf in the Bellingshausen Sea has increased the near-surface ocean heat content (Martinson et al., 2008), with considerable repercussions for the declining SIE. Increased westerly winds away from the coast could enhance the upwelling of CDW, with clear implications for increases in the ocean heat flux in the sea-ice zone (Ferreira et al., 2015).

There is evidence of anthropogenic signals in some of the atmospheric modes that affect sea ice. For the summer and autumn seasons when sea-ice trends are strongest, the SAM trend can be attributed to a combination of greenhouse gas forcing and ozone depletion (Arblaster and Meehl, 2006; Gillett et al., 2013;

Christidis and Stott, 2015). However, as seen in Figure 10.4, SAM alone is not sufficient to explain the observed SIE trends (Lefebvre et al., 2004; Liu et al., 2004; Yu et al., 2011). The reduction of SIE in the Bellingshausen Sea is thought to be related to the deepening of the ASL, which is directly related to the SAM. The observed ASL deepening is consistent with the expected response to anthropogenic forcing, in particular stratospheric ozone depletion (Fogt and Zbacnik, 2014; Fogt and Wovrosh, 2015). However, the observed deepening is found to be the range of internal variability of global coupled climate models (Turner et al., 2015). Nevertheless, there is some evidence, though uncertain, which would indicate that we might expect to see a sea-ice response to anthropogenic forcing. We must continue monitoring the sea-ice changes over the come years, as more data will help to elucidate the climate signals in the sea-ice variability.

10.5 CONCLUSIONS

We analyzed SIE over the 1979–2018 period using satellite-based observations of sea-ice concentrations. Trends were computed for each decade. The WS showed a negative trend in SIE for all seasons during 2009–2018 and 1979–1988 (except for spring), and positive trends during 1999–2008 and 1989–1998 (except for summer). The IO sector exhibited negative trends for all seasons during 1979–1988 (except for spring), 1989–1998, and 2009–2018, while positive trends were observed during 1999–2008. The SIE trend was significant at 99% confidence for the autumn period (−20.12%/decade). With negative trends during 1979–2088, the SIE in the PO sector showed positive trends during 1989–1998, 1999–2008 (except for autumn and winter), and 2009–2018 (except for spring). We encountered a significant SIE trend at 95% confidence in winter during 1989–1998 (16.38%/decade). During 1979–1988 and 1989–1998, the SIE showed positive trends in the RS sector, and negative trends during 1999–2008 (except for spring) and 2009–2018. Trends significant at 99% confidence were encountered in the RS in summer (74.48%/decade) and spring (21.26%/decade) during 1979–1988, and in winter (−11.89%/decade; 95% confidence) during 2009–2018. With negative trends during 1979–1988 and 1999–2008, we detected positive trends during 1989–1998 (except for spring) and 2009–2018 (except for autumn and spring). Overall, the SH exhibited negative SIE trends during 1979–1988 (−0.9%/decade) and 2009–2018 (−9.87%/decade), and positive trends during 1989–1998 (2.78%/decade) and 1999–2008 (2.22%/decade).

Among the local drivers, WS, TA, and SST were negatively correlated to SIE, and their association was significant at 95% confidence during most of the seasons for the WS. In IO, we observed a highly negative correlation (significant at 95% confidence) between TA, SST, and SIE during 1989–1998, 1999–2008, and 2009–2018. A negative correlation between TA and SIE was also observed to be significant for the PO sector. For the RS sector, WS in summer and SST in winter, while TA in summer played a key role in the modulation of SIE during 1979–1988. In winter TA during 1989–1998, while HFLUX in winter and WS and MSLP in spring induced SIE changes during 1999–2008; the SST in autumn and spring influenced SIE during 2009–2018. As for the BAS sector, TA negatively influenced

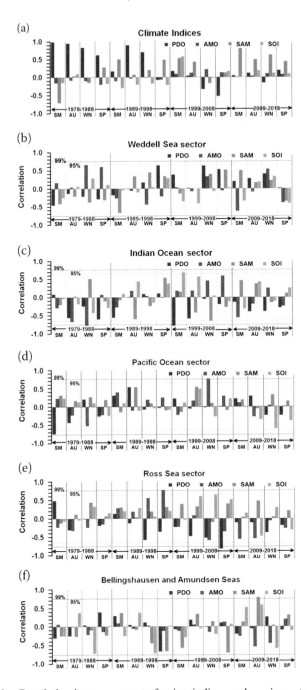

FIGURE 10.4 Correlation between remote forcing indices and sea-ice extent. The factors include Pacific Decadal Oscillation (PDO), Atlantic Meridional Oscillation (AMO), Southern Annular Mode (SAM), and Southern Oscillation Index (SOI). Abbreviations are SM: summer, AU: autumn, WN: winter, and SP: spring season.

SIE in autumn and winter seasons; higher MSLP was found to be facilitating a sea-ice increase in winter during 1989–1998.

As for the remote forcing variables, we found AMO and SIE were positively and significantly correlated in winter during 1979–1988, while SAM was negatively and significantly correlated to SIE in summer during 1989–1998. In the IO sector, AMO and SIE were negatively correlated to SIE in autumn and winter during 1979–1988. The PDO (SOI) exhibited a negative (positive) correlation with SIE during 1999–2008. The PDO was also negatively correlated to SIE during 1979–1988, while AMO was positively correlated to SIE during 2009–2018 for PO. For the RS sector, the AMO was positively correlated to SIE in spring during 1989–1998, while SOI (PDO) was significantly and positively (negatively) to SIE in winter (spring) during 1999–2008. For BAS, SOI was negatively correlated with SIE in winter during 1979–1988 and 1989–1998, while PDO was negatively correlated with SIE in spring during 1989–1998; springtime SAM negatively influenced SIE during 1989–1998 and 1999–2008. In contrast, SAM was positively correlated to SIE during 2009–2018.

We have consolidated different studies that correlated different drivers with SIE for different seasons. There is a superimposition of these signals in different seasons and years, so their interaction with sea ice is complicated. It is prudent to use model simulations such as CMIP version 5 to infer the trends and the role of the forcing parameters. However, Turner et al. (2013c) argued that the models have very large differences in SIE over 1860–2005, as compared to the satellite observations. Most of the control runs have statistically significant trends in SIE over their full time span, and all of the models have a negative trend in SIE since the mid-19th century. Shu et al. (2015) pointed out that multi-model ensemble mean results can give good climatology of SIE, but the linear trend is incorrect. In most of the model runs, Turner et al. (2013c) reported negative trends in the SIE for 1979–2005 that were inferred as the continuation of an earlier decline, suggesting that the processes responsible for the observed increase over the last 30 years are not being simulated correctly. So there is a need for a review of the model physics and better parameterization for the physical mechanisms.

ACKNOWLEDGMENTS

We thank NCPOR Director, Dr. M. Ravichandran, for his encouragement. The SAM index was downloaded from https://climatedataguide.ucar.edu/climate-data/marshall-southern-annular-mode-sam-index-station-based, PDO from http://research.jisao.washington.edu/pdo/PDO.latest.txt, and AMO from https://www.esrl.noaa.gov/.

REFERENCES

Agnew, T. A., and Howell, S. (2002). Comparison of digitized Canadian ice charts and passive microwave sea-ice concentrations. Geoscience and Remote Sensing Symposium, 2002. IGARSS '02. 2002 IEEE International1, 231–233. https://doi.org/10.1109/IGARSS.2002.1024996.

Agnew, T., and Howell, S. (2003). The use of operational ice charts for evaluating passive microwave ice concentration data. *Atmosphere-Ocean*, 41(4), 317–331. https://doi.org/10.3137/ao.410405.

Arblaster, J. M., and Meehl, G. A. (2006). Contributions of external forcings to southern annular mode trends. *J. Clim.*, 19, 2896–2905.

Arrigo, K. R., Lizotte, M. P., and Mock, T. (2009). Sea ice algae. In: Thomas, D. N., and Dieckmann, G. S. (Eds.). *Sea Ice*, 2nd ed. Wiley-Blackwell, Oxford, UK.

Baines, P. G., and Fraedrich, K. (1989). Topographic effects on the mean tropospheric flow patterns around Antarctica. *J. Atmos. Sci.*, 46, 3401–3415.

Barker, S., and Ridgwell, A. (2012). Ocean acidification. *Nature Education Knowledge*, 3(10), 21.

Bintanja, R., van Oldenborgh, G. J., Drijfhout, S. S., Wouters, B., and Katsman, C. A. (2013). An important role for ocean warming and increased ice-shelf melt in Antarctic sea-ice expansion. *Nature Geosci*, 6, 376–379.

Bracegirdle, T. J., and Marshall, G. J. (2012). The reliability of Antarctic tropospheric pressure and temperature in the latest global reanalyses. *J. Clim*, 25, 7138–7146.

Bromwich, D. H., Nicolas, J. P., and Monaghan, A. J. (2011). An assessment of precipitation changes over Antarctica and the Southern Ocean since 1989 in contemporary global reanalyses. *J. Clim*, 24, 4189–4209.

Cavalieri, D. J., Gloersen, P., and Campbell, W. J. (1984). Determination of sea ice parameters with the Nimbus 7 SMMR. *J. Geophys. Res.*, 89, 5355–5369.

Cavalieri, D. J., Parkinson, C. L., Gloersen, P., Comiso, J. C., and Zwally, H. J. (1999). Deriving long-term time series of sea ice cover from satellite passive-microwave multisensor datasets. *J. Geophys. Res.*, 104, 15803–15814.

Cavalieri, D. J., Parkinson, C. L., DiGirolamo, N., and Ivanoff, A. (2012). Intersensor calibration between F13 SSMI and F17 SSMIS for global sea ice data records. *IEEE Geosci. Remote Sens. Lett.*, 9, 233–236.

Christidis, N., and Stott, P. A. (2015). Changes in the geopotential height at 500 hPa under the influence of external climatic forcings. *Geophys. Res. Lett.*, 42 (2015), 10, 798–810. https://doi.org/10.1002/2015gl066669

Clem, K. R., and Renwick, J. A. (2015). Austral spring Southern Hemisphere circulation and temperature changes and links to the SPCZ. *J. Clim*, 28, 7371–7384.

Comiso, J. C., Gersten, R. A., Stock, L. V., Turner, J., Perez, G. J., and Cho, K. (2017). Positive trend in the Antarctic Sea ice cover and associated changes in surface temperature. *J. Clim.*, 30, 2251–2267.

Comiso, J. C., Kwok, R., Martin, S., and Gordon, A. L. (2011). Variability and trends in sea ice extent and ice production in the Ross Sea. *J. Geophys. Res.*, 116, C04021. https://doi.org/10.1029/2010JC006391

Deppeler, S. L., and Davidson, A. T. (2017). Southern Ocean phytoplankton in a changing Climate. *Front. Mar. Sci.*, 4, 40. https://doi.org/10.3389/fmars.2017.00040

Ding, Q. H., Steig, E. J., Battisti, D. S., and Kuttel, M. (2011). Winter warming in West Antarctica caused by central tropical Pacific warming. *Nat. Geosci.*, 4, 398–403. https://doi.org/10.1038/NGEO1129.

Durack, P. J., Wijffels, S. E., and Matear, R. J. (2012). Ocean salinities reveal strong global water cycle intensification during 1950 to 2000. *Science*, 336(6080), 455–458.

Dutrieux, P., de Rydt, J., Jenkins, A., Holland, P. R., Ha, H. K., Lee, S. H., Steig, E. J., Ding, Q., Abrahamsen, E. P., and Schröder, M. (2014). Strong sensitivity of Pine Island ice-shelf melting to climatic variability. *Science*, 343, 174–178.

Emery W. J., Fowler C., and Maslanik J. (1994). Arctic sea ice concentrations from special sensor microwave imager and advanced very high-resolution radiometer satellite data. *J. Geophys. Res.*, 99, 18329–18342.

Ferreira, D., Marshall, J., Bitz, C. M., Solomon, S., and Plumb, A. (2015). Antarctic Ocean and sea ice response to ozone depletion: A two-time-scale problem. *J. Clim.*, 28, 1206–1226.

Fetterer, F., Knowles, K., Meier, W. N., Savoie, M., and Windnagel, A. K. (2017). Sea Ice Index, ver 3. Product No. G02135, Boulder, Colorado USA. *NSIDC: National Snow and Ice Data Center*. https://doi.org/10.7265/N5K072F8. Accessed on 1st March 2020.

Fogt, R. L., and Bromwich, D. H. (2006). Decadal variability of the ENSO teleconnection to the high-latitude South Pacific governed by coupling with the southern annular mode. *J. Clim.*, 19, 979–997. https://doi.org/10.1175/Jcli3671.1

Fogt, R. L., Bromwich, D. H., and Hines, K. M. (2010). Understanding the SAM influence on the South Pacific ENSO teleconnection. *Clim. Dyn.*, 36, 1555–1576. https://doi.org/10.1007/s00382-010-0905-0

Fogt, R. L., and Wovrosh, A. J. (2015). The relative influence of tropical sea surface temperatures and radiative forcing on the Amundsen Sea low. *J. Clim.*, 28, 8540–8555.

Fogt, R. L., Wovrosh, A. J., Langen, R. A., and Simmonds, I. (2012). The characteristic variability and connection to the underlying synoptic activity of the Amundsen-Bellingshausen Seas low. *J. Geophys. Res.*, 117, D07111. https://doi.org/10.1029/2011JD017337

Fogt, R. L., and Zbacnik, E. A. (2014). Sensitivity of the Amundsen Sea low to stratospheric ozone depletion. *J. Clim.*, 27, 9383–9400.

Gillett, N. P., Fyfe, J. C., and Parker, D. E. (2013). Attribution of observed sea level pressure trends to greenhouse gas, aerosol, and ozone changes. *Geophys. Res. Lett.*, 40, 2302–2306.

Goosse, H., Lefebvre, W., de Montety, A., Crespin, E., and Orsi, A. H. (2009). Consistent past half-century trends in the atmosphere, the sea ice, and the ocean at high southern latitudes. *Clim. Dyn*, 33, 999–1016.

Goosse, H., and Zunz, V. (2014). Decadal trends in the Antarctic sea ice extent ultimately controlled by ice-ocean feedback. *Cryosphere*, 8, 453–470.

Gordon, A. L. (1981). Seasonality of Southern-Ocean sea ice. *J. Geophys. Res.-Oc. Atm.*, 86, 4193–4197.

Hall, A., and Visbeck, M. (2002). Synchronous variability in the southern hemisphere atmosphere, sea ice, and ocean resulting from the annular mode. *J. Clim.*, 15, 3043–3057. https://doi.org/10.1175/1520-0442

Harangozo S. A. (2006). Atmospheric circulation impacts on winter maximum sea ice extent in the West Antarctic Peninsula region (1979–2001). *Geophys. Res. Lett.*, 33(2), L02502. https://doi.org/10.1029/2005GL024978

Haumann, F. A., Gruber, N., Münnich, M., Frenger, I., and Kern, S. (2016). Sea-ice transport driving Southern Ocean salinity and its recent trends. *Nature*, 537(7618), 89–92.

Haumann, F. A., Notz, D., and Schmidt, H. (2014). Anthropogenic influence on recent circulation-driven Antarctic sea ice changes. *Geophys. Res. Lett.*, 41, 8429–8437

Hobbs W. R., Massom R., Stammerjohn S., Reid, P., Williams, G., and Meier, W. (2016). A review of recent changes in Southern Ocean sea ice, their drivers and forcings. *Global Planet Change*, 143, 228–250. https://doi.org/10.1016/j.gloplacha.2016.06.008

Holland, M. M., Bitz, C. M., and Hunke, E. C. (2005). Mechanisms forcing an Antarctic dipole in simulated sea ice and surface ocean conditions. *J. Clim.*, 18, 2052–2066.

Holland, P. R., and Kwok R. (2012). Wind-driven trends in Antarctic sea-ice drift. *Nature Geosci*, 5(12), 872–875. https://doi.org/10.1038/ngeo1627

Hosking, J. S., Orr, A., Marshall, G. J., and Turner, J. Phillips, T. (2013). The influence of the Amundsen-Bellingshausen seas low on the climate of West Antarctica and its representation in coupled climate model simulations. *J. Clim.*, 26, 6633–6648. https://doi.org/10.1175/JCLI-D-12-00813.1

Jacobs, S. S., and Comiso, J. C. (1997). Climate variability in the Amundsen and Bellinghausen Seas. *J. Clim.*, 10(4), 697709.

Jacobs, S. S., Giulivi, C. F., and Mele, P. A. (2002). Freshening of the Ross Sea during the late 20th century. *Science*, 297(5580), 386–389.

Jena, B., Kumar, A., Ravichandran, M., and Kern, S. (2018). Mechanism of sea-ice expansion in the Indian Ocean sector of Antarctica: Insights from satellite observation and model reanalysis. *Plos One*, 13(10). https://doi.org/10.1371/journal.pone.0203222

Jin, D., and Kirtman, B. P. (2009). Why the southern hemisphere ENSO responses lead ENSO. *J. Geophys. Res.*, 114, D23101. https://doi.org/10.1029/2009jd012657

Kimura, N., and Wakatsuchi, M. (2011). Large-scale processes governing the seasonal variability of the Antarctic sea ice. *Tellus A*, 63, 828–840.

Kumar, A., J. Yadav, and R. Mohan (2021). Seasonal sea-ice variability and its trend in the Weddell Sea sector of West Antarctica. *Environ. Res. Lett.*, 16(2), 024046.

Kwok, R., and Comiso, J. C. (2002). Southern Ocean climate and sea ice anomalies associated with the southern oscillation. *J. Clim.*, 15, 487–501.

Lefebvre, W., and Goosse, H. (2005). Influence of the Southern Annular Mode on the Sea Ice–ocean system: The role of the thermal and mechanical forcing. *Ocean Science*, 2, 299–329.

Lefebvre, W., Goosse, H., Timmermann, R., and Fichefet, T. (2004). Influence of the Southern Annular Mode on the sea ice-ocean system. *J. Geophys. Res-Oceans*, 109, C09005. https://doi.org/10.1029/2004jc002403

Li, X., Holland, D. M., Gerber, E. P., and Yoo, C. (2014). Impacts of the north and the tropical Atlantic Ocean on the Antarctic Peninsula and sea ice. *Nature*, 505, 538–542.

Liu, J. P., and Curry, J. A. (2010). Accelerated warming of the Southern Ocean and its impacts on the hydrological cycle and sea ice. *Proc. Natl. Acad. Sci. U.S.A.*, 107(14), 987–14, 992.

Liu J., Curry, J. A., and Martinson, D. G. (2004). Interpretation of recent Antarctic sea ice variability. *Geophys. Res. Lett.*, 31, L02205. https://doi.org/10.1029/2003GL018732

Liu, J., Tingyin, X., and Chen, L. (2011). Intercomparisons of air-sea heat fluxes over the Southern Ocean. *J. Clim.* 24, 1198–1211. https://doi.org/10.1175/2010jcli3699.1

Mantua, N. J., Hare, S. R., Zhang, Y., Wallace, J. M., Francis, R. C., Mantua, N. J., Hare, S. R., Zhang, Y., and Francis, R. C. (1997). A Pacific interdecadal climate oscillation with impacts on salmon production. *Bull. Ame. Meteorol. Soc.*, 78(6), 1069–1079.

Marshall, G. J. (2003). Trends in the Southern annular mode from observations and reanalyses. *J. Clim.*, 16, 4134–4143.

Martinson, D. G., Stammerjohn, S. E., Iannuzzi, R. A., Smith, R. C., and Vernet, M. (2008). Western Antarctic peninsula physical oceanography and Spatio-temporal variability, *Deep-Sea Res II*, 55, 1964–1987.

Meehl, G. A., Arblaster, J. M., Bitz, C. M., Chung, C. T. Y., and Teng, H. (2016). The Antarctic sea-ice expansion between 2000 and 2014 was driven by tropical Pacific decadal climate variability. *Nature Geosci*, 9, 590–595.

Meredith, M. P., and King, J. C. (2005). Rapid climate change in the ocean west of the Antarctic Peninsula during the second half of the 20th century. *Geophys. Res. Lett.*, 32(2005), L19604. https://doi.org/10.1029/2005gl024042

Mo, K. C. (2000). Relationships between low-frequency variability in the southern hemisphere and sea surface temperature anomalies. *J. Clim.*, 13, 3599–3610.

Mo, K. C., and Higgins, R. W. (1998). The Pacific-South American modes and tropical convection during the southern hemisphere winter. *Mon. Wea. Rev.*, 126, 1581–1596.

Mo, K. C., and Paegle, J. N. (2001). The Pacific-South American modes and their downstream effects. *Int. J. Clim.*, 21, 1211–1229.

Okumura, Y. M., Schneider, D., Deser, C., and Wilson, R. (2012). Decadal-interdecadal climate variability over Antarctica and linkages to the tropics: analysis of ice Core, instrumental, and tropical proxy data. *J. Clim.*, 25, 7421–7441. https://doi.org/10.1175/Jcli-D-12-00050.1

Paolo, F. S., Fricker, H. A., and Padman, L. (2015). Ice sheets. Volume loss from Antarctic ice shelves is accelerating. *Science*, 348(6232), 327–331.

Parkinson, C. L. (2014). Global sea ice coverage from satellite data: Annual cycle and 35-Yr trends. *J. Clim.*, 27, 9377–9382.

Parkinson, C. L. (2019). A 40-y record reveals gradual Antarctic sea ice increases followed by decreases at rates far exceeding the rates seen in the Arctic. *Proc. Natl. Acad. Sci USA*, 116(29), 14414–14423. https://doi.org/10.1073/pnas.1906556116.

Parkinson, C. L., and Cavalieri, D. J. (2012). Antarctic sea ice variability and trends, 1979–2010. *Cryosphere*, 6, 871–880. https://doi.org/10.5194/tc-6-871-2012

Purich, A., Cai, W., England, M. H., and Cowan, T. (2016). Evidence for link between modelled trends in Antarctic sea ice and underestimated westerly wind changes. *Nature Comm*, 7, 10409. https://doi.org/10.1038/ncomms10409

Rahmstorf, S. (2003). The concept of thermohaline circulation. *Nature*, 421, 699. https://doi.org/10.1038/421699a

Raphael, M. N., and Hobbs, W. (2014), The influence of the large-scale atmospheric circulation on Antarctic sea ice during ice advance and retreat seasons. *Geophys. Res. Lett.*, 41, 5037–5045. https://doi.org/10.1002/2014GL060365

Scafetta, N., and Mazzarella A. (2015). The Arctic and Antarctic Sea-Ice Area index records versus measured and modeled temperature data. *Hindawi*, 8. https://doi.org/10.1155/2015/481834

Schmidtko, S., Heywood, K. J., Thompson, A. F., and Aoki, S. (2014). Multidecadal warming of Antarctic waters. *Science*, 346, 1227–1231.

Schneider, D. P., and Steig, E. J. (2008). Ice cores record significant 1940s Antarctic warmth related to tropical climate variability. *Proc. Natl. Acad. Sci USA*, 105, 12154–12158. https://doi.org/10.1073/pnas.0803627105

Screen, J. A., and Simmonds, I. (2010). The central role of diminishing sea ice in recent Arctic temperature amplification. *Nature*, 464, 1334–1337. https://doi.org/10.1038/nature09051

Shu, Q., Song, Z., and Qiao, F. (2015). Assessment of sea ice simulations in the CMIP5 models. *The Cryosphere*, 9, 399–409.

Sigmond, M., and Fyfe, J. C. (2014). The Antarctic sea ice response to the ozone hole in climate models. *J. Clim.*, 27(3), 1336–1342. https://doi.org/10.1175/JCLI-D-13-00590.1

Simpkins, G. R., Ciasto, L. M., Thompson, D. W. J., and England, M. H. (2012). Seasonal relationships between large-scale climate variability and Antarctic sea ice concentration. *J. Clim.*, 25(16), 5451–5469. https://doi.org/10.1175/JCLI-D-11-00367.1

Simpkins, G. R., McGregor, S., Taschetto, A. S., Ciasto, L. M., and England, M. H. (2014). Tropical connections to climatic change in the extratropical southern hemisphere: the role of Atlantic SST trends. *J. Clim.*, 27, 4923–4936. https://doi.org/10.1175/Jcli-D-13-00615.1

Stammerjohn, S. E., Martinson, D. G., Smith, R. C., Yuan, X., and Rind, D. (2008). Trends in Antarctic annual sea ice retreat and advance and their relation to El Niño–Southern Oscillation and Southern Annular Mode variability. *J. Geophys. Res.*, 113, C03S90. https://doi.org/10.1029/2007jc004269

Steffen, K., and Schweiger, A. (1991). NASA team algorithm for sea ice concentration retrieval from Defense Meteorological Satellite Program special sensor microwave imager: comparison with Landsat satellite imagery. *J. Geophys. Res.*, 96, 21971–21987.

Steig, E. J., Ding, Q., Battisti, D. S., and Jenkins, A. (2012). Tropical forcing of circumpolar deep water inflow and outlet glacier thinning in the Amundsen Sea embayment, West Antarctica. *Ann. Glaciol.*, 53, 19–28.

Stroeve, J., and Notz, D. (2018). Changing state of Arcticsea ice across all seasons. *Environ. Res. Lett.*, 13, 103001.

Swart, N. C., and Fyfe, J. C. (2013). The influence of recent Antarctic ice sheet retreat on simulated sea ice area trends. *Geophys. Res. Lett.*, 40, 4328–4332. https://doi.org/10.1 002/grl.50820

Thoma, M., Jenkins, A., Holland, D., and Jacobs, S. (2008). Modelling circumpolar deep water intrusions on the Amundsen Sea continental shelf, Antarctica. *Geophys. Res. Lett.*, 35, L18602. https://doi.org/10.1029/2008gl034939

Tréguer, P., and Pondaven, P. (2002). Climatic changes and the cycles of carbon in the Southern Ocean: A step forward (II). *Deep-Sea Res II*, 49(16), 3103–3104.

Turner, J. (2004). The El Nino-southern oscillation and Antarctica. *Int. J. Climatol.*, 24, 1–31. https://doi.org/10.1002/joc.965

Turner, J., Bracegirdle, T. J., Phillips, T., Marshall, G. J., and Hosking, J. S. (2013c). An initial assessment of Antarctic Sea ice extent in the CMIP5 Models. *J. Clim.*, 26(5), 1473–1484.

Turner, J., Comiso, J. C., Marshall, G. J., Lachlan-Cope, T. A., Bracegirdle, T., Maksym, T., Meredith, M. P., Wang, Z., Orr, A. (2009). Non-annular atmospheric circulation change induced by stratospheric ozone depletion and its role in the recent increase of Antarctic sea ice extent. *Geophys. Res. Lett.*, 36, L08502. https://doi.org/10.1029/2 009GL037524

Turner, J., Guarino, M. V., Arnatt, J., Jena, B., Marshall, G. J., Phillips, T., Bajish, C. C., Clem, K., Wang, Z., Andersson, T., Murphy, E. J. , and Cavanagh, R. (2020). The recent decrease of summer sea ice in the Weddell Sea, Antarctica. *Geophys. Res. Lett*, 47, e2020GL087127. https://doi.org/10.1029/2020GL087127

Turner, J., Hosking, J. S., Marshall, G. J., Phillips, T., and Bracegirdle, T. J. (2015). Antarctic sea ice increase consistent with intrinsic variability of the Amundsen Sea low. *Clim. Dyn.* https://doi.org/10.1007/s00382-015-2708-9

Turner, J., Hosking, J. S., Phillips, T., and Marshall, G. J. (2013a). Temporal and spatial evolution of the Antarctic sea ice prior to the September 2012 record maximum extent. *Geophys. Res. Lett.*, 40, 5894–5898. https://doi.org/10.1002/2013GL058371

Turner, J., Lu, H., King, J., Marshall, G. J., Phillips, T., Bannister, D., and Colwell, S. (2021). Extreme temperatures in the Antarctic. *J. Clim*, 34, 2653–2668.

Turner, J., Phillips, T., Hosking, J. S., Marshall, G. J., and Orr, A. (2013b). The Amundsen Sea low. *Int. J. Climatol.*, 33, 818–1829.

Turner, J., Phillips, T., Marshall, G. J., Hosking, J. S., Pope, J. O., Bracegirdle, T. J., and Deb, P. (2017). Unprecedented springtime retreat of Antarctic sea ice in 2016. *Geophys. Res. Lett.*, 44, 6868–6875. https://doi.org/10.1002/2017GL07365

van den Broeke, M., Reijmer, C., and van de Wal, R. (2004). Surface radiation balance in Antarctica as measured with automatic weather stations. *J. Geophys. Res.*, 109, D09103. https://doi.org/10.1029/2003JD004394

Walsh, J. E. (2013). Melting ice: What is happening to Arctic sea ice, and what does it mean for us? *Oceanography*, 26, 171–181. https://doi.org/10.5670/oceanog.2013.19

Yu, L. J., Zhang, Z., Zhou, M., Zhong, S., Lenschow, D. H., Gao, Z., Wu, H., Li, N., and Sun, B. (2011). Interpretation of recent trends in Antarctic sea ice concentration. *J. of Applied Remote Sensing*, 5(1), 053557. https://doi.org/10.1117/1.3643691

Yu, L., Zhong, S., Winkler, J. A., Zhou, M., Lenschow, D. H., Li, B., Wang, X., and Yang, Q. (2017). Possible connections of the opposite trends in Arctic and Antarctic sea-ice cover. *Sci. Rep*, 7, 45804. https://doi.org/10.1038/srep45804

Yuan, N., Ding, M., Ludescher, J., and Bunde, A. (2017). The increase of the Antarctic sea ice extent is highly significant only in the Ross Sea. *Sci Rep*, 7, 41096. https://doi.org/ 10.1038/srep41096

Yuan, X. J. (2004). ENSO-related impacts on Antarctic sea ice: A synthesis of phenomenon and mechanisms. *Antarct. Sci.*, 16, 415–425. https://doi.org/10.1017/S0954102004002238

Yuan, X. J., and Li, C. H. (2008). Climate modes in southern high latitudes and their impacts on Antarctic sea ice. *J. Geophys. Res.*, 113, C06S91. https://doi.org/10.1029/2006jc004067

Zhang Y., Wallace J. M., and Battisti D. S. (1997). ENSO-like interdecadal variability: 1900–93. *J. Clim.*, 10, 1004–1020.

Zhang J. (2007). Increasing Antarctic sea ice under warming atmospheric and oceanic conditions. *J. Clim.*, 20, 2515–2529. https://doi.org/10.1175/JCLI4136.1

Zwally, H. J., Comiso, J. C., Parkinson, C. L., Cavalieri, D. J., and Gloersen, P. (2002). Variability of Antarctic sea ice 1979–1998. *J. Geophys. Res.*, 107 (C5), 3041. https://doi.org/10.1029/2000JC000733

11 Southern Hemispheric Climate Change, Interhemispheric Teleconnection, and the Observed Trends in the Seasonal Mean Monsoon Rainfall over the Indian Region in the Last Century (1871–2004)

Rajib Chattopadhyay and A.K. Sahai
Indian Institute of Tropical Meteorology, Pune, India

CONTENTS

DOI: 10.1201/9781003203742-11

11.1 CLIMATE CHANGE IN THE SOUTHERN HEMISPHERE AND ANTARCTICA

In the last few decades, after several long-term reanalyses, observational as well as satellite data are available for the Southern Hemisphere; several studies have quantified the cause and impact of Southern Hemispheric climate change using these data sets. The data sets like near-surface temperatures, winds, sea level pressure, sea-ice change, change in Antarctic ice mass, etc. reflect the Southern Hemispheric warming. As a result of this, it is now recognized that several long-term to seasonal to sub-seasonal meteorological time-series indicators show a substantial shift in the last hundred years, and this long-term trend is attributed to the rising temperature trend in the hemispheric as well as global scale. The poleward extension of the Hadley cell, the poleward shift of the storm tracks, and change in the frontal activity are discussed in several papers as a manifestation of this warming trend (Gastineau et al., 2009; Fyfe et al., 2012; Solman and Orlanski, 2013, 2015). As a result of this climatic shift, in addition to the hemispheric change, several regional impacts are also documented. Changes in rainfall patterns over different parts of Australia, New Zealand, southern Africa, and South America are now well known.

Quantification of the impact of Southern Hemispheric climate change is important in understanding the changes in general circulation. Southern Hemispheric oceans south of 50°S and the Antarctica continent, in general, play a significant role in the maintenance of the overturning and eddy-driven circulation through heat balance. The melting or freezing of the ice sheets in Antarctica and the adjoining ocean surface can control or support the overall southern oceanic and atmospheric flux exchange. This exchange is a multi-scale spatial-temporal climate feedback process through physical processes like the ice-albedo feedback, Ekman transport, and the eddy-mediated heat transfer by the zonal temperature gradients, in which Antarctica acts as a heat sink. The integrated effects of this feedback affect the general circulation of the Southern Hemispheric climatic features. Another aspect of the Southern Hemispheric climate change is related to the development and healing of the ozone holes. Many modeling studies (e.g. Son 2010; Polvani et al., 2011b; Hu et al., 2013) suggested that severe Antarctic ozone depletion caused cooling over the high latitudes that extend to the troposphere. It has led to enhanced meridional temperature gradients between the tropospheric polar region and the extratropics. Consequently, the thermal wind balance leads to a poleward shift of westerly winds. The net effect on the Hadley cell is the weakening and widening over the extra-tropical to high-latitude region (Hu and Fu, 2007; Hu et al., 2018). The impact of healing of ozone holes (Kuttippurath and Nair, 2017) though uncertain, some studies suggest a weakening of the climate change trends during Austral summer in the southward shifting of the Southern Hemispheric westerlies (Polvani et al., 2011a; Fogt et al., 2017; Fogt and Marshall, 2020).

Studies based on empirical orthogonal functions (EOFs) using sea-level pressure data in the Southern Hemisphere have found a zonally symmetric annular mode in the atmosphere: The Southern annular mode (SAM) and similarly analyses in the Northern Hemisphere isolate the northern annular mode (NAM) (Rogers and van Loon, 1982; Thompson and Wallace, 2000). The SAM explains a dominant part of the week-to-week, month-to-month, and year-to-year variance in the Southern Hemispheric extratropical atmospheric flow than any other climate phenomenon. Like its Northern Hemispheric counterpart or NAM, the SAM explains about ~20–30% of the total variance in the geopotential height and wind fields of their respective hemispheres, depending on the level and timescale considered. The SAM is also known as Antarctic Oscillations (AAO) in the Southern Hemisphere. Studies based on observations and numerical modeling experiments suggest the annular modes have played and will continue to play a role in determining the variations of weather patterns and weather regimes (Pohl and Fauchereau, 2012). The annular modes are also defined by defining the indices derived from the sea level pressures (Gong and Wang, 1999; Marshall, 2003; Meneghini et al., 2007). Both methods provide similar information, and several studies use this pressure-based index to study the climate trends in the Southern Hemisphere.

The atmospheric and climatic response of Southern Hemispheric warming is popularly studied through the impact on the variability of the southern annular mode (SAM). Variability and trends in SAM provide crucial information on the impact of climate change in the Southern Hemisphere and its teleconnection to the different belts of Southern Hemispheric latitudes. The positive phase of the SAM (positive SAM), i.e. a positive value of the index, is associated with a poleward shift of the Southern Hemispheric mid-latitude eddy-driven jet. This jet typically is located around a latitude band ~50°S. As a result, there is an equator-ward extension of the cyclones. This shifting of the jet is a complex process mediated by the coupled SAM and the Antarctic sea-ice extent relation, that effectively controls the equator-ward or poleward propagation of eddies (cyclones) and the shifting of the jet position (Hartmann and Lo 1998; Pezza et al., 2008). Hence, anomalies of reduced rainfall/higher temperature in the mid-latitude and enhanced rainfall/lower temperature in the high latitudes are observed with positive SAM phases, and vice versa with negative SAM whose regional implications would be found in several pieces of literature (Silvestri and Vera, 2003; Gillett et al., 2006; Sen Gupta and England, 2006; Hendon et al., 2007; Meneghini et al., 2007). The appearance of high pressure over mid-latitude and high sea surface temperature trends associated with global warming over low southern latitude close to the equator would generate increased weather disturbance, resulting in increased precipitation over low southern latitudes, for example, northern Australia (Gillett et al., 2006; Meneghini et al., 2007). A sufficient indication of this mechanism would also be manifested in the increase in cyclonic potential vorticity and its trend, as discussed later in Section 5.

Several studies investigated how the SAM or AAO changes in the future climate change scenario. The observation-based analysis suggests that SAM shows an increasing trend towards the positive phases from the mid-1950s (Kidson, 1999; Gong and Wang, 1999). Several studies have also shown that the SAM phase index change has influenced the circulation over the Antarctic Peninsula. Changes in the SAM phase have been attributed to causing substantial warming in the Antarctic Peninsula and cooling over much of East Antarctica.

Over the last 40 years, there has been a continuous measurement of Antarctic sea ice using satellite-based passive microwave remote sensing instruments. Observations indicate that there is a small but statistically significant increment in the sea ice in the Southern Ocean sea ice coverage in the long term, which is followed by a decrease at a very high rate (Hobbs et al., 2016; Parkinson, 2019; Parkinson and Cavalieri, 2012). This contrasts with the situation in the Northern Hemisphere where the sea in the Arctic is strongly and monotonically declining in the recent decades. The sea-ice concentration also shows positive trends though there are regional distribution and seasonal variations. During summer and autumn, large positive trends are seen in the Weddel and the Western Ross Sea, while negative trends are seen in the Amundsen and the Bellingshausen sea, and recent studies indicate slow growth and fast melt season (Hobbs et al., 2016; Eayrs et al., 2019).

The Antarctic continental land ice is another crucial Southern Hemispheric feature that contrasts with the Arctic continental feature, which is mostly ocean. The Antarctic land ice holds enough ice to raise the sea level by 58 m. A recent Ice Sheet Mass Balance Inter-comparison Exercise (IMBIE) team (Shepherd, 2018) report shows that it lost 2,720 ±1,390 billion tons of ice between 1992 and 2017, which is likely to increase the mean sea level by 7.6 ±3.9 millimeters (errors are one standard deviation). This trend is consistent with long-term analysis, which also shows similar longer-term trends (Rignot et al., 2019). The melting of land ice would create other surface thermal effects besides increasing sea level. Over the Antarctic continent, it would change the ice-albedo feedback to a considerable extent, affecting the tropospheric and stratospheric temperature balance.

The brief review done above suggests that the Southern Hemispheric climate change is clear from the long-term and short-term data. Despite several uncertainties related to data availability in the Antarctic subcontinent, a consistent trend is clear in terms of warming of the surface air temperature and melting of land ice and sea ice. Such trends affect the Southern Hemispheric weather and climate features, as documented in several studies.

11.2   SOUTHERN HEMISPHERIC CLIMATE CHANGE AND THE INTERHEMISPHERIC TELECONNECTION

Interhemispheric teleconnections are known to impact the climate response in the opposite hemisphere. Studies, for example, have shown that monsoon diabatic heating can have a role in the strengthening of the Southern Hemispheric subtropical anticyclone (Lin, 2009; Lee et al., 2013). Interhemispheric teleconnection originating from the Southern Hemisphere has recently received a lot of attention in literature due to its potential role in influencing the Northern Hemispheric precipitation and its predictability (Song et al., 2009; Nan et al., 2009; Zhao et al., 2015; Mamalakis et al., 2018). Simple interhemispheric teleconnection indices have been shown to efficiently differentiate the global warming signatures as compared to the natural variability and are sometimes used for detection and attribution studies related to global warming (Karoly and Braganza, 2001; Drost and Karoly, 2012). As a result of climate change, some studies have shown that these inter-hemispheric teleconnections can change significantly in the warming scenarios (Karoly and Braganza, 2001; Friedman et al., 2013; Drost and Karoly, 2012). Since the northern and Southern Hemispheres are different in terms of land-water mass distribution, these studies show that the interhemispheric indicators like Northern and Southern Hemispheric temperature contrast are often very robust and reflect several large-scale manifestations of climate change.

In the context of global warming and Southern Hemispheric climate change scenarios presented previously, how does the interhemispheric teleconnection modulate the monsoon? What is the impact of Southern Hemispheric climate change on the Northern Hemispheric tropics and monsoonal region in general? Monsoon plays a significant role in the land hydrology over the boreal summer monsoon zone over the Indian Subcontinent. Since the hemispheric branch of the Hadley cell is related to convective updrafts originating over the equatorial region, the question of whether any change in interhemispheric teleconnection would be reflected in the change in the convective activity over the convergence zones (land and ocean) during boreal summer has remained unanswered. The boreal summer monsoon, especially the Indian summer monsoon, is driven by an interhemispheric migrating outflow pattern originating from the Southern Hemispheric Mascarene high-pressure belt. Strong cross-equatorial flows as well as low-level jets, which are signatures of interhemispheric teleconnection, are embedded in the definition of monsoon. Hence, any change in monsoon rainfall could be attributed to the change in the convergence of moisture over land and ocean and hence change in rainfall could be associated with a change in interhemispheric teleconnection. In recent years, some studies have shown the weakening of the low-level jet (Joseph and Simon, 2005) and change in the Mascarene high-pressure belts due to the global warming hiatus (Vidya et al., 2020). Thus, the role of interhemispheric tele-connection, as well as Southern Hemispheric climate change, should be explored. It would be a tricky question to explore, though. Multiple teleconnections influence tropical rainfall, especially monsoon rainfall. The El-Niño southern oscillations noticeably contribute to the development of the major tropical and extratropical teleconnection patterns (e.g. PNA pattern) and the well-known ENSO Indian summer monsoon teleconnection. It is well known that these teleconnections have been affected by climate change. While several papers have reported the ENSO-monsoon teleconnection in a warming scenario, the role of the interhemispheric change in teleconnection pattern on the monsoon interannual variability is not clear in the century-scale timespan. In this chapter, we will focus on the probable trends and changes and variability of monsoon system because of the change in the in-terhemispheric teleconnection due to the climate change in the Southern Hemisphere in the past century or more. Before taking up this issue, in order to have a broader horizon, a brief review of the existing evidence of change in monsoon rainfall pattern and monsoon teleconnections associated with other global circula-tion features would be done in the next two sections.

11.3   CENTURY SCALE TRENDS IN THE INDIAN SUMMER MONSOON RAINFALL

Several studies have documented a change in the summer monsoon climate over the Indian region because of anthropogenic warming. One of the most sensitized results is that the extreme rainfall events over the Indian region show increasing trends (Goswami et al., 2006; Roxy et al., 2017; Goswami et al., 2019; Mishra, 2019). In the last century, the seasonal mean all-India averaged summer monsoon rainfall has decreased for many meteorological subdivisions with a change in seasonality, and also it has increased in some other subdivision (Guhathakurta and Rajeevan, 2008; Naidu et al., 2009; Kumar et al., 2010; Guhathakurta et al., 2015; Rai and Dimri, 2020; Viswambharan, 2019; Praveen et al., 2020). Guhathakurta et al. (2015) also concluded that the July–August all-India mean monsoon rainfall has shown a significant decreasing trend for most parts of India. This is also confirmed by many other studies (Turner and Annamalai, 2012; Jin and Wang, 2017). We plot the "*All-India*" area-averaged monthly mean rainfall (***AISMR***) trend and the six yearly smoothed trends in Figure 11.1 for the years 1871–2004. The plot is based on publicly available homogeneous rainfall data for the Indian region (Kothawale and Rajeevan, 2017) from the Indian Institute of Tropical Meteorology (IITM) data repository. The data set used in the study is described in Table 11.1. The plot shows that there is a statistically significant long-term trend in the monthly rainfall, which is clearer in the six yearly smoothed (running averaged) data. The statistical significance for this plot or all the plots is assessed by computing the *t-statistic* and the "*t-value*", which is mentioned for each plot. The plot shows that the monthly rainfall has decreased for June, July, and September and has increased in August. The increase in rainfall is also mild. This mild increase in rainfall during August and decreasing trend in rainfall during other monsoon months is in contradiction to the global warming scenario in which it is hypothesized that evaporation would strongly increase as a response to the warming and hence rainfall would increase in a homogeneous manner assuming the *Clasius-Clayperon* relation controlling the moisture availability over the atmosphere. It is already noted in many studies that this amplification may not happen homogeneously for globally averaged precipitation (Stephens and Ellis, 2008; Skliris et al., 2016; Pall et al., 2011) as well as for monsoon precipitation (Asharaf and Ahrens, 2015; Lal et al., 1994) in a global warming scenario. Normally, a monsoon is known to be significantly controlled by El-Niño conditions in the Pacific. However, the ENSO-monsoon correlation is a week in the latter half of the last century (Kumar et al., 1999). Then how, is this general weakening trend of rainfall could be explained when it is *expected* that it would increase based on *Clasius-Clayperon* conditions? To understand this, we note that a monsoon can be explained as a system driven by asymmetric heating over land during boreal summer (Zhang and Krishnamurti, 1996). A study by Jin and Wang (2017) have explored this weakening trend based on the relative gradient of heating over the Indian Ocean and Indian landmass. This conclusion of land-sea thermal contrast control of monsoon rainfall is further verified by other studies (Gnanaseelan et al., 2017; Chou, 2003; Roxy, 2017).

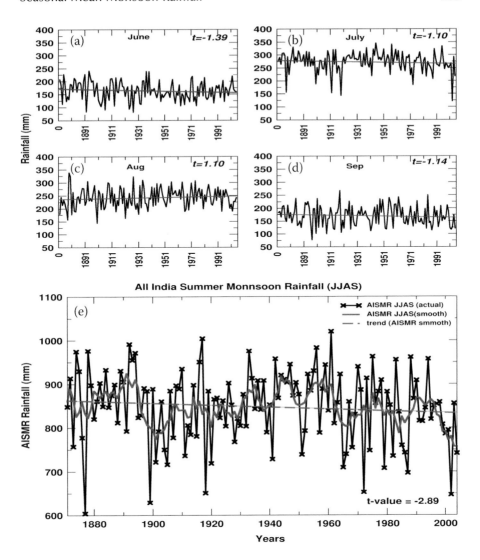

FIGURE 11.1 (a)–(d) Monthly (Jun, July, August, and September) rainfall over the Indian region using IITM(REF) data. (e) AISMR (JJAS) yearly variation (black curve) and a six yearly smoothed value of the same rainfall (red curve). The red line in each plot shows a linear trend and the t-values of the plotted trends are mentioned in each plot panel.

Most of these studies demonstrate the thermal contrast change and its impact on monsoon circulation in terms of equatorial to tropical land-ocean warming. How much of this thermal contrast would be controlled by southern extratropical to polar (or entire) Southern Hemispheric climate change is not yet clear. In this chapter, we will take up this issue and explain this weakening trend in the context of the role of entire Southern Hemispheric warming in maintaining the monsoonal thermal contrast. Mascarene high, for example, is known to be a significant source of monsoon

TABLE 11.1
Data Used in This Study

Data abbreviation	Reference	Download link	Description
All-India rainfall data	(Kothawale and Rajeevan, 2017)	https://tropmet.res.in/static_pages.php?page_id=53 **(last accessed: 20/10/2020) Based on 306 station data where data is available for the full period.**	**Primary data source: IMD(MoES).**
HADSST1 HADSeaIce	(Rayner et al., 2003)	https://www.metoffice.gov.uk/hadobs/ **(last accessed: 20/10/2020)**	**Hadley, Centre, UK data**
HADSLP	(Allan and Ansell, 2006)	https://www.metoffice.gov.uk/hadobs/hadslp2/ **(last accessed: 20/10/2020)**	**-do-**
HADCRU	(Morice et al., 2012)	https://crudata.uea.ac.uk/cru/data/temperature/(last accessed **20/10/2020)**	**Climatic Research Unit (University of East Anglia) and Hadley Centre (UK Met Office)**
ERSSTv5	(Huang, 2017)	https://www1.ncdc.noaa.gov/pub/data/cmb/ersst/v5/netcdf/ **(last accessed: 20/10/2020)**	**NCDC/NOAA Global gridded SST data**
20Cv2 Reanalysis	(Compo, 2011)	https://www.psl.noaa.gov/data/20thC_Rean/	
SAM Index	(Gong and Wang, 1999)	https://www.psl.noaa.gov/data/20thC_Rean/timeseries/monthly/SAM/sam.20crv2.long.data **(last accessed: 20/10/2020) Acknowledgment statement:** The current study uses the SAM index created from 20CV2 reanalysis data. Support for the Twentieth Century Reanalysis Project data set is provided by the U.S. Department of Energy, Office of Science Innovative and Novel Computational Impact on Theory and Experiment (DOE INCITE) program, and Office of Biological and Environmental Research (BER), by the National Oceanic and Atmospheric Administration Climate Program Office, and by the National Oceanic and Atmospheric Administration	**20cV2Reanalysis** (Compo 2011)

(Continued)

TABLE 11.1 (Continued)
Data Used in This Study

Data abbreviation	Reference	Download link	Description
NCEP-NCAR Reanalysis-1	(Kalnay, 1996)	Climate Program Office, and by the NOAA Physical Sciences Laboratory. https://psl.noaa.gov/data/gridded/data.ncep.reanalysis.html **(last accessed: 20/10/2020) Acknowledgment Statement:** NCEP Reanalysis data provided by the NOAA/OAR/ESRL PSL, Boulder, Colorado, USA.	**The PV index is computed based on this data**

circulation. Because of a change in the Southern Hemispheric climate pattern, we hypothesize that the Southern Hemispheric control on monsoon flow has weakened, which leads to a change (weakening) in monsoon circulation or rainfall. This proposition will be elaborated in Section 5.

Many studies have noted significant decadal variability, which may be a manifestation of natural monsoon variability as such longer period secular or sea surface temperature forced variability is reported for a monsoon (Kucharski et al., 2006) can also cause the trend. Whether it is forced by natural or forced by anthropogenic warming, overall, for most of the subdivisions as well as in the all-India scale, a decreasing trend is seen in the last century. Based on the evidence presented in the earlier sections, it is clear that the natural variability and warming planet both would contribute to the trend. The weakening of the monsoon wind also supplements the weakening of the rainfall. The low-level *Findlater Jet* is weakened (Joseph and Simon, 2005), tropical easterly jet at the upper level has weakened (Sathiyamoorthy, 2005; Abish et al., 2013) or the overturning (Hadley cell) has weakened (Krishnan, 2013). One of the dynamic mechanisms being argued for the weakening of the monsoon flow over the Indian region is the weakening of the synoptic systems and monsoon depressions over Bengal's Bay (Sandeep et al., 2018; Vishnu et al., 2016). Since most trends are attributed to anthropogenic warming, such trends indicate strong monsoon variability in the future climate change scenario (Sharmila et al., 2015).

11.4 MONSOON AND TROPICAL TELECONNECTIONS (IOD AND ENSO) MODES IN A CLIMATE CHANGE SCENARIO

The change in the monsoon climate is also associated with the shift in the teleconnection patterns. The ENSO-monsoon teleconnection revealed by the Nino index, for example, has undergone a strong to weak correlation phase (Kumar et al., 1999). However, modeling studies in future warming scenarios do not show a consistent response in the future, with studies arguing that the ENSO-monsoon relation weakens or strengthens (Annamalai et al., 2007; Roy et al., 2019). Similarly, other research has shown that there could be a shift in ENSO's spectral peak, leading to ENSO's frequent appearance; hence, more drought-like situations over the Indian region (Azad and Rajeevan, 2016). Thus, several studies indicate that the ENSO cycle and ENSO-monsoon teleconnection would change shortly.

The Indian Ocean Dipole mode (Saji et al., 1999; Webster et al., 1999) is another important mode of variability over the Indian Ocean region that is known to moderate the monsoon rainfall distribution and amplitude (Behera and Ratnam, 2018; Ashok et al., 2001; Bala and Singh, 2008). Studies show that the frequency of the positive dipole mode events is changed in the recent decade. The reason for this shift is attributed to the climate change impacts over the Indian Ocean region (Cai et al., 2009). Although the effect on rainfall variability is often conflicting in a warming scenario due to *dynamic* and *thermodynamic* effects are often counter-acting (Huang et al., 2019), the increase in the positive dipole mode frequency would likely impact the dipole mode and monsoon teleconnection.

Thus, it is clear that the tropical teleconnection pattern, evidenced from AISMR and the NINO/IOD index correlation, also controls the monsoon (refer to Table 11.2a) and in a warming scenario, it would provide important feedback to monsoon rainfall. However, as Table 11.2a shows, the ENSO-monsoon tele-connection though is the dominant teleconnection pattern; it explains only ~25% of the monsoon interannual variability, and as correlation with other indices suggest, other teleconnection patterns could also be important. The hemispheric temperature gradient for the monsoon region (gradT, defined in Section 6), for example, would have a significant correlation with a monsoon. This interhemispheric temperature gradient would be taken up next for further analysis.

11.5 SOUTHERN (INTERHEMISPHERIC) TELECONNECTION AND INDIAN SUMMER MONSOON

Southern Hemispheric influence on the monsoon climate and interhemispheric teleconnection is still a less studied affair than other teleconnection effects. Although the monsoon flow has its dynamic and thermodynamic origin over the Southern Hemisphere, with Mascarene High playing a significant role in its evolution and asymmetric heating can generate monsoon circulation (Zhang and Krishnamurti, 1996), it is not at once clear how the monsoon climate would be impacted by a climate change in the Southern Hemisphere. To get some idea in this regard, it would be useful to review some of the past studies that document the Southern Hemispheric influence on the monsoon system.

A study by Rodwell (1997) has documented the role of Southern Hemispheric synoptic systems in the modulation of monsoon flow. From the conservation of potential vorticity principle, he showed that the traveling synoptic disturbances in the Southern Hemisphere could induce a break monsoon situation if they enter the Northern Hemisphere. Such an induction can take place under several general conditions, and hence it can be treated as a potential mechanism for affecting the sub-seasonal to seasonal variability of monsoon. Under climate change scenarios, intensification and development of Southern Hemispheric circulations and storms are projected to be changed due to a change in the SAM variability (Grieger et al., 2018; Screen et al., 2018). Similarly, a direct descending SAM-induced circulation mechanism is reported in other studies, which says SAM can impact the rainfall over other regions, e.g. the Korean Peninsula (Prabhu et al., 2017) and precipitation over the South China Sea (Liu et al., 2018; Wu et al., 2015; Liu et al., 2015). Similarly, the variability of Mascarene High and the Australian High also impact the rainfall over East Asia during the summer monsoon season (Xue et al., 2004). A study by Fan and Wang (2006) has shown that zonal mean wind at Southern Hemispheric high latitudes has a significant positive correlation with that of Eurasia during boreal winter and has a negative correlation with the Pacific North American teleconnection (PNA) pattern during boreal spring, showing some sort of meridional teleconnections existing between both hemispheres. Thus, a pathway can exist in which the impact of Southern Hemispheric systems into the monsoon flow over the Northern Hemisphere is possible. Such mechanisms can induce a reduction in monsoonal rainfall over the Indian region.

Some studies have well documented the role of SAM in the modulation of monthly rainfall over the Indian region. Viswambharan and Mohanakumar (2013) suggest that a positive SAM index impacts the monsoonal flow during the onset time. Also, humidity and precipitation are affected during the monsoon time in this period. They also suggested that the SAM and July–August rainfall are anti-correlated, with the high SAM index not favoring the rain over the Indian region. The same result is also discussed in another study, which further elaborated on the mechanism through the role of warm sea surface temperature during positive SAM phases (Pal et al., 2017). A statistical analysis by Dugam and Kakade (2004) showed that the Antarctic sea-ice extent in the December–January season is inversely associated with the monsoon rainfall in the next season. Their analysis also showed regional variation with the southern peninsula does not have much association. An indirect effect of the Southern Hemisphere on tropical climate is proposed by Terray (2011). The study has suggested that after the 1976–1977 climate shift, the ENSO (and hence tropical climate) is more strongly influenced by extratropical SST modes.

11.6 THE TREND IN THE SEASONAL MONSOON RAINFALL ASSOCIATED WITH THE TREND IN THE SOUTHERN HEMISPHERIC TELECONNECTION

Based on the previous review, it is clear that the change in monsoon rainfall and circulation can largely be attributed to a change in the teleconnection patterns in a warming scenario. The interhemispheric differential warming impacts indeed cannot be neglected in a climate change scenario (Friedman et al., 2013). In this study, we will use the indicators based on interhemispheric gradients (Karoly and Braganza, 2001) to understand the impact of the Southern Hemisphere on the monsoon climate. Two features from the previous review, i.e. change in the low-level *Findlater* jet intensity and change in the monthly rainfall and break days, are noteworthy in terms of interhemispheric impacts (Joseph and Simon, 2005). Based on these features, two mechanisms based on Southern Hemispheric

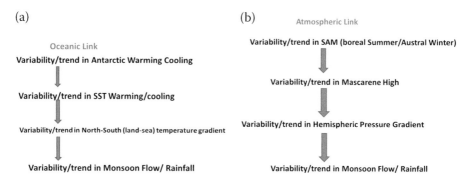

FIGURE 11.2 (a) The probable oceanic mechanism of interhemispheric trend. (b) The atmospheric mechanism of rainfall trend.

teleconnections are proposed, which can impact the monsoon rainfall and northern tropical circulations. The first one is the large-scale Indian Ocean control or surface temperature control on monsoon interannual variability (Webster et al., 2002; Webster, 2006), a hypothesis based on interhemispheric heat transport can affect the monsoon rainfall in a warming scenario, as shown in the sketch diagram presented in Figure 11.2a. The second mechanism is based on the Southern Hemispheric teleconnection associated with potential vorticity induction associated with Southern Hemispheric weather disturbances, as proposed in Rodwell (1997), which is shown in Figure 11.2b. There can be other indirect mechanisms. As mentioned previously, these mechanisms would suggest that the Southern Hemispheric teleconnection, which can be linked to the variability of SAM and interhemispheric heat transport, can impact the observed monsoon trend. These links are explored in this section.

11.6.1 Mechanisms of Monsoonal Trend Related to Indian Ocean/ Southern Hemispheric Warming Affecting the Interhemispheric Gigantic Land-Sea Breeze Structure

The Southern Hemisphere has undergone a large change in the sea surface temperature (SST), and a monthly plot would show a largely positive trend in the SST (*HADSST1;* Rayner et al., 2003), as shown in Figure 11.3a. The *HADSST1* is the grid-interpolated monthly SST data that is available for the study period (1871–2004). The data source is described in Table 11.1. The spatial pattern shows strong regional gradients, with the Eastern Hemisphere showing more warming than the Western Hemisphere, which is also seen in *ERSSTv5* data (Huang 2017; plot not shown). The sea-ice concentration plot using the *Hadley Centre, UK,* data (*HADSeaIce;*Rayner et al., 2003) is shown in Figure 11.3b; however, it does not show a uniform warming trend. The Eastern Hemisphere shows a decreasing trend, and in the Western hHmisphere, it shows an increasing trend. This trend is consistent with the SST trend, which shows strong warming in the Eastern Hemisphere.

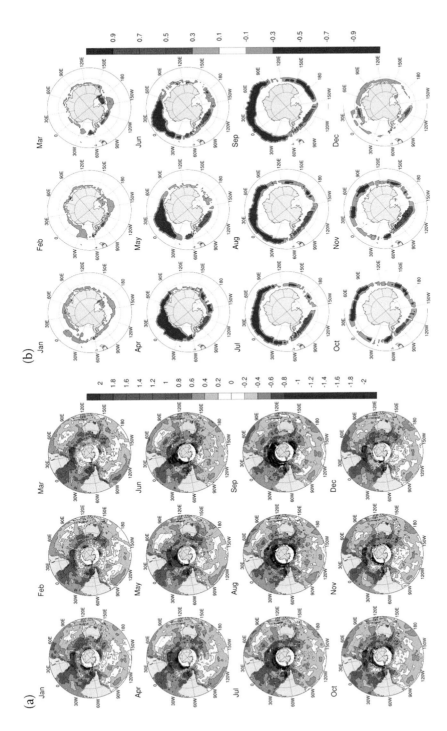

FIGURE 11.3 (a) Monthly sea surface temperature (SST) trend (unit °C per 100 years) based on HadSST and (b) monthly sea-ice concentration trend (unit: fractional change per 100 years) from HadSeaIce data. Data used to calculate trend is for the period 1871–2004. For data, source refer Table-1.

How would a hemispheric temperature change affect the monsoon? In this re-gard, we would like to point out the well-known self-regulatory mechanism of monsoon interannual variability (Webster et al., 2002; Webster, 2006). This theory states that for the ocean-atmosphere coupled system like the south Asian monsoon, the amplitude of the interannual variability measured by its precipitation indices and its interannual variability (as compared to the seasonal mean) is smaller than the variability exhibited in other climate systems of the tropics. It is argued that the reason for this stability is related to the ocean-atmosphere interaction in which the Indian Ocean and Indian subcontinent act as a heat source with a land-sea contrast maintained on an interannual scale, and an imbalance in the land-sea heat transport could lead to a self-regulated biennial variability (Fig. 1.33 of Webster, 2006). This gigantic monsoonal land-sea breeze hypothesis is essentially linked to the oceanic heat-transport process and sea surface temperature, creating the land-sea heat contrast. Naturally, the temperature trend would change in circulation, which would lead to a change in monsoon flow. This can be visualized as follows using the well-known circuit diagram from the dynamic meteorology book of Holton (2004, fourth edition; Fig. 4.3) related to land-sea breeze (Figure 11.4).

The criteria that monsoonal ocean to land circulation would be generated (similar to a direct or reverse monsoon Hadley cell) can be obtained, assuming a monsoon is a thermal contrast-driven system with the circulation is actively supported by the Mascarene outflow and high-pressure system. At the start of the season, the contrast would be well reflected by vertically averaged air temperature (representative of vertically integrated diabatic heating) and after it is established and the rainfall is spread all over the region, it can be well represented by the upper-tropospheric temperature gradients created due to deep heating (Xavier et al., 2007). The rate of change of monsoon circulation $C_a > 0$ to a crude order can be derived from the land-sea breeze formula (Holton 2004) and is given as:

$$\frac{dC_a}{dt} = R. \ \ln\left(\frac{p_0}{p_1}\right)(\bar{T}_L - \bar{T}_o) > 0 \qquad [11.1]$$

where $\bar{T}_L$ denotes the mean vertically averaged temperature over land (i.e. Indian subcontinent) and $\bar{T}_o$ denotes the mean vertically averaged temperature over the ocean. This *means* variation implies any time scale and any vertical level of interest. Although a monsoon is not a purely land-sea breeze driven system, rather it also can be explained by the migratory *Intertropical Convergence Zone* (ITCZ) and latent heating-based theory (Gadgil, 2018), the land-sea thermal contrast essentially generates the circulation, whether it is due to surface warming or is due to latent heating. Thus, both $\overline{T_L}$ and $\overline{T_O}$ can be represented by vertically integrated heating. The monsoon Hadley cell, to a first order, is driven by this circuit flow. Since the change in circulation is related to the shift in tangential velocity along the circuit, the previous equation can be written in terms of mean tangential velocity. For monsoon flow, it can be the mean meridional velocity (Hadley cell flow). This is written as:

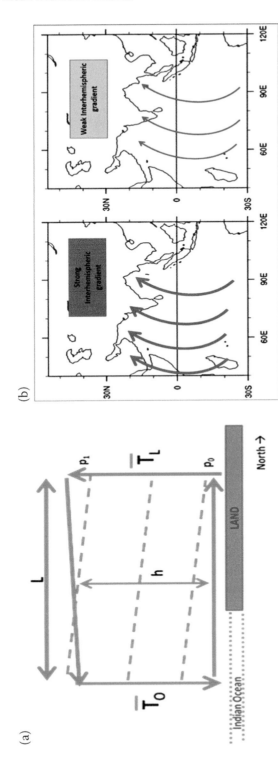

FIGURE 11.4 (a) The schematic mechanism of monsoon as a land-sea breeze, and (b) a mechanism of monsoon interannual variability as shown in Webster et al. (2006).

$$\frac{d\bar{v}}{dt} = R. \ \ln\left(\frac{p_0}{p_1}\right)(\bar{T}_L - \bar{T}_o) > 0 \qquad [11.2]$$

Taking $(\bar{T}_L - \bar{T}_o) = \Delta T_{LO}$ (x, y, t) and denoting it as a land-ocean gradient term. Thus, we see that:

$$\frac{d\bar{v}}{dt} \propto \Delta T_{LO} \ (x, y, t)$$

To the first order, since the atmosphere above land is warmer than the ocean (due to sensible or latent heating), the monsoon flow (i.e. land-sea breeze) is generated depending on the strength of the vertically integrated temperature gradient. From the moisture budget equation, precipitation P can be differentiated as a local component and a large-scale component: $P = P_{local} + P_{Large\ Scale}$ and can be written as:

$$P = P_{local} + P_{Large\ Scale\ dynamics} = E - \int \nabla. \ q\vec{v} \ . \qquad [11.3]$$

E is the *local* contribution coming from local surface fluxes, evaporation, etc. and q is the available moisture and v is the wind. Thus, from Equations 11.2 and 11.3, it should be noted that, if the land is cooler or the land-ocean contrast is weaker, circulation (i.e. dv/dt) decrease, then $+v$ *(southerly)* decrease; hence, large-scale moisture convergence decreases (i.e. moisture divergence increase) and precipitation would decrease and vice-versa. Similarly, the "E" term that represents the source term evaporation flux from various processes is also dependent on local temperature, Bowen ratio, as well as the water-holding capacity of the atmosphere, as can be easily seen from the *Clasius-Clayperon* relation. Thus, the precipitation response to an increased temperature (I.e. warming) scenario is a complex process and is shown that a part of it is temperature forced and is tied to the thermodynamic energy budget and distribution of latent heating. As depicted by Trenberth and Shea (2005), land surface temperature is inversely correlated to precipitation with the hot and dry or wet and cold combination as the most likely scenarios in tropics, when E dominates the large-scale dynamics term over the land. However, when the circulation, i.e. large-scale dynamics dominates, e.g. in northern mid-latitudes, precipitation and surface temperature are directly correlated. On a global scale, as a result, the change of precipitation in the past century is shown to not directly follow warming temperature trends (Allen and Ingram, 2002; Skliris et al., 2016; Hegerl, 2015). For the monsoon zone, the monsoonal moisture convergence is also not proper in the reanalysis (Trenberth et al., 2007) and hence E, P, and E-P estimates (thus, relationship with land or ocean temperature) is uncertain. The same is true for monsoon precipitation. Monsoon precipitation shows a decreasing trend rather than a linear formula predicted pathway: an increase due to increased evaporation.

As the monsoon progresses, Indian land gets cooler and wetter due to increased rain during the July–August month. Thus, over Indian land and Indian ocean, increased rainfall reduces surface temperature and hence is inversely correlate, e.g. Fig. 4 of Naidu et al. (2020) and Fig. 2 of Trenberth and Shea (2005). However, as shown by Xavier et al. (2007), the vertically integrated upper tropospheric temperature gradient can be defined, which would include the warming effect over the land region. Thus, the gradient (difference) term on the right-hand side of Equation 11.2 is valid to a first order during the monsoon season and proportional to the vertically integrated heat transport and hence monsoon circulation. Precipitation would also be inversely proportional to the local (surface) term, as discussed in the previous paragraph. If the trend in the vertical temperature gradient as defined in Equation 11.1 is available, the previous equation can be used to qualitatively explain the slow warming i.e. long-term change (or the trend) in the circulation component associated with the trend. Then, using Equation 11.3, a long-term trend of precipitation (forced due to southern to Northern Hemispheric temperature gradient) could be defined quantitatively.

To demonstrate the observed temperature trend (both surface and vertically integrated), we use the *HADCRUT4* data set (Morice et al., 2012) and *20Cv2* reanalysis (Compo, 2011) (Table 11.1, downloaded from https://crudata.uea.ac.uk/cru/data/temperature/) and create a gradient index data based on box area-averaged surface temperature data by defining two regions: the northern region N (50°–110°E; 5°–45°N) and the southern region S (50°–110°E; 45°S–5°N). The gradient index is then defined by taking the difference, i.e. N minus S (referred to as gradT index). The index is computed for the period 1871–2004 and the plot for seasonally averaged surface temperature and vertically averaged temperature gradient values is shown in Figure 11.5. In this study, we define June–July–August averaged rainfall as the "JJA" season and July–August–September rainfall as the "JAS" season. The plots for the two seasons JJA and JAS are shown separately as it is now known that the monsoon season shows early (July) and late (August) peaks. The early peak years would be represented by the *JJA* index, and the late peak years would be represented by the *JAS* index. The gradient index clearly shows two things for both seasons: (a) there is a clear trend that the surface and vertically averaged gradient is decreasing for this period, implying that the monsoonal circulation should reduce provided other terms do not change significantly (refer to eq,1 and eq,2), and (b) the gradient shows a clear year-to-year low-frequency variability including the decadal variability consistent with the self-regulatory theory of Webster et al. (2002). Thus, it is clear that the vertically integrated gradT during the monsoon season and hence north-south circulation has decreased in recent years due to ocean-land differential imbalance warming trends. A reduction in circulation reduced the convergence (increased divergence); hence, there is a reduction in the precipitation in the monsoon (JJA or JAS) season. A reduction in the temperature gradient (gradT, ΔT_{LO}) implies more oceanic warming compared to an earlier part of the century or previous century as there is no slowdown in land warming. Thus, the result showed a decrease in gradT due to the warming of the Southern Ocean, making the monsoon season dry and hot.

Is it possible that changing any other teleconnection pattern could have contributed to this trend? To answer this, we plot the correlation of ISMR rainfall with the newly defined indices and compare them with other existing indices (ENSO,

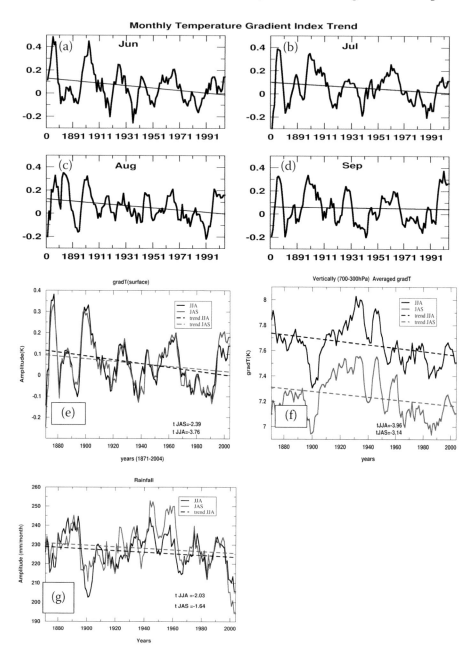

FIGURE 11.5   (a)–(d): Monthly surface temperature gradient index (gradT) trend. For the definition of temperature, gradient refers to text. (e) Surface gradT index for JJA and JAS season (f) vertically (700–300 hPa) averaged gradT trend using 20cV2 reanalysis (g) rainfall during JJA and JAS season. For the (e), (f), and (g) panel, the six yearly runnings mean is removed.

IOD) for the JJA and JAS season in Table 11.2a. It may be seen that the correlation of the newly defined indices is comparable to the correlation of other known indices. Thus, a similar order of associativity indicates that a monsoon would be influenced by the north-south gradient indices as much as other indices. Also, the gradT index (Table 11.2b), is inversely correlated to the southern annular mode (~ **–0.2**), which implies that in a warming scenario, when the SAM index strength increases, the gradT index strength decreases, causing a weakening of the monsoon. Finally, we also note that the southern box temperature is linked to the SAM variability in the sense that the southern box surface temperature index and SAM index show a significant positive correlation, along with a Mascarene High index, as shown in Table 11.2c. Thus, the change in the surface temperature gradient is driven by Southern Hemispheric climate change as the southern box is located in the south Indian Ocean with a significant impact of SAM and sea surface temperature over the region. Thus, as given in the schematic diagram in Figure 11.2, Southern Hemispheric climate change over the south Indian Ocean is a significant driver for monsoon rainfall and since the warming trend has impacted the southern Indian Ocean deep in the Southern Hemisphere, the temperature gradient, as well as

TABLE 11.2A
Correlation of Teleconnection Indices with All India Summer Monsoon Rainfall (AISMR) for the Period 1871–2004

AISMR	Nino3.4[*]	IOD[*]	PDO[#]	AMO[#]	gradT[*](surface)	gradT[++](vertically integrated)
JJA	–0.51	0.25	–0.12	0.24	–0.42	0.45
JAS	–0.50	0.10	–0.18	0.21	–0.15	0.39

Notes
[*] Nino3.4, IOD, and gradT indices are computed using HADCRU surface temperature data
[#] PDO index is based on ERSST and is downloaded from: https://www.ncdc.noaa.gov/teleconnections/pdo/
[#] AMO index is based on Kaplan SST and is downloaded from: http://www.psl.noaa.gov/data/timeseries/AMO/
[++] vertically integrated gradT is calculated from 20Cv2 reanalysis

TABLE 11.2B
Correlation of gradT (Temperature Gradient) Index with Other Northern and Tropical Teleconnection Indices for the Period 1871–2004

	Nino3.4	PDO	AMO
JJA,gradT	0.16	0.11	0.23
JAS,gradT	0.19	0.06	0.29

TABLE 11.2C
Correlation of Southern Ocean Box Temperature Indices with Other Southern Hemispheric Teleconnection Indices for the Period 1871–2004

Southern Ocean Box Surface Temperature (S Box)	SAM^	MAS (Mascarene High)	gradT
JJA	0.75	0.36	−0.42
JAS	0.74	0.33	−0.48

Note
^ SAM index is based on 20Cv2 reanalysis and is downloaded from: https://psl.noaa.gov/data/20thC_Rean/timeseries/monthly/SAM/

monsoon rainfall over the Indian region, show a decreasing trend. The gradT index bears the footprint of this interhemispheric teleconnection and brings out the trend. Also, it may be seen that the gradT index is associated with other teleconnection indices, e.g. ENSO, PDO, AMO, and IOD (Table 11.2b). Any warming trend in these indices would also thus potentially impact the gradT index and thus monsoon rainfall. However, this is not explored in this chapter.

11.6.2 Mechanisms of Monsoonal Trend Associated with Southern Hemispheric Weather Variability

The variability of SAM is plotted in Figure 11.6 for individual months and JJA and JAS seasons. The SAM index is calculated from *20Cv2 reanalysis* (Compo, 2011) and is based on the definition of Gong and Wang (1999) as given in Table 11.1. The data is downloaded from https://www.psl.noaa.gov/data/20thC_Rean/timeseries/monthly/SAM/sam.20crv2.long.data.

Studies show that a positive SAM increases the rainfall in southern low latitudes (Gillett et al., 2006). It shows that there is an increase in the high pressure in the mid-latitude belt (Fig. 1 of Gillett et al., 2006) and an increase in low Southern Hemispheric easterlies, leading to an increase in the frequency of weather disturbances over northern Australia. How would this impact the monsoon? To influence the monsoon, it would require that the SAM would impact the southern low latitudes, and especially it would impact the outflow from the Mascarene High. It is seen that there is a strong positive trend of Mascarene High in the last century. Therefore, the monsoon should increase. However, as discussed here, it is not so. The apparent paradox is not very hard to understand, as we know monsoon flow depends on the pressure gradient force rather than pressure over a location itself. We plot the pressure gradient for the same box as defined for temperature gradient in Figure 11.7a. The pressure gradient (gradP) index is calculated from the Hadley center, UK mean sea level pressure data (*HADSLP2 data;* Allan and Ansell, 2006; also refer to Table 11.1). It can be seen that the pressure gradient decreases between the north and south boxes. A reduction in the pressure gradient would increase divergence (decrease monsoonal convergence) and hence monsoon rainfall would

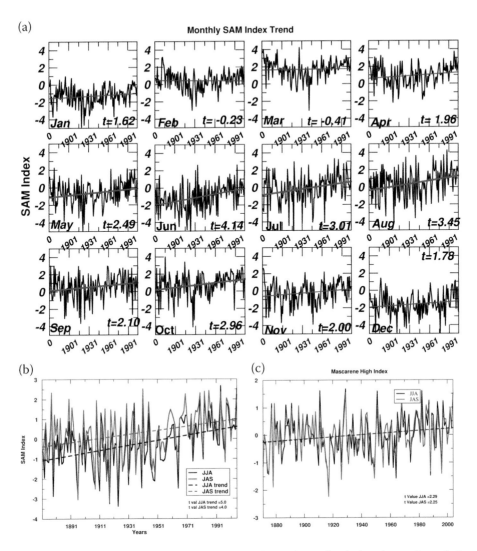

FIGURE 11.6 (a) Trend in SAM index for individual months during the study period (1871–2004). May to October shows a highly significant trend with t-values mentioned at the bottom. (b) Smoothed SAM index during JJA and JAS along with its linear trendline and (c) Mascarene High index during JJA and JAS along with its linear trendline.

decrease. Although it is somewhat intuitive that pressure change and temperature change should be linked by gas laws, the changes are induced by land-sea warming (i.e. thermodynamic reason), or the changes are induced by Southern Hemispheric potential vorticity insertion (dynamic reason) is not immediately clear. So we would check the potential vorticity aspect a little closer.

Since the *NCEP-NCAR reanalysis-1 data* (Kalnay 1996) is available from 1948 onwards, we plot the 850hPa potential vorticity trend computed from this data in the

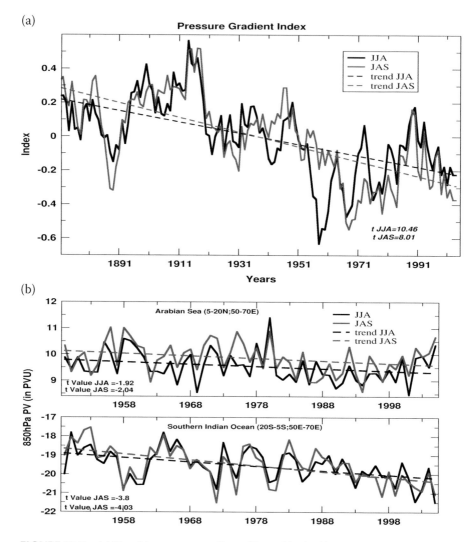

FIGURE 11.7 (a) Trend in pressure gradient, (b) trend in Arabian sea PV, and (c) southern Indian Ocean potential vorticity (PV). The PV values are calculated from monthly zonal wind data from NCEP-NCAR reanalysis for the period 1948–2004.

next figure. It is evident in this plot (Figure 11.7b) that indeed there is a decreasing trend of cyclonic (positive in Northern Hemisphere) potential vorticity (PV) since 1948 over the Arabian Sea where monsoon flow enters first before reaching the land region. This cyclonic PV can be decreased via injection of Southern Hemispheric negative PV. If the PV is conserved, an increased value of Southern Hemispheric negative PV, if it is injected in the Northern Hemisphere, would reduce the Northern Hemispheric positive PV. The trend of Southern Hemispheric cyclonic (negative) PV is shown in Figure 11.7c. It shows an increase in negative PV in the southern Indian Ocean, consistent with an increase in the positive phase of SAM. This increase in cyclonic (negative) PV in the Southern Hemisphere means an increase in weather disturbances in the positive phase of SAM that gives increased rainfall over the southern low latitude. Increased weather disturbances could be generated due to an increase in low-latitude easterlies associated with an increase in the SAM index.

11.7 INDIAN MONSOON RAINFALL, MULTIDECADAL VARIABILITY OF THE INDICES, AND THEIR RELATIONSHIP WITH MONSOONS

In this study, we have not discussed the decadal variability. We also have not seen whether the relation of the land temperature or its gradient to the Southern Ocean has multidecadal variability or not. Multidecadal variability of an upper tropospheric temperature gradient can significantly contribute to the trend shown by Xavier et al. (2007). To get a complete picture, the 31-year running mean of some of the indices is shown in Figure 11.8. The first panel shows the running correlation of surface temperature gradient index (gradT index), the second panel shows the

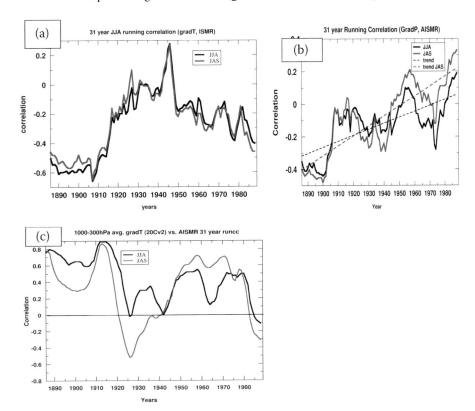

FIGURE 11.8 Thirty-one-year running correlation of AISMR with (a) gradT (surface), (b) gradP, and (c) gradT (1,000–300 hPa average from 20Cv2 reanalysis) index.

running correlation of pressure gradient (gradP) index, and the third panel shows the vertically integrated temperature gradient index from 20Cv2 reanalysis data. The plot shows the weakening and strengthening of the relation in different decades. The surface gradT index shows that the relationship has weakened during the 1930–1960 decades, which then again strengthened. The reason for this is not explored and it is speculated that since a monsoon is controlled by multiple tele-connection processes, such changes could be expected and would require further examination. The gradP index on the other hand shows systematic strengthening with decadal variability embedded in it. In recent years, it shows higher pressure (i.e. positive index value) over land, indicating a weakening of monsoons during the 1980–2000 decades. The vertically averaged gradT index shows a positive corre-lation with monsoons, implying a vertically averaged thermal gradient is the dia-batic forcing primarily arising from the latent heating. It is not, however, clear why the correlation or association with a vertically averaged temperature shows a strong decadal variability, with a JJA correlation showing a strong decreasing trend (trend line not shown). Weak diabatic heating would indicate a weak monsoon, which can partly explain the weakening trend in the last century. However, the detailed im-plications of the figure are not explored here and it is not clear if the Southern Hemispheric teleconnection has any role in this or if it is an artifact of other tel-econnection or a reanalysis data issue.

11.8 CONCLUSIONS

The Southern Hemispheric climate has undergone an extensive modulation due to the global warming trend in the last century. Several studies have documented the Southern Hemispheric climate change in different contexts. In this study, we have used the available climate data from 1871–2004 to understand and explore the role of Southern Hemispheric climate change on the monsoonal variability based on the definition of regional interhemispheric gradient indices. It is already known that interhemispheric gradients are representative of interhemispheric teleconnections and are showing strong temperature anomaly asymmetry as a result of global warming (Friedman et al., 2013). Since Indian monsoons can be explained as a response of asymmetric heating over the monsoonal zone (Zhang and Krishnamurti, 1996) and can be efficiently represented as a system responding to tropospheric temperature gradients (Xavier et al., 2007; Jin and Wang, 2017), we hypothesize that the Southern Hemispheric warming trend would reflect in interhemispheric teleconnection patterns represented by the north-south temperature gradients con-trolling the monsoonal rainfall and circulation.

Based on this, we have demonstrated that the monsoon rainfall has not increased in the past hundred years in response to global warming, as a simple theory using an increase in moisture availability (e.g. perceptible water) and warming ocean (the Bay of Bengal and the Arabian Sea) would suggest (e.g. Fig. 3 of Mishra, 2019). Instead, the three-month seasonal mean has decreased in the past century and shows a gradual decreasing trend in the last century, especially rainfall during the *JJA,* and the *JAS* season has reduced with a certain statistical level of significance. This chapter shows how the Southern Hemispheric climate change can explain this decreasing trend in a simplistic framework if we assume that the monsoon flow, to a first order, is a land-sea breeze circulation with the interhemispheric link, which known for a long time (for example, Mascarenes High and Findlater jet–induced monsoonal flow). It is shown that the seasonal decrease in rainfall could be linearly related to a change in the interhemispheric temperature gradient and pressure gradient, which will explain the seasonal rainfall decrease to the first order. This interhemispheric temperature gradient is inversely correlated with the SAM index (Table 11.2b). Thus, an increase in the SAM index would weaken the land-sea temperature gradient, as actually seen in the plots (Figures 11.5 and 11.6). Table 11.2b also shows that the gradT index is also correlated with AMO or ENSO indices. This suggests that multiple teleconnections control the land-sea thermal contrast. It would require a further modeling study to see how the active participation of Southern Hemispheric conditions and other teleconnection over the tropics and the Northern Hemisphere would modulate the interannual variability of monsoons and the interhemispheric teleconnection in the context of global warming. We have not described how the trend in the other teleconnection processes (e.g. IOD or ENSO) would change the temperature gradient or the monsoon rainfall. To minimize IOD or ENSO effect, we have done a six yearly smoothing. Since the strength of correlation of the gradP and gradT index with AISMR is of the same order as the ENSO-ISMR and stronger than IOD-ISMR values in the 134 years of data that is used (Tables 11.2a, 11.2b, and 11.2c), we conclude that the trend in the temperature gradient indices can explain the weakening (or not strengthening) of monsoons, despite increased evaporation due to global warming. The weakening of rainfall is linked to the weakening of meridional flow, which, in addition to other teleconnections, is also linked to the Southern Hemispheric warming.

ACKNOWLEDGMENTS

IITM is an autonomous institute funded by the Ministry of Earth Sciences (MoES), Govt. of India. Data used in this analysis are downloaded from several freely available data sources (Tables 11.1, 11.2a, 11.2b, and 11.2c). All of the plots are prepared using the freely available NCAR Command Language (NCL) and the XMGRACE software.

REFERENCES

Abish, B., Joseph, P. V., and Johannessen, O. M. (2013). Weakening trend of the tropical easterly jet stream of the Boreal Summer monsoon season 1950–2009. *Journal of Climate*, 26, 9408–9414, https://doi.org/10.1175/JCLI-D-13-00440.1.

Allan, R., and Ansell, T. (2006). A new globally complete monthly historical gridded mean sea level pressure data set (HadSLP2): 1850–2004. *Journal of* , 19, 5816–5842, https://doi.org/10.1175/JCLI3937.1.

Allen, M. R., and Ingram, W. J. (2002). Constraints on future changes in climate and the hydrologic cycle. *Nature*, 419, 228–232, https://doi.org/10.1038/nature01092.

Annamalai, H., Hamilton, K., and Sperber, K. R. (2007). The South Asian summer monsoon and its relationship with ENSO in the IPCC AR4 simulations. *Journal of Climate*, 20, 1071–1092, https://doi.org/10.1175/JCLI4035.1.

Asharaf, S., and Ahrens, B. (2015). Indian summer monsoon rainfall processes in climate change scenarios. *Journal of Climate*, 28, 5414–5429, https://doi.org/10.1175/JCLI-D-14-00233.1.

Ashok, K., Guan, Z., and Yamagata, T. (2001). Impact of the Indian Ocean dipole on the relationship between the Indian monsoon rainfall and ENSO. *Geophysical Research Letters*, 28, 4499–4502, https://doi.org/10.1029/2001GL013294.

Azad, S., and Rajeevan, M. (2016). Possible shift in the ENSO-Indian monsoon rainfall relationship under future global warming. *Scientific Reports*, 6, 20145, https://doi.org/10.1038/srep20145.

Bala, I., and Singh, O. P. (2008). Teleconnections to the Indian summer monsoon under changing climate. *Mausam*, 59(2), 167–172.

Behera, S. K., and Ratnam, J. V. (2018). Quasi-asymmetric response of the Indian summer monsoon rainfall to opposite phases of the IOD. *Scientific Reports*, 8, 123, https://doi.org/10.1038/s41598-017-18396-6.

Cai, W., Sullivan, A., and Cowan, T. (2009). Climate change contributes to more frequent consecutive positive Indian Ocean Dipole events. *Geophysical Research Letters*, 36, https://doi.org/10.1029/2009GL040163.

Chou, C. (2003). Land–sea heating contrast in an idealized Asian summer monsoon. *Climate Dynamics*, 21, 11–25, https://doi.org/10.1007/s00382-003-0315-7.

Compo, G. P., and Coauthors (2011). The twentieth century reanalysis project. *Quarterly Journal of the Royal Meteorological Society*, 137, 1–28, https://doi.org/10.1002/qj.776.

Drost, F., and Karoly, D. (2012). Evaluating global climate responses to different forcings using simple indices. *Geophysical Research Letters*, 39, https://doi.org/10.1029/2012GL052667.

Dugam, S. S., and Kakade, S. B. (2004). Antarctic sea-ice and monsoon variability. *Indian Journal of Radio and Space Physics*, 33, 306–309.

Eayrs, C., Holland, D., Francis, D., Wagner, T., Kumar, R., and Li, X. (2019). Understanding the seasonal cycle of Antarctic sea ice extent in the context of longer-term variability. *Reviews of Geophysics*, 57, 1037–1064, https://doi.org/10.1029/2018RG000631.

Fan, K., and Wang, H. (2006). Interannual variability of Antarctic Oscillation and its influence on East Asian climate during boreal winter and spring. *Science in China Series D*, 49, 554–560, https://doi.org/10.1007/s11430-006-0554-7.

Fogt, R. L., and Marshall, G. J. (2020). The southern annular mode: Variability, trends, and climate impacts across the Southern Hemisphere. *WIREs Climate Change*, 11, e652, https://doi.org/10.1002/wcc.652.

Fogt, R. L., Marshall, G. J., Goergens, C. A., Jones, J. M., Schneider, D. P., Nicolas, J. P., Bromwich, D. H., and Dusselier, H. E. (2017). A twentieth-century perspective on summer Antarctic pressure change and variability and contributions from tropical SSTs and ozone depletion. *Geophysical Research Letters*, 44, 9918–9927, https://doi.org/10.1002/2017GL075079.

Friedman, A. R., Hwang, Y.-T., Chiang, J. C. H., and Frierson, D. M. W. (2013). Interhemispheric temperature asymmetry over the twentieth century and in future projections. *Journal of Climate*, 26, 5419–5433, https://doi.org/10.1175/JCLI-D-12-00525.1.

Fyfe, J. C., Gillett, N. P., and Marshall, G. J. (2012). Human influence on extratropical Southern Hemisphere summer precipitation. *Geophysical Research Letters*, 39, https://doi.org/10.1029/2012GL054199.

Gadgil, S. (2018). The monsoon system: Land–sea breeze or the ITCZ? *Journal of Earth System Science*, 127, 1, https://doi.org/10.1007/s12040-017-0916-x.

Gastineau, G., Li, L., and Le Treut, H. (2009). The Hadley and Walker circulation changes in global warming conditions described by idealized atmospheric simulations. *J. Climate*, 22, 3993–4013, https://doi.org/10.1175/2009JCLI2794.1.

Gillett, N. P., Kell, T. D., and Jones, P. D. (2006). Regional climate impacts of the southern annular mode. *Geophysical Research Letters*, 33, https://doi.org/10.1029/2006GL 027721.

Gnanaseelan, C., Roxy, M. K., and Deshpande, A. (2017). Variability and trends of sea surface temperature and circulation in the Indian Ocean. *Observed Climate Variability and Change over the Indian Region*, M.N. Rajeevan and S. Nayak, Eds., Springer Singapore, pp. 165–179.

Gong, D., and Wang, S. (1999). Definition of Antarctic oscillation index. *Geophysical Research Letters*, 26, 459–462, https://doi.org/10.1029/1999GL900003.

Goswami, B. N., Venugopal, V., Sengupta, D., Madhusoodanan, M. S., and Xavier, P. K. (2006). Increasing trend of extreme rain events over India in a warming environment. *Science*, 314, 1442–1445, https://doi.org/10.1126/science.1132027.

Goswami, B. N., Venugopal, V., and Chattopadhyay, R. (2019). Chapter 2 - South Asian monsoon extremes. *Tropical Extremes*, V. Venugopal, J. Sukhatme, R. Murtugudde, and R. Roca, Eds., Elsevier, pp. 15–49.

Grieger, J., Leckebusch, G. C., Raible, C. C., Rudeva, I., and Simmonds, I. (2018). Subantarctic cyclones identified by 14 tracking methods, and their role for moisture transports into the continent. *Tellus A: Dynamic Meteorology and Oceanography*, 70, 1–18, https://doi.org/10.1080/16000870.2018.1454808.

Guhathakurta, P., and Rajeevan, M. (2008). Trends in the rainfall pattern over India. *International Journal of Climatology*, 28, 1453–1469, https://doi.org/10.1002/joc. 1640.

Guhathakurta, P., Rajeevan, M., Sikka, D. R., and Tyagi, A. (2015). Observed changes in southwest monsoon rainfall over India during 1901–2011. *International Journal of Climatology*, 35, 1881–1898, https://doi.org/10.1002/joc.4095.

Hartmann, D. L., and Lo, F. (1998). Wave-driven zonal flow vacillation in the Southern Hemisphere. *Journal of the Atmospheric Sciences*, 55, 1303–1315, https://doi.org/ 10.1175/1520-0469(1998)055<1303:WDZFVI>2.0.CO;2.

Hegerl, G. C., and Coauthors (2015). Challenges in quantifying changes in the global water cycle. *Bulletin of the American Meteorological Society*, 96, 1097–1115, https:// doi.org/10.1175/BAMS-D-13-00212.1.

Hendon, H. H., Thompson, D. W. J., and Wheeler, M. C. (2007). Australian rainfall and surface temperature variations associated with the Southern Hemisphere annular mode. *Journal of Climate*, 20, 2452–2467, https://doi.org/10.1175/JCLI4134.1.

Hobbs, W. R., Massom, R., Stammerjohn, S., Reid, P., Williams, G., and Meier, W. (2016). A review of recent changes in Southern Ocean sea ice, their drivers and forcings. *Global and Planetary Change*, 143, 228–250, https://doi.org/10.1016/j.gloplacha.2016. 06.008.

Holton, J. R., (2004). Chapter 4.1: The circulation theorem, *An Introduction to Dynamic Meteorology*, J. R. Holton, Ed. International Geophysics Series, Elsevier, USA, pp. 90–91.

Hu, Y., and Fu, Q. (2007). Observed poleward expansion of the Hadley circulation since 1979. *Atmos. Chem. Phys.*, 7, 5229–5236, https://doi.org/10.5194/acp-7-5229-2007.

Hu, Y., Tao, L., and Liu, J. (2013). Poleward expansion of the Hadley circulation in CMIP5 simulations. *Advances in Atmospheric Sciences*, 30, 790–795, https://doi.org/10.1007/ s00376-012-2187-4.

Hu, Y., Huang, H., and Zhou, C. (2018). Widening and weakening of the Hadley circulation under global warming. *Science Bulletin*, 63, 640–644, https://doi.org/10.1016/j.scib. 2018.04.020.

Huang, B., and Coauthors (2017). Extended reconstructed sea surface temperature, Version 5 (ERSSTv5): Upgrades, validations, and intercomparisons. *J. Climate*, 30, 8179–8205, https://doi.org/10.1175/JCLI-D-16-0836.1.

Huang, P., Zheng, X.-T., and Ying, J. (2019). Disentangling the changes in the Indian Ocean Dipole–Related SST and rainfall variability under global warming in CMIP5 Models. *Journal of Climate*, 32, 3803–3818, https://doi.org/10.1175/JCLI-D-18-0847.1.

Jin, Q., and Wang, C. (2017). A revival of Indian summer monsoon rainfall since 2002. *Nature Climate Change*, 7, 587–594, https://doi.org/10.1038/nclimate3348.

Joseph, P. V., and Simon, A. (2005). Weakening trend of the southwest monsoon current through peninsular India from 1950 to the present. *Current Science*, 89, 687–694.

Kalnay, E., and Coauthors (1996). The NCEP/NCAR 40-Year reanalysis project. *Bulletin of the American Meteorological Society*, 77, 437–472, https://doi.org/10.1175/1520-0477(1996)077<0437:TNYRP>2.0.CO;2.

Karoly, D. J., and Braganza, K. (2001). Identifying global climate change using simple indices. *Geophysical Research Letters*, 28, 2205–2208, https://doi.org/10.1029/2000GL011925.

Kidson, J. W. (1999). Principal modes of Southern Hemisphere low-frequency variability obtained from NCEP–NCAR Reanalyses. *J. Climate*, 12, 2808–2830, https://doi.org/10.1175/1520-0442(1999)012<2808:PMOSHL>2.0.CO;2.

Kothawale, D. R., and Rajeevan, M. (2017). *Seasonal and Annual Rainfall Time Series for All India, Homogeneous Regions and Meteorological Subdivisions:1871-2016*. IITM, https://tropmet.res.in/static_pages.php?page_id=53 (Accessed March 1, 2020).

Krishnan, R., and Coauthors (2013). Will the South Asian monsoon overturning circulation stabilize any further? *Climate Dynamics*, 40, 187–211, https://doi.org/10.1007/s00382-012-1317-0.

Kucharski, F., Molteni, F., and Yoo, J. H. (2006). SST forcing of decadal Indian Monsoon rainfall variability. *Geophysical Research Letters*, 33, https://doi.org/10.1029/2005GL025371.

Kumar, K. K., Rajagopalan, B., and Cane, M. A. (1999). On the weakening relationship between the Indian Monsoon and ENSO. *Science*, 284, 2156, https://doi.org/10.1126/science.284.5423.2156.

Kumar, V., Jain, S. K., and Singh, Y. (2010). Analysis of long-term rainfall trends in India. *null*, 55, 484–496, https://doi.org/10.1080/02626667.2010.481373.

Kuttippurath, J., and Nair, P. J. (2017). The signs of Antarctic ozone hole recovery. *Scientific Reports*, 7, 585, https://doi.org/10.1038/s41598-017-00722-7.

Lal, M., Cubasch, U., and Santer, B. D. (1994). Effect of global warming on Indian monsoon simulated with a coupled ocean-atmosphere general circulation model. *Current Science*, 66, 430–438.

Lee, S.-K., Mechoso, C. R., Wang, C., and Neelin, J. D. (2013). Interhemispheric influence of the northern summer monsoons on southern subtropical anticyclones. *Journal of Climate*, 26, 10193–10204, https://doi.org/10.1175/JCLI-D-13-00106.1.

Lin, H. (2009). Global extratropical response to diabatic heating variability of the Asian summer monsoon. *Journal of the Atmospheric Sciences*, 66, 2697–2713, https://doi.org/10.1175/2009JAS3008.1.

Liu, T., Li, J., and Zheng, F. (2015). Influence of the Boreal autumn southern annular mode on winter precipitation over land in the Northern Hemisphere. *Journal of Climate*, 28, 8825–8839, https://doi.org/10.1175/JCLI-D-14-00704.1.

Liu, T., Li, J., Li, Y., Zhao, S., Zheng, F., Zheng, J., and Yao, Z. (2018). Influence of the May Southern annular mode on the South China Sea summer monsoon. *Climate Dynamics*, 51, 4095–4107, https://doi.org/10.1007/s00382-017-3753-3.

Mamalakis, A., Yu, J.-Y., Randerson, J. T., AghaKouchak, A., and Foufoula-Georgiou, E. (2018). A new interhemispheric teleconnection increases predictability of winter precipitation in the southwestern US. *Nature Communications*, 9, 2332, https://doi.org/10.1038/s41467-018-04722-7.

Marshall, G. J. (2003). Trends in the Southern annular mode from observations and re-analyses. *Journal of Climate*, 16, 4134–4143, https://doi.org/10.1175/1520-0442(2003) 016<4134:TITSAM>2.0.CO;2.

Meneghini, B., Simmonds, I., and Smith, I. N. (2007). Association between Australian rainfall and the Southern annular mode. *International Journal of Climatology*, 27, 109–121, https://doi.org/10.1002/joc.1370.

Mishra, A. K. (2019). Quantifying the impact of global warming on precipitation patterns in India. *Meteorological Applications*, 26, 153–160, https://doi.org/10.1002/met.1749.

Morice, C. P., Kennedy, J. J., Rayner, N. A., and Jones, P. D. (2012). Quantifying uncertainties in global and regional temperature change using an ensemble of observational estimates: The HadCRUT4 data set. *Journal of Geophysical Research: Atmospheres*, 117, https://doi.org/10.1029/2011JD017187.

Naidu, C. V., Durgalakshmi, K., Muni Krishna, K., Ramalingeswara Rao, S., Satyanarayana, G. C., Lakshminarayana, P., and Malleswara Rao, L. (2009). Is summer monsoon rainfall decreasing over India in the global warming era? *Journal of Geophysical Research: Atmospheres*, 114, https://doi.org/10.1029/2008JD011288.

Naidu, P. D., Ganeshram, R., Bollasina, M. A., Panmei, C., Nürnberg, D., and Donges, J. F. (2020). Coherent response of the Indian monsoon rainfall to Atlantic multi-decadal variability over the last 2000 years. *Scientific Reports*, 10, 1302, https://doi.org/10.1038/s41598-020-58265-3.

Nan, S., Li, J., Yuan, X., and Zhao, P. (2009). Boreal spring Southern Hemisphere annular mode, Indian Ocean sea surface temperature, and East Asian summer monsoon. *Journal of Geophysical Research: Atmospheres*, 114, https://doi.org/10.1029/2008JD010045.

Pal, J., Chaudhuri, S., Roychowdhury, A., and Basu, D. (2017). An investigation of the influence of the southern annular mode on Indian summer monsoon rainfall. *Meteorological Applications*, 24, 172–179, https://doi.org/10.1002/met.1614.

Pall, P., Aina, T., Stone, D. A., Stott, P. A., Nozawa, T., Hilberts, A. G. J., Lohmann, D., and Allen, M. R. (2011). Anthropogenic greenhouse gas contribution to flood risk in England and Wales in autumn 2000. *Nature*, 470, 382–385, https://doi.org/10.1038/nature09762.

Parkinson, C. L. (2019). A 40-y record reveals gradual Antarctic sea ice increases followed by decreases at rates far exceeding the rates seen in the Arctic. *Proceedings of the National Academy of Sciences of the United States of America*, 116, 14414, https://doi.org/10.1073/pnas.1906556116.

Parkinson, C. L., and Cavalieri, D. J. (2012). Antarctic sea ice variability and trends, 1979–2010. *The Cryosphere*, 6, 871–880, https://doi.org/10.5194/tc-6-871-2012.

Pezza, A. B., Durrant, T., Simmonds, I., and Smith, I. (2008). Southern Hemisphere synoptic behavior in extreme phases of SAM, ENSO, sea ice extent, and Southern Australia rainfall. *Journal of Climate*, 21, 5566–5584, https://doi.org/10.1175/2008JCLI2128.1.

Pohl, B., and Fauchereau, N. (2012). The Southern annular mode seen through weather regimes. *J. Climate*, 25, 3336–3354, https://doi.org/10.1175/JCLI-D-11-00160.1.

Polvani, L. M., Previdi, M., and Deser, C. (2011a). Large cancellation, due to ozone recovery, of future Southern Hemisphere atmospheric circulation trends. *Geophysical Research Letters*, 38, https://doi.org/10.1029/2011GL046712.

Polvani, L. M., Waugh, D. W., Correa, G. J. P., and Son, S.-W. (2011b). Stratospheric ozone depletion: The main driver of twentieth-century atmospheric circulation changes in the Southern Hemisphere. *Journal of Climate*, 24, 795–812, https://doi.org/10.1175/201 0JCLI3772.1.

Prabhu, A., Kripalani, R., Oh, J., and Preethi, B. (2017). Can the Southern annular mode influence the Korean summer monsoon rainfall? *Asia-Pacific Journal of Atmospheric Sciences*, 53, 217–228, https://doi.org/10.1007/s13143-017-0029-0.

Praveen, B., Talukdar, S., Shahfahad, S., Mahato, J., Mondal, P., Sharma, A. R., Islam, Md. T., and Rahman, A. (2020). Analyzing trend and forecasting of rainfall changes in India using non-parametrical and machine learning approaches. *Scientific Reports*, 10, 10342, https://doi.org/10.1038/s41598-020-67228-7.

Rai, P., and Dimri, A. P. (2020). Changes in rainfall seasonality pattern over India. *Meteorological Applications*, 27, e1823, https://doi.org/10.1002/met.1823.

Rayner, N. A., Parker, D. E., Horton, E. B., Folland, C. K., Alexander, L. V., Rowell, D. P., Kent, E. C., and Kaplan, A. (2003). Global analyses of sea surface temperature, sea ice, and night marine air temperature since the late nineteenth century. *Journal of Geophysical Research: Atmospheres*, 108, https://doi.org/10.1029/2002JD002670.

Rignot, E., Mouginot, J., Scheuchl, B., van den Broeke, M., van Wessem, M. J., and Morlighem, M. (2019). Four decades of Antarctic Ice Sheet mass balance from 1979–2017. *Proceedings of the National Academy of Sciences of the United States of America*, 116, 1095, https://doi.org/10.1073/pnas.1812883116.

Rodwell, M. J. (1997). Breaks in the Asian Monsoon: The influence of Southern Hemisphere weather systems. *Journal of the Atmospheric Sciences*, 54, 2597–2611, https://doi.org/10.1175/1520-0469(1997)054<2597:BITAMT>2.0.CO;2.

Rogers, J. C., and van Loon, H. (1982). Spatial variability of sea level pressure and 500 MB height anomalies over the Southern Hemisphere. *Monthly Weather Review.*, 110, 1375–1392, https://doi.org/10.1175/1520-0493(1982)110<1375:SVOSLP>2.0.CO;2.

Roxy, M. K. (2017). Land warming revives monsoon. *Nature Climate Change*, 7, 549–550, https://doi.org/10.1038/nclimate3356.

Roxy, M. K., Ghosh, S., Pathak, A., Athulya, R., Mujumdar, M., Murtugudde, R., Terray, P., and Rajeevan, M. (2017). A threefold rise in widespread extreme rain events over central India. *Nature Communications*, 8, 708, https://doi.org/10.1038/s41467-017-00744-9.

Roy, I., Tedeschi, R. G., and Collins, M. (2019). ENSO teleconnections to the Indian summer monsoon under changing climate. *International Journal of Climatology*, 39, 3031–3042, https://doi.org/10.1002/joc.5999.

Saji, N. H., Goswami, B. N., Vinayachandran, P. N., and Yamagata, T. (1999). A dipole mode in the tropical Indian Ocean. *Nature*, 401, 360–363, https://doi.org/10.1038/43854.

Sandeep, S., Ajayamohan, R. S., Boos, W. R., Sabin, T. P., and Praveen, V. (2018). Decline and poleward shift in Indian summer monsoon synoptic activity in a warming climate. *Proceedings of the National Academy of Sciences of the United States of America*, 115, 2681, https://doi.org/10.1073/pnas.1709031115.

Sathiyamoorthy, V. (2005). Large scale reduction in the size of the Tropical Easterly Jet. *Geophysical Research Letters*, 32, https://doi.org/10.1029/2005GL022956.

Screen, J. A., Bracegirdle, T. J., and Simmonds, I. (2018). Polar climate change as manifest in atmospheric circulation.*Current Climate Change Reports*, 4, 383–395, https://doi.org/10.1007/s40641-018-0111-4.

Sen Gupta, A., and England, M. H. (2006). Coupled ocean-atmosphere–ice response to variations in the southern annular mode. *Journal of Climate*, 19, 4457–4486, https://doi.org/10.1175/JCLI3843.1.

Sharmila, S., Joseph, S., Sahai, A. K., Abhilash, S., and Chattopadhyay, R. (2015). Future projection of Indian summer monsoon variability under climate change scenario: An assessment from CMIP5 climate models. *Global and Planetary Change*, 124, 62–78, https://doi.org/10.1016/j.gloplacha.2014.11.004.

Shepherd, A., and Coauthors (2018). Mass balance of the Antarctic Ice Sheet from 1992 to 2017. *Nature*, 558, 219–222, https://doi.org/10.1038/s41586-018-0179-y.

Silvestri, G. E., and Vera, C. S. (2003). Antarctic Oscillation signal on precipitation anomalies over southeastern South America. *Geophysical Research Letters*, 30, https://doi.org/10.1029/2003GL018277.

Skliris, N., Zika, J. D., Nurser, G., Josey, S. A., and Marsh, R. (2016). Global water cycle amplifying at less than the Clausius-Clapeyron rate. *Scientific Reports*, 6, 38752, https://doi.org/10.1038/srep38752.

Solman, S. A., and Orlanski, I. (2013). Poleward shift and change of frontal activity in the Southern Hemisphere over the last 40 years. *Journal of the Atmospheric Sciences*, 71, 539–552, https://doi.org/10.1175/JAS-D-13-0105.1.

Solman, S. A., and Orlanski, I. (2015). Climate change over the extratropical Southern Hemisphere: The tale from an ensemble of reanalysis data sets. *Journal of Climate*, 29, 1673–1687, https://doi.org/10.1175/JCLI-D-15-0588.1.

Son, S.-W., and Coauthors (2010). Impact of stratospheric ozone on Southern Hemisphere circulation change: A multimodel assessment. *Journal of Geophysical Research: Atmospheres*, 115, https://doi.org/10.1029/2010JD014271.

Song, J., Zhou, W., Li, C., and Qi, L. (2009). Signature of the Antarctic oscillation in the Northern Hemisphere. *Meteorology and Atmospheric Physics*, 105, 55–67, https://doi.org/10.1007/s00703-009-0036-5.

Stephens, G. L., and T. D. Ellis, (2008): Controls of global-mean precipitation increases in global warming GCM experiments. *Journal of Climate*, 21, 6141–6155, https://doi.org/10.1175/2008JCLI2144.1.

Terray, P., (2011): Southern Hemisphere extra-tropical forcing: A new paradigm for El Niño-Southern oscillation. *Climate Dynamics*, 36, 2171–2199, https://doi.org/10.1007/s00382-010-0825-z.

Thompson, D. W. J., and Wallace, J. M. (2000). Annular modes in the extratropical circulation. Part I: Month-to-month variability. *Journal of Climate*, 13, 1000–1016, https://doi.org/10.1175/1520-0442(2000)013<1000:AMITEC>2.0.CO;2.

Trenberth, K. E., and Shea, D. J. (2005). Relationships between precipitation and surface temperature. *Geophysical Research Letters*, 32, https://doi.org/10.1029/2005GL022760.

Trenberth, K. E., Shea, D. J., Smith, L., Qian, T., Dai, A., and Fasullo, J. (2007). Estimates of the global water budget and its annual cycle using observational and model data. *Journal of Hydrometeorology*, 8, 758–769, https://doi.org/10.1175/JHM600.1.

Turner, A. G., and Annamalai, H. (2012). Climate change and the South Asian summer monsoon. *Nature Climate Change*, 2, 587–595, https://doi.org/10.1038/nclimate1495.

Vidya, P. J., Ravichandran, M., Subeesh, M. P., Chatterjee, S., Murukesh, N. (2020). Global warming hiatus contributed weakening of the Mascarene High in the Southern Indian Ocean. *Scientific Reports*, 10, 3255, https://doi.org/10.1038/s41598-020-59964-7.

Vishnu, S., Francis, P. A., Shenoi, S. S. C., and Ramakrishna, S. S. V. S. (2016). On the decreasing trend of the number of monsoon depressions in the Bay of Bengal. *Environmental Research Letters*, 11, 014011, https://doi.org/10.1088/1748-9326/11/1/014011.

Viswambharan, N. (2019). Contrasting monthly trends of Indian summer monsoon rainfall and related parameters. *Theoretical and Applied Climatology*, 137, 2095–2107, https://doi.org/10.1007/s00704-018-2695-y.

Viswambharan, N., and Mohanakumar, K. (2013). Signature of a Southern Hemisphere extratropical influence on the summer monsoon over India. *Climate Dynamics*, 41, 367–379, https://doi.org/10.1007/s00382-012-1509-7.

Webster, P. J. (2006). The coupled monsoon system, *The Asian Monsoon*, B. Wang, ed., Springer Berlin Heidelberg, pp. 3–66.

Webster, P. J., Moore, A. M., Loschnigg, J. P., and Leben, R. R. (1999). Coupled ocean-atmosphere dynamics in the Indian Ocean during 1997–98. *Nature*, 401, 356–360, https://doi.org/10.1038/43848.

Webster, P. J., Clark, C., Cherikova, G., Fasullo, J., Han, W., Loschnigg, J., and Sahami, K. (2002). The monsoon as a self-regulating coupled ocean-atmosphere system, in*Meteorology at the Millennium*, edited by R.P. Pearce, International Geophysics Series, Vol. 83. Academic Press, pp. 198–219.

Wu, Z., Dou, J., and Lin, H. (2015). Potential influence of the November–December Southern Hemisphere annular mode on the East Asian winter precipitation: A new mechanism. *Climate Dynamics*, 44, 1215–1226, https://doi.org/10.1007/s00382-014-2241-2.

Xavier, P. K., Marzin, C., and Goswami, B. N. (2007). An objective definition of the Indian summer monsoon season and a new perspective on the ENSO–monsoon relationship. *Quarterly Journal of the Royal Meteorological Society*, 133, 749–764, https://doi.org/10.1002/qj.45.

Xue, F., Wang, H., and He, J. (2004). Interannual variability of Mascarene high and Australian high and their influences on East Asian Summer Monsoon. *Journal of the Meteorological Society of Japan. Ser. II*, 82, 1173–1186, https://doi.org/10.2151/jmsj.2004.1173.

Zhang, Z., and Krishnamurti, T. N. (1996). A generalization of Gill's heat-induced tropical circulation. *Journal of the Atmospheric Sciences*, 53, 1045–1052, https://doi.org/10.1175/1520-0469(1996)053<1045:AGOGHI>2.0.CO;2.

Zhao, S., Li, J., and Li, Y. (2015). Dynamics of an interhemispheric teleconnection across the critical latitude through a southerly duct during Boreal Winter*. *Journal of Climate*, 28, 7437–7456, https://doi.org/10.1175/JCLI-D-14-00425.1.

12 Spatial and Temporal Variability of Physical Parameters in the Prydz Bay for Climate Change

S.M. Pednekar
National Centre for Polar and Ocean Research, Ministry of
Earth Science, Goa, India

CONTENTS

12.1 INTRODUCTION

Antarctica is the coldest, windiest, and driest continent on the Earth. The Antarctic weather condition is highly variable and it can change dramatically in short periods. The atmospheric temperature in summer can exceed +10°C near the coast and fall to below −40°C in winter. At times, the resulting wind speeds can exceed 100 km/hrs. Antarctica has just two seasons: 6 months of daylight in its summer and 6 months of darkness in its winter. The seasons are caused by the tilt of the Earth's axis with the sun. The average annual temperature on the Antarctic coast ranges from −10°C to −60°C in the interior region. The inland temperature rises to about −30°C in summer but falls below −80°C in winter. The lowest temperature recorded was −89.2°C at the Vostok station on 21st July 1983. The polar night occurs from

DOI: 10.1201/9781003203742-12

late May to late July, and the polar day from late November to early February, with the lowest monthly mean air temperature in July (−18.6°C) and the highest in January (0.4°C) (Lei and Quot, 2010).

Climate change refers to variations in physical parameters at a considerably longer time duration and place. According to the IPCC (2007) report, climate change is defined as any change in climate over time, whether due to natural variability or as a result of human activities. The Intergovernmental Panel on Climate Change (IPCC (2007)) pointed out that, in the past (50 years) the western coastline of the Antarctic Peninsula has undergone some severest arming world-wide. The warming over the eastern coastline of the Antarctic Peninsula is fairly slow, with the faster warming seasons being summer and autumn. The southern Indian Ocean is written in the IIOE (International Indian Ocean Expedition) Atlas by Wyrtki (1971) and the Soviet Antarctic Atlas (Tolstikov, 1966). The thermo-haline circulation taking place on the continental shelf of Antarctica is on a large scale and the most active region of the world.

Prydz Bay is one of the regions that is undergoing the revolution of a climate change problem and is an important area because of its V-shape and unique nature in the polar region. It is the third-largest embayment in the Antarctic continent (Figure 12.1) and lies in the Indian Ocean sector. The southeastern coast of Prydz Bay is associated with the group of islands known as the Larsemann Hills, and the

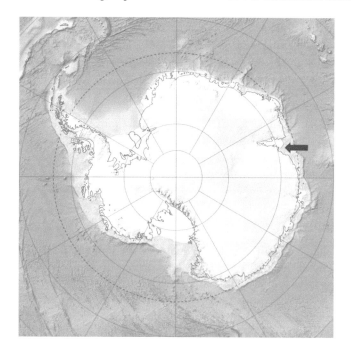

FIGURE 12.1 Antarctica is a landmass surrounded by the Southern Ocean. The red arrow indicates the Prydz Bay. Magnification of Prydz Bay in the circle is shown in Fig 12.4. The ocean seafloor topography is plotted. Image adopted from www.grida.no/resources/7162.

Amery Ice Shelf, Princess Elizabeth Land, and East Antarctica (Chauhan et al., 2015). The annual mean temperature is increasing by 0.28°C/10a and the winter mean temperature by nearly 0.3°C/10a. The lowest monthly mean air temperature occurred in August and the highest in January. In summer, the surface radiation budget directly affects the change in physical parameters of the surface water. The open surface can absorb more energy from downward shortwave radiation compared to that gained by the ice-covered surface. The hydrological balance is maintained by the variability of ice conditions temporally and spatially, particularly regarding the surface structure (Kupetskii, 1959; Denisov and Myznhikova, 1978). The Prydz Bay continental shelf water is warmer and colder as it gets deeper (Zverev, 1959, 1963; Izvekov, 1959).

North of the Antarctic continental slope environment, it is extremely stormy with wind stress on the ocean toward the east and thus the transport in the surface Ekman layer is toward the north. To conserve mass, the water column below the Ekman layer moves to the south, slowly and steadily driving deep Circumpolar Deep Water (CDW) up the continental slope and has been in the deep ocean for a long time (~1,000 years), and it is relatively salty, warm, low in oxygen, and rich in nutrients compared to the Antarctic coastal waters.

Surface ocean currents are the main sources of deepwater masses and the deepwater masses also feed one another. However, the surface and deep ocean currents are an integrated system known as the Global Conveyor Belt (Figure 12.2). It explains heat transport, bottom water aging, and nutrient supply in the oceans. It also redistributes heat taken up from the Pacific through thermohaline circulation. Through thermohaline circulation, ocean heat transports over large geographic distances and it takes approximately 1,000 years to complete one round-trip.

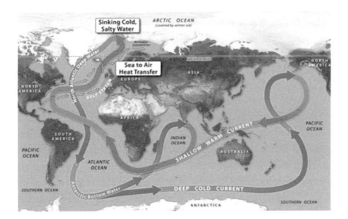

FIGURE 12.2 Diagram demonstrates the major surface and deep water circulation component of the ocean that combined to form the Global Conveyor Belt. Blue arrows represent deep-cold water currents, while red arrows represent surface warm currents. Credit to the Smithsonian Institution websites for providing the image.

In the recent decade, the major challenges among researchers are to understand a change in the climate system, which is one of the major physical processes that contributes to the world's oceans. Little is known about Prydz Bay compared to the Weddell Sea and the Ross Sea in terms of spatial and temporal variability due to lack of data. In this chapter, spatial and temporal variability of the physical parameters in Prydz Bay is explained using long-term in-situ climatology data collected by seal since 2004. The following sections are described:

    i. Background
   ii. Data and methodology
  iii. Classification of water masses
  iv. Spatial distribution of mCDW
   v. Annual spatial and temporal variability of mCDW
  vi. Summary

12.2   BACKGROUND

The Prydz Bay topographical features described are located in the Indian sector of the Southern Ocean, occupying about an 80,000 km^2 area that occupies between 65°E and 80°E (Taylor and McMinn, 2002). It has a direct connection with the open water in the north adjacent to the Indian Ocean sector of the Southern Ocean. Towards the south, the borders are up on the ice shelf and the Antarctic continent to its east and west. The water stretches more inland (southward) compared to the Weddell Sea and the Ross Sea. The Prydz Bay is narrow in the southwest and wide in the northeast, with the farthest eastern end at 70°S, 76°E near Four Ladies Bank, and the farthest western end at 68°S, 69°E near the Fram Bank. The bank's performs as a barrier to water exchange between the bay and the deep ocean (Smith and Treguer, 1994). The water on the continental shelf of the bay is shallow, with a depth from 400 m to 600 m (Alberts, 1995).

The majority of the area of the bay is covered by ice about 2 m thick in winter. The Lambert glaciers extend from the inland to the bay (Christie et al., 1990), and connect with the Amery Ice Shelf. The ice becomes broken and partly molten, but the ice amount is still great and the floe distributed area is changeable in summer (Dong et al., 1984). The water becomes sharply deeper in the north of the bay and keeps the flat bottom to the far north on the continental shelf, which extends to the slope break at about 67°S. The depth of the water column is as deep as 3,000 m or deeper, and is distributed to the north of the continental slope in the Southern Ocean. The Antarctic Coastal Current (CoC) and Antarctic Slope Current (ASC) flow westward around Prydz Bay and the Antarctic Circumpolar Current flows eastward. The cyclonic Gyre in the Prydz Bay and the Antarctic Divergence Zone (ADZ) in the vicinity rotates anticlockwise (Smith et al., 1984; Vaz and Lennon, 1996; Taylor and McMinn, 2002; Yabuki et al., 2006; Williams et al., 2016). Seasonal sea ice formed in the Prydz Bay increases up to ~58°S in Austral winter and withdraws back to the continental shelf in Austral summer, though some of the previous year ice may present in coastal areas (Smith and Treguer, 1994; Worby et al., 1998).

The geographical condition in the Prydz Bay region has special features that make the shelf water of the bay to keep the memory of the extremely cold events that occur on the continent in winter for a longer time. Such memory can be evidenced by the bottom water with low temperature and high salinity on the shelf of the bay. The formation of the bottom waters are explained as follows. In late autumn or winter, the cold air cools the surface water and the strong wind above the sea surface causes the deep convection in the shelf water so that the cooled and dense water near the sea surface subsides downward to the bottom and at the same time takes part in the coastal circulation that includes the wind drive current by polar easterly. After the sea surface has become frozen, the ice-covered water can still be further cooled by surface ice rather than the wind mixing. Furthermore, the salt-release process starts to affect and make the salinity and the density of the ice-covered water increase when the surface water becomes frozen. Therefore, the extremely cold, salty, and dense water will go down to the bottom of the bay to form the shelf water, which bears the memory of the extreme events in the winter. When it is summer, the density of the upper layer of water decreases by the effect of the seasonal warming and the freshwater flux coming from the molten ice in the seasonal ice zone around Prydz Bay. The seasonal thermoclines are formed, and the vertical stability of the water column is increased. However, these surface changes cannot penetrate through the thermoclines and reach the lower layer because the wind force seasonally decreases and wind mixing becomes weaker. Therefore, the winter water or the bottom water still bear the memory of the coldest events from the previous winter (Figure 12.3).

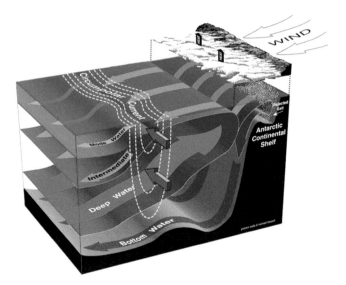

FIGURE 12.3 Simplified view of the Antarctic polar system showing subglacial hydrology along the Antarctic continental shelf and Southern Ocean circulation. Antarctic Bottom Water (AABW), Antarctic Circumpolar Current (ACC), Antarctic Slope Current (ASC), and Circumpolar Deep Water (CDW) (Figure adopted from the website https://www.csiro.au/en/).

12.3 DATA AND METHODOLOGY

This chapter explains spatial and temporal variability of the physical parameters using long-term data and was possible only because of the data collected by the new techniques of instrumented elephant seals. Hydrographic profiles obtained with instrumented elephant seals analyze spatial and temporal variability of the physical parameters in the Prydz Bay. Seal-derived data are making a growing contribution to climatologists in building upon existing oceanographic databases (Fabien et al., 2014), such as the World Ocean Database. The climatic factor is particularly responsible for the distribution of temperature and salinity near the ocean surface as a result of complicated air-sea-ice interactions. The hydrographic temperature and salinity parameters obtained from instrumented elephant seals were used to illustrate the variability in Prydz Bay (Figure 12.4). The calibration processing procedure for the seal CTD data is described in Roquet et al. (2014), with the temperature accuracy estimated to be within ±0.03°C and salinity within ±0.05. CTDSRDL (CTD as Conductivity, Temperature, and Depth; SRDL as Satellite, Relayed, Data, and Loggers) for the instrumental southern elephant seals measure vertical profiles of temperature and salinity during their foraging trips on the continental slope and shelf regions of Antarctica. The data processing steps involved calibration using delayed mode techniques and cross-comparison with the existing in-situ profiles to establish similar protocols within the Argo community. All stations located in the

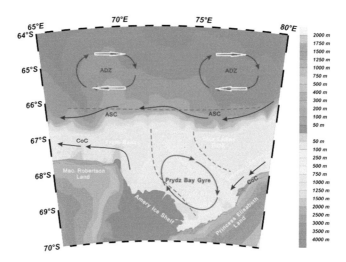

FIGURE 12.4   Map boundaries are 65°E to 80°E, 64°S to 70°S, and the circulation process involved in Prydz Bay, East Antarctica. The blue lines represent the Antarctic Coastal Current (CoC) and Antarctic Slope Current (ASC). The Prydz Bay Gyre and Antarctic Divergence Zone (ADZ) are denoted as red circles. Following Cooper and O'Brien (2004) and Wua et al. (2017), the currents are traced. The light gray dotted lines indicate the inner bay, banks, and Deep Ocean (Haozhuang et al., 2015). CD: Cape Darnley; FLB: Four Ladies Bank. Locations of the Mackenzie Polynya (MP) and Davis Polynya (DP) are shown as cyan lines following Williams et al. (2016).

region bounded 66°–70°S, 65°–85°E have been utilized since 2004. The data format is provided following the Argo netCDF format so that it can be easily processed.

12.4 CLASSIFICATIONS OF WATER MASSES

The spatial distribution of various water masses identified based on the horizontal distribution of potential temperature (T) and salinity (S) diagram are shown in Figure 12.5 as the scattered plot during Austral summer and Austral winter. Instrumented seals climatological profiles obtained during Austral summer (November to April, Figure 12.5a) and Austral winter (May to October, Figure 12.5b) seasons have been used since 2004. Oceanic water mass south of the Polar Frontal Zone comprises three major source masses: Antarctic Surface Water (AASW), Circumpolar Deep Water (CDW), and Antarctic Bottom Water (AABW). Based on these primary sources, water masses identified in the Prydz Bay and the nearby offshore region are AASW and subdivided as Antarctic Summer Surface Water (AASSW) and Winter Water (WW); Shelf Water (SW) is subdivided as High Salinity Shelf Water (HSSW) and Low Salinity Shelf Water (LSSW), modified SW (mSW), Ice Shelf Water (ISW), CDW, modified CDW (mCDW), and AABW. The TS characteristics of the different water masses have been explained by various previous studies (Orsi and Wiederwohl, 2009; Smith et al., 1984; Whitworth et al., 1998; Wong et al., 1998). In this section, all the water mass TS characteristics are explained briefly.

12.4.1 AASW

AASW is identified above the thermocline with a warmer surface water in summer and the colder WW in winter with a temperature nearer to the freezing point. The

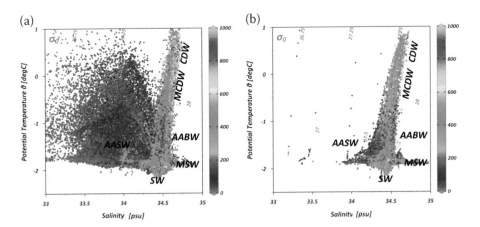

FIGURE 12.5 Climatology of potential temperature and salinity scatters plot with colors bar indicating depth: (a) left panel for summer and (b) right panel for winter. Major water masses are labeled as Antarctic Surface Water (AASW), Modified Circumpolar Deep Water (mCDW/CDW), Modified Shelf Water (mSW/SW), and Antarctic Bottom Water (AABW).

WW formed during the previous winter is found below the seasonal thermocline. The WW mass is not well defined near the continental shelf. The AASW (Guo et al., 2019) is subdivided into AASSW (-1.8 °C < T < 2.1 °C, 30.6 < S < 34.2 psu) and WW.

12.4.2 SW, ᴍSW, ᴀɴᴅ ISW

Mosby (1934) segregated continental SW as cold continental SW with low salinity and high salinity and near-freezing ISW. SW is close to near-freezing temperatures but varying salinity in the water mass in the shelf domain is described as LSSW and HSSW. Dense SW is referred to as HSSW, with a temperature range from -1.95°C < T < -1.85°C and salinity S ≥ 34.5. Whitworth et al. (1998) redefined the LSSW in Prydz Bay. LSSW and HSSW are usually separated at a salinity of 34.6 psu (Smith et al., 1984). Smith and Treguer (1994) draw the boundary between these two types at a salinity of 34.5 psu. Nunes Vazand Lennon (1996) investigated the HSSW with a salinity of more than 34.6 psu. When the salinity of the SW is over 34.6 psu, the dense bottom water that is formed due to the mixing process can descend the slope of the continental shelf, termed *Prydz Bay Bottom Water* (Middleton and Humphries, 1989).

The water in the bottom layer, with a temperature above -1.85°C, has revealed the mixing of SW with the warmer water above. The SW has a near-freezing temperature of -1.89°C, salinity 34.4 psu, and density equal zero and is constantly mixing vertically with the relatively warm inflow of mCDW above, producing transitional dense waters known as mSW. In the Weddell Sea, mSW produces AABW through a classic mechanism (Gill, 1973; Foster and Carmack, 1976). At the mouth of the Drygalski Trough, mSW was produced during the strongest spring tidal currents (Whitworth and Orsi, 2006; Muench et al., 2009; Padman et al., 2008).

The SW colder than -1.95°C interacts with the ice shelf from below and adopts this temperature as an upper limit to ISW (Sverdrup, 1940; Lusquinos, 1963). The signature of ISW at the bottom of the troughs is traced toward the sills as an intermediate temperature minimum core (Orsi and Wiederwohl, 2009). The spatial distribution of ISW was studied by Zheng et al. (2011) in the west near the Amery Ice Shelf and it is suggested that its origin is near the Amery Ice Shelf as dense water. The distribution of ISW is restricted within the western Amery Basin. The ISW in the eastern part of the bay is probably related to the West Ice Shelf (Yabuki et al., 2006). The ISW close to the Amery Ice Shelf is characterized as salty and cool due to loss of heat and rejection of salt below the ice shelf. Prydz Bay topography depression, together with Fram Bank and Four Ladies Bank, controlled the exchange of water with the ocean, due to which increases the effect of the deep waters within the bay. The deeper depth next to the Amery Ice Shelf leads to freezing below the shelf and the existence of cold and salty water (Morgan, 1972; Budd et al., 1982). The ISW was found to be less in volume (about 2.4% of the total volume) and appears to have only a less marginal influence on the large-scale oceanography in terms of thermodynamic property (Carmack, 1977).

12.4.3   CDW AND mCDW

CDW is the most abundant water mass in the Southern Ocean and contributes 55% of the water from the Indian Ocean (Carmacks, 1977). According to Callahan (1972), the North Atlantic Deep Water converted into CDW through a complicated process along isopycnal surfaces. Water columns are divided into three neutral density layers according to Jackett and McDougall (1997) and Orsi and Wiederwohl (2009). Whitworth et al. (1998) considered a density of 28.00 kg/m^3 as the lower limit of AASW since it is present near the warmest subsurface temperature maximum of CDW and, according to Orsi et al. (1999), the density of 28.27 kg/m^3 separates CDW from the AABW. Therefore, the middle-density layer is occupied by CDW and mCDW in the north and south of the shelf break (Figures 12.5a,b). The characteristics of the CDW in the Prydz Bay, according to Callahan (1972), are the relatively warm temperature (>0.5°C) and salinity (>34.65 psu) in deep water and modified CDW (mCDW) as the cold shelf water in the shelf region (Nunes Vaz and Lennon, 1996).

   CDW was identified at a depth of ~400 m, with the warmest layer temperature > 1.2°C (Whitworth et al., 1998). Separated from mCDW by a marked alongshore front (Jacobs, 1991), the Antarctic slope front, SW is cooled almost down to the surface freezing point on the continental shelf. The mixing processes between CDW, SW, and AASW result in the production of a transition water mass, mCDW. The mixing of CDW, SW, and AASW produces mCDW, with a temperature range from −1.85°C < T < −0.5°C and potential density between 27.72 kg/m^3 and 27.85 kg/m^3 (Herraiz-Borreguero et al., 2015, 2016; Wong et al., 1998).

12.4.4   AABW

AABW is generally recognized by temperatures below 0°C and salinity in the range of 34.60–34.72 psu (Gordon, 1971a; Carmack, 1977). The key element in the formation of AABW is the possibility of CDW rising and extending onto the shelf (Gill, 1973; Foster and Carmack, 1976) and may be responsible for breaking pack ice (Jacobs et al., 1970). AABW is formed due to the mixing of mCDW with salty shelf water (Foster and Carmack, 1976) and spread over the abyssal layer of the world ocean. Jacobs and Georgi (1977) identified cold and high oxygenated water near the bottom of the continental slope at 60°E, which is the source of AABW coming from the Enderby Land/Prydz Bay. Orsi et al. (1999) inferred the AABW formation based on the large-scale flow pattern and chlorofluorocarbon (CFC) distribution and inferred its formation not only in the Weddell Sea and the Ross Sea but also in the Prydz Bay region. Jacobs et al. (1970) discovered regions with temperatures near the freezing point at the base of the Ross Ice Shelf and suggested that the formation of AABW in summer is due to the mixing of Ross Sea ISW and CDW with low salinity. Mosby (1934) showed that if continental shelf waters become sufficiently salty due to the sea-ice formation, it will sink the continental slope due, to greater density to form bottom water (Deacon, 1937). The mixing processes are complex and have been undertaken in the Weddell Sea, the principal source region for AABW (e.g. Foster et al., 1987).

12.5   SPATIAL DISTRIBUTION OF mCDW

This section explains spatial and temporal variability of the mCDW in the Prydz Bay region. Spatial distribution of mCDW is perceived using a potential temperature and salinity diagram. The Austral summer (November to April) and the Austral winter (May to October) seasons are demonstrated in Figure 12.5a and Figure 12.5b, respectively. Foster and Carmack (1976) considered $-1.7°C$ isotherm as the lower limit of temperature for mCDW. The range of temperature varies between $-1.7°C$ and $0.2°C$ and the neutral density between 28.00 kg/m^3 and 28.27 kg/m^3. The spatial display of mCDW on average neutral density $\gamma N = 28.135$ kg/m^3 is the intermediate value. Therefore, potential temperature and salinity are chosen on the average neutral density isopycnal surface $\gamma N = 28.135$ kg/m^3 to analyze horizontal distribution during Austral summer and Austral winter. The distribution of potential temperature and salinity for Austral summer is shown in Figure 12.6 (left panel) and for Austral winter is shown in Figure 12.6 (right panel). The distribution of data in the Prydz Bay region in Austral winter is less as compared to Austral summer. In both seasons (summer and winter), mCDW is identified north of 67°S across a zonal section near the continental slope of the Prydz Bay. The mCDW water mass is warmer and saltier, which enters the southeastern inner shelf region, as demonstrated by Lina et al. (2016) between 72°E and 75°E, adjacent to 67.25°S.

The vertical sections of potential temperature for two zonal transects are shown in Figure 12.7 along 66.3°S (Figure 12.7a) and 67°S (Figure 12.7b) across Prydz Bay between 65° and 85°E. The two transects are randomly selected, occupying the slope and shelf region of Prydz Bay. The variability in the vertical structure of potential temperature was noticed in both transects. Along the 66.3°S zonal transect, the slope of the Prydz Bay occupies the CDW below 200 m between 65°E and 75°E below 28.00 kg/m^3 isopycnals. The shelf region of Prydz Bay is occupied by the mCDW along 67°S (Figure 12.7b). The presence of mCDW flows onto Prydz Bay between 28.00 kg/m^3 and 28.27 kg/m^3 isopycnal surface between 66°E–70°E and 72°E–79°E, the slightly modified core of warm water below 200 m. The mCDW occurs further south below 100 m in pockets in the transect of 67°S between 72°E–78.5°E and 74°E (Lina et al., 2016). The warmer core identified near the surface at 75°E could be due to ocean ice-atmosphere interaction and the floe concentration at the sea surface (Shuzhen et al., 2010). Thus, the intrusion of CDW is depicted in the vertical column of zonal transects.

12.6   ANNUAL SPATIAL AND TEMPORAL VARIABILITY OF mCDW

In this section, the annual time scale variability is established using 2011 and 2012 profiles obtained using instrumental seal climatology. The horizontal distribution of the potential temperature and salinity is shown in Figure 12.8 for 2011 and in Figure 12.9 for 2012 consecutive periods on selected isopycnal surface of $\gamma N = 28.00$ kg/m^3 (Figures 12.8a,b and Figures 12.9a,b) and $\gamma N = 28.27$ kg/m^3 (Figures 12.8c,d and Figures 12.9c,d). The isotherm $-1.7°C$ superimposed over the spatial distribution on potential temperature and salinity to highlight the extent of the mCDW signal to the interior of the bay, as depicted in Figures 12.8 and 12.9,

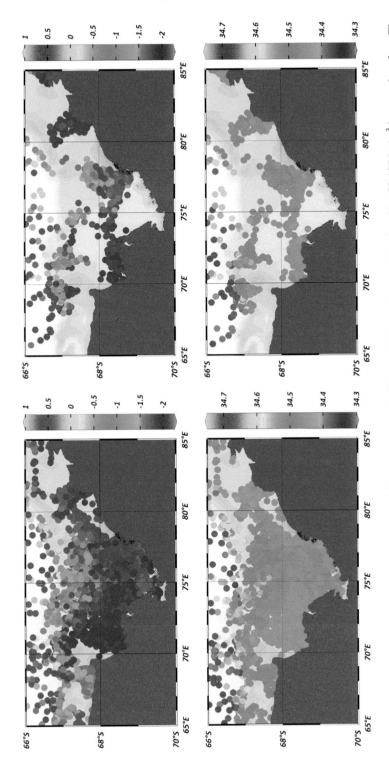

FIGURE 12.6 Climatological distribution of potential temperature and salinity on horizontal neutral density $\gamma N = 28.135$ kg/m^3 isopycnal surface. The left panel is for summer and the right panel is for winter. Top raw for potential temperature, and down raw for salinity. The $-1.7°C$ is considered the lower temperature limit of mCDW according to Foster and Carmack (1976).

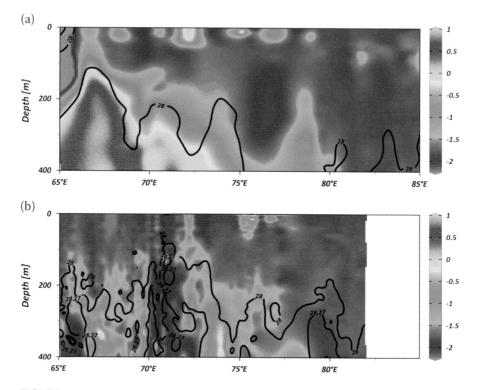

FIGURE 12.7   Vertical sections of potential temperature for two zonal transects 66.3°S (a) and 67°S (b) during summer (November to April). The solid contour lines represent the γN = 28.00 kg/m^3 and γN = 28.27 kg/m^3 isotherms defining the boundary of mCDW.

and there exists variability in the extension of mCDW each year in both isopycnal surfaces. In 2011, warm and saline water identifies north of −1.7°C isotherm that is north of 67°S. The mCDW resides in the slope section between 67°E and 72°E. The mCDW occupies the shelf region between 73°E and 78°E north of 68°S. In 2012, warm and saline water occupied a more interior region of Prydz Bay as −1.7°C isotherm extended more southward 68.5°S with small pockets up to 69°S, whereas in Figures 12.9c,d the mCDW extended up to 68°S, which is less compared to Figures 12.9a,b, approaching towards shelves and disappearing faster.

12.7   SUMMARY

CTD data was collected in the region using instrumented seals south of the polar front zone covered by the sea ice during the Austral winter season. Instrumental seal data supported the scientific community to understand better the climate change process and fill important data gaps in the polar region, in studying the role of Antarctica in the global climate. In this chapter, the attempt has been made to show the spatial and temporal distribution of physical parameters and how they vary with space and time in Prydz Bay. The process seal data is analyzed to find the existence

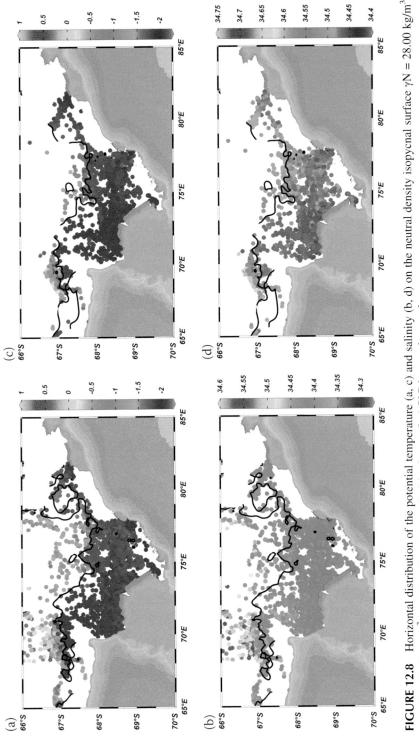

FIGURE 12.8 Horizontal distribution of the potential temperature (a, c) and salinity (b, d) on the neutral density isopycnal surface γN = 28.00 kg/m³ and γN = 28.27 kg/m³ for 2011. The thick dashed isotherm of −1.7°C is superimposed.

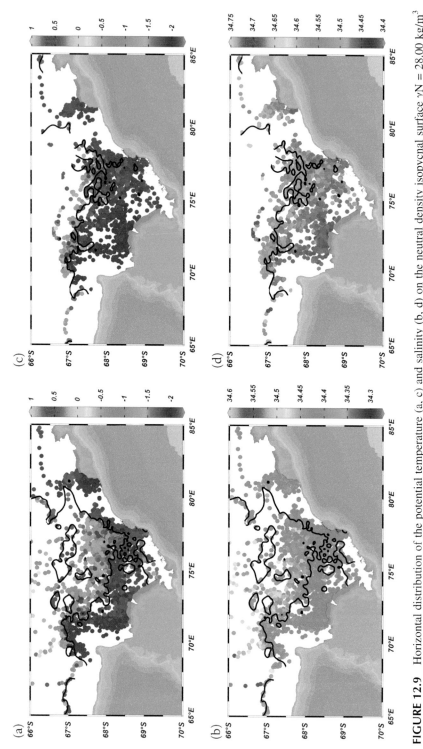

FIGURE 12.9 Horizontal distribution of the potential temperature (a, c) and salinity (b, d) on the neutral density isopycnal surface $\gamma N = 28.00$ kg/m^3 and $\gamma N = 28.27$ kg/m^3 for 2012. The thick dashed isotherm of $-1.7°$C is superimposed.

of the various water masses. The major water masses are Antarctic Surface Water (AASW), Circumpolar Deep Water (CDW), modified CDW (mCDW), Shelf Water (SW), modified SW (mSW), Ice SW, and Antarctic Bottom Water (AABW), identified during summer and winter periods based on potential temperature and salinity diagrams.

Further, the spatial distribution of mCDW is perceived based on the horizontal distribution of the potential and salinity diagram. The analyses have shown the entrance of CDW into Prydz Bay near the shelf break during the summer, with warmer and saltier water characteristics identified. The vertical section of potential temperature along two zonal transects has shown CDW below 200 m between 65°E and 75°E. The presence of mCDW flows onto Prydz Bay between 28.00 kg/m^3 and 28.27 kg/m^3 isopycnal surface between 66°E and 70°E and 72°E–79°E and the slightly modified core of warm water below 200 m. The mCDW occurs farther south below 100 m in pockets in the transect of 67°S between 72°E and 78.5°E and 74°E.

An annual timescale variability was established using 2011 and 2012 profiles. The isotherm −1.7°C distribution on potential temperature and salinity highlights the extent of the mCDW signal to the interior of the bay in 2011. In 2012, warm and saline water occupied a more interior region of Prydz Bay as −1.7°C isotherm extended more southward 68.5°S with small pockets up to 69°S. There exists variability in the extension of mCDW each year in both isopycnal surfaces on a spatial and temporal scale.

ACKNOWLEDGEMENTS

The author is thankful to the secretary, Ministry of Earth Sciences, and director, National Centre for Polar and Ocean Research for their continuous support. Thanks to Schlitzer, Reiner, Ocean Data View, 2020 for providing a package for the interactive exploration, analysis, and visualization of oceanographic data. The marine mammal data were collected by the International MEOP Consortium and the national programs that contribute to it are gratefully acknowledged.

REFERENCES

Alberts, F. G. (1995). *Geographic Names of the Antarctic*. 2nd ed, United States Board on Geographic Names.

Budd, W. F., Corry, M. J., and Jacka, T. H. (1982). Results from the Amery Ice Shelf Project. *Annals of Glaciology*, 3, 36–41.

Callahan, L. E. (1972). The structure and circulation of deep water in the Antarctic. *Deep-Sea Research*, 19, 563–575.

Carmack, E. C. (1977). Water characteristics of the southern ocean south of the polar front. In: Angel M. ed., *A Voyage of Discovery, Deacon 70th* , Pergamon Press, Oxford, pp. 15–61.

Chauhan, A., Bharti, P. K., Goyal, P., Varma, A., & Jindal, T. (2015). Psychrophilic pseudomonas in antarctic freshwater lake at stornes peninsula, larsemann hills over east Antarctica. *SpringerPlus*, 4, 582. https://doi.org/10.1186/s40064-015-1354-3.

Christie, J., Bartholomew, J. C., Jones, R., Lewis Obe, H. A. G., Lippard, S., Rothery, D., and Whitehouse, D. (1990). *The Concise Atlas of the World*, Times Books Press, London.

Cooper, A. K., and O'Brien, P. E. (2004). Leg-188 synthesis: Transitions in the glacial history of the Prydz Bay region, East Antarctica, from ODP drilling. A. K. Cooper, P. E. O'Brien, and C. Richter (Eds). Proceedings of the Ocean Drilling Program, Scientific Results. Available from Ocean Drilling Program, Texas A&M University, College Station, TX, Vol. 188 [CD-ROM], 1–42.

Deacon, G. E. R. (1937). The hydrography of the Southern Ocean. *Discovery Report*, 15, 124.

Denisov, A. S., and Myznhikova, M. N. (1978). Osobennosti gidrologicheskogo rezhima v zalive Priuds (fevral' 1973 g.) (Features of the oceanographic regime in Prydz Bay, February 1973), *Trudy Sovetskoi Antarticheskoi Ekspeditsii*, 68(1978), 100–105.

Dong, Z. Q., Smith, N. R., Kerry, K. R., and Wright, S. (1984). Water masses and circulation in the region of Prydz Bay, Antarctica. *A collection of Antarctica Scientific Exploration*, No. 2, Ocean Press, Beijing, pp. 1–24 (in Chinese).

Fabien, R. et al. (2014). A southern Indian Ocean database of hydrographic profiles obtained with instrumental elephant seal. *Scientific Data*. https://doi.org/10.1038/sdata.2014.28.

Foster, T. D., and Carmack, E. C. (1976). Frontal zone mixing and Antarctic Bottom Water formation in the southern Weddell Sea. *Deep-sea Res*, 23, 301–317.

Foster, T. D., Foldwk, A., and Middleton, J. H. (1987). Mixing and bottom water formation in the shelf break region of the southern Weddell Sea. *Deep-Sea Res.*, 34, 1771–1794.

Gill, A. E. (1973). Circulation and bottom water production in the Weddell Sea. *Deep-Sea Res.* 20, 111–140.

Gordon, A. L. (1971a). Recent physical oceanography studies of Antarctic waters. In: *Research in the Antarctic*, L. Quam, ed., American Association for the Advancement of Science, Washington, DC, pp. 609–629.

Gordon, A. L. (1971b). Oceanography of Antarctic waters. In: *Antarctic oceanography I: Antarctic research series*, Vol. 15, J. L. Reid, ed., American Geophysical Union, pp. 169–203.

Guo, G., Shi, J., Gao, L., Tamura, T., and Williams, G. D. (2019). Reduced Sea Ice production due to upwelled oceanic heat flux in Prydz Bay, East Antarctica, *Geophys. Res. Lett.*, 46(9), 4782–4789.

Haozhuang, W., Chen, Z., Wang, K., Liu, H., Tang, Z., and Huang, Y. (2015). Characteristics of heavy minerals and grain size of surface sediments on the continental shelf of Prydz Bay: Implications for sediment provenance *Antarctic Science*, 28, 103–114.

Herraiz-Borreguero, L., Coleman, R., Allison, I., Rintoul, S. R., Craven, M., and Williams, G. D. (2015). Circulation of modified Circumpolar Deep Water and basal melt beneath the Amery Ice Shelf, East Antarctica. *J. Geophys Res., Oceans*, 120, 3098–3112.

Herraiz-Borreguero, L., Lannuzel, D., van der Merwe, P., Treverrow, A., and Pedro, J. B. (2016). Large flux of iron from the Amery Ice Shelf marine ice to Prydz Bay, East Antarctica. *J. Geophys Res., Oceans*, 121(8), 6009–6020.

IPCC (2007). Climate Change 2007 The Physical Science Basis. In: *Contribution of Working Group I to the Fourth Assessment Report of the IPCC*. S. Solomon, D. Qin, M. Manning, Z. Chen, M. Marquis, K. B. Averyt, M. Tignor and H. L. Miller, eds., Cambridge University Press, Cambridge, UK.

Izvekov, M. V. (1959). Results of observations on currents in the region of the West ice Shelf. *Soviet Antarctic Expedition*, 2, 9–93.

Jackett, D. R., and McDougall, T. J. (1997). A neutral density variable for the world's oceans. *Journal of Physical Oceanography*, 27(2), 237–263.

Jacobs, S. S., Amos, A. F., and Bruchhausen, P. M. (1970). Ross Sea oceanography and Antarctic Bottom Water formation. *J. Geophys. Res.*, 17, 935–962.

Jacobs, S. S., and Georgi, D. T. (1977). Observation on the southwest Indian Antarctic ocean In a voyage of discovery. In: *Supplement to Deep-Sea Research*, M. V. Angel, ed., Pergamon Press, Oxford, pp. 43–84.

Jacobs, S. S. (1991). On the nature and significance of the Antarctic Slope Front. *Marine Chemistry*, 35(1–4), 9–24.

Kupetskii, V. N. (1959). O prichinakh anomalii gidrologicheskikh uslovii zaliva Olaf Priuds (On the causes of anomalous hydrological conditions in Prydz Bay). *Izvestiya Vsesoyuznogo Geographicheskogo Obshchestvo (VGO)*, 91, 356–357.

Lei, R., and Quot (2010). Annual cycle of land-fast sea ice in Prydz Bay East Antarctica, *J. Geophy. Res.*, 115(C2).

Lina, L., Hongxia, C., and Na, L. (2016). The characteristics of warm water inflowing and its temporal and spatial variation on the Prydz Bay continental shelf. *Antarctic Acta Oceanol. Sin.*, 35(9), 51–57

Lusquinos, A. J. (1963). Extreme temperatures in the Weddell Sea. Arbok for Universitet: Bergen, *Mathemetisk-naturvitenskapelig serie*, 23, 19.

Middleton, J. H., and Humphries, S. E. (1989). Thermohaline structure and mixing in the region of Prydz Bay Antarctica. *Deep-Sea Research*, 36, 1255–1266.

Morgan, V. L. (1972). Oxygen isotope evidence for bottom freezing on the Amery Ice Shelf. *Nature, London*, 238, 393–394.

Mosby, H. (1934). *The waters of the Atlantic Antarctic Ocean. Scientific Results of the Norwegian Antarctic Expeditions, 1927–1928*. Oslo, 1, 131.

Muench, R. D., Padman, L., Gordon, A. L., and Orsi, A. H. (2009). Mixing of a dense water outflow from the Ross Sea, Antarctica: the contribution of tides. *Journal of Marine Systems*, accepted. https://doi.org/10.1016/j.jmarsys.2008.11.003.

Nunes Vaz, R. A., and Lennon, G. W. (1996). Physical oceanography of the Prydz Bay region of Antarctic waters. *Deep-Sea Research Part* I :*Oceanographic Research Papers*, 43(5), 603–641.

Orsi, A. H., Johnson, G. C., and Bullister, J. L. (1999). Circulation, mixing, and production of Antarctic Bottom Water. *Progress in Oceanography*, 43(1), 55–109.

Orsi, A. H., and Wiederwohl, C. L. (2009). A recount of Ross Sea waters. *DeepSea Res. Part* II : *Topical Studies in Oceanography*, 56(13–14), 778–795.

Padman, L., Howard, S. L., Orsi, A. H., and Muench, R. D. (2008). Tides of the northwestern Ross Sea and their impact on dense outflows of Antarctic Bottom Water. *Deep-Sea Res. II*, 56, 818–834. https://doi.org/10.1016/jdsr2.2008.10.026.

Roquet, F., Williams, G., Hindell, M. A., Harcourt, R., McMahon, C. R., Guinet, C., Charrassin, J. B., Reverdin, G., Boehme, L., Lovell, P., and Fedak, M. A. (2014). A Southern Indian Ocean database of hydrographic profiles obtained with instrumented elephant seals. *Nature Scientific Data*, 1, 140028.

Shuzhen, P., Renfeng, G., and Zhaoqian, D. (2010). Variability of marine hydrological features at the northern margin of Amery Ice Shelf. *Chinese Journal of Polar Research* (in Chinese), 22(3), 244–253.

Smith, N. R., Zhaoqian, D., Kerry, K. R. and Wright, S. (1984). Water masses and circulation in the region of Prydz Bay Antarctica. *Deep-Sea Res.*, 31, 1121–1147.

Smith, N. R., and Treguer, P. (1994). *Physical and chemical oceanography in the vicinity of Prydz Bay*, Cambridge University Press, Cambridge, Antarctica.

Sverdrup, H. U. (1940). Hydrology, Section 2, Discussion. Reports of the B.A.N.Z. Antarctic Research Expedition 1921–1931, Series A, 3, Oceanography, Part 2, Section 2, pp. 88–126.

Taylor F. and McMinn, A. (2002). Late quaternary diatom assemblages from Prydz Bay, Eastern Antarctica. *Quaternary Research*, 57, 151–161.

Tolstikov, E. E. (1966). Atlas Antarktiki, Vol. I, G.U.C.K., Moscow (English translation, Soviet Geography: Reviews and Translations, 8, Nos 5-6, 225 pp, American Geographical Society, New York, 1967).

Vaz, R. A. N., and Lennon, G. W. (1996), Physical oceanography of Prydz Bay region of Antarctic waters. *Deep-Sea Research (I)*, 43(5), 603–641.

Whitworth, T., and Orsi, A. H. (2006). Antarctic bottom water production and export by tides in the Ross Sea. *Geophysical Research Letters*, 33(12), 1–4.

Whitworth, T., Orsi, A. H., Kim, S.-J., Nowlin Jr., W. D., and Locarnini, R. A. (1998). Water masses and mixing near the Antarctic slope front. In: *Ice, and Atmosphere: Interactions at the Antarctic Continental Margin*, S. S. Jacobs, R. F. Weiss, eds., American Geophysical Union, Washington, DC, pp. 1–27.

Williams, G. D., Herraiz-Borreguero, L., Roquet, F., Tamura, T., Ohshima, K. I., Fukamachi, Y., Fraser, A. D., Gao, L., Chen, H., McMahon, C. R., Harcourt, R. and Hindell, M. (2016). The suppression of Antarctic bottom water formation by melting ice shelves in Prydz Bay, *Nature Communications*, 7, 1–9.

Wong, A. P. S., Bindoff, N. L., and Forbes, A. (1998). Ocean-ice shelf interaction and possible bottom water formation in Prydz Bay, Antarctica. In: *Ocean, Ice, and Atmosphere: Interactions at the Antarctic Continental Margin*, S. S. Jacobs and R. F. Weiss, eds., American Geophysical Union, Washington, DC, pp. 173–187.

Worby, A. P., Massom, R. A., Allison, I., Lytle, V. I., and Heil, P. (1998). East Antarctic sea ice: A review of its structure, properties and drift Antarctic Sea Ice Properties, *Processes and Variability*, 74, 41–67.

Wua, L., Wanga, R., Xiaoa, W., Geb, S., Chenb, Z., and Krijgsmanc, W. (2017). Productivity-climate coupling recorded in Pleistocene sediments off Prydz Bay (East Antarctica). *Palaeogeography, Palaeoclimatology, Palaeoecology*, 485: 260–270.

Wyrtki, K. (1971). *Oceanographic Atlas of the International Indian Ocean Expedition*. National Science Foundation, Washington, DC, p. 531.

Yabuki, T., Suga, T., Hanawa, K., Matsuoka, K., Kiwada, H., and Watanabe, T. (2006). Possible source of the Antarctic Bottom Water in Prydz Bay region. *Journal of Oceanography*, 62(5), 649–655.

Zheng, S., Shi, J., Jiao, Y., and Ge, R. (2011). Spatial distribution of Ice Shelf Water in front of the Amery Ice Shelf, Antarctica in summer. *Chinese Journal of Oceanology and Limnology*, 29(6), 1325–1338.

Zverev, A. A. (1959). Anomalous seawater temperatures in Olaf Prydz Bay. *Soviet Antarctic Expedition*, I, 269–271.

Zverev, A. A. (1963). Currents in the Indian, sector of the Antarctic. *Trudy Sovetskoi Antarkticheskoi Ekspeditsii*, 17, 144–155.

Index